Developments in Earth Surface Processes, 18

Series Editor – J.F. Shroder Jr

For previous volumes refer
http://www.sciencedirect.com/science/bookseries/09282025

Developments in Earth Surface Processes

Volume 18

Geomorphological Fieldwork

Edited by

Mary J. Thornbush

School of Geography, Earth and Environmental Sciences
University of Birmingham, Edgbaston, Birmingham, UK

Casey D. Allen

Department of Geography and Environmental Sciences,
University of Colorado Denver, Colorado, USA

Faith A. Fitzpatrick

U.S. Geological Survey, Wisconsin Water Science Center,
Middleton, Wisconsin, USA

ELSEVIER

AMSTERDAM • BOSTON • HEIDELBERG • LONDON
NEW YORK • OXFORD • PARIS • SAN DIEGO
SAN FRANCISCO • SINGAPORE • SYDNEY • TOKYO

Elsevier
Radarweg 29, PO Box 211, 1000 AE Amsterdam, Netherlands
The Boulevard, Langford Lane, Kidlington, Oxford OX5 1GB, UK
225 Wyman Street, Waltham, MA 02451, USA

ISBN: 978-0-444-63402-3
ISSN: 0928-2025

For information on all Elsevier publications
visit our website at http://store.elsevier.com/

Contents

List of contributors xi
Foreword xiii
Preface xvii

1. Introduction 1

Mary J. Thornbush, Casey D. Allen, Faith A. Fitzpatrick

References 7

2. Why Fieldwork? 11

Casey D. Allen

2.1 Introduction 11
2.2 Wither Fieldwork? The Necessity of "Ground Truthing" and the Importance of Place 12
2.3 Teaching and Learning Through Fieldwork 15
2.4 Fieldwork Enhances Subfield and Interdisciplinary Relationships 18
2.5 Fieldwork Challenges Established Paradigms 20
2.5.1 Global-Scale: Continental Drift 20
2.5.2 Regional-Scale: The Channeled Scablands 21
2.5.3 Local-Scale: Yosemite Valley, CA (USA) 21
2.5.4 Hand Sample-Scale: Rock Decay Science 22
2.6 Why *Not* Fieldwork? 23
References 25

3. Getting into the Field 31

Mary J. Thornbush, Casey D. Allen, Faith A. Fitzpatrick

3.1 Preparing for Fieldwork 33

Mick Day

3.1.1 Introduction 33
3.1.2 Consulting the Existing Information 35
3.1.3 Time 36
3.1.4 Reconciling Academic and Logistic Considerations 39
3.1.5 Institutional and Personal Support 39

3.1.6 Field Techniques and Equipment 40
3.1.7 Preparing for Problems 42
3.1.8 Personnel Considerations 43
3.1.9 Budgets, Expenses, Funding, and Authorization 46
3.1.10 Reconciling Academic and Logistic Preparation 47
3.1.11 Transportation 48
3.1.12 Accommodation 49
3.1.13 Provisions 51
3.1.14 Considering the Physical Environment 52
3.1.15 Considering the Human Environment 55
3.1.16 Conclusion 59
Acknowledgements 60
References 61

3.2 Field Safety: Principles, Practice, and Culture 65

Sam Alpay, Roger C. Paulen

3.2.1 Introduction 65
3.2.2 Principles 65
3.2.2.1 Preservation of Life, Health, Property, and the Environment 66
3.2.2.2 Culture and Conscience 66
3.2.2.3 Approach 67
3.2.3 Practice 68
3.2.3.1 Standards 68
3.2.3.2 Processes 69
3.2.3.3 Training 75
3.2.3.4 Field Parties 80
3.2.3.5 Communications 82
3.2.3.6 Equipment 83
3.2.4 Leadership of Safety Culture 84
3.2.5 Conclusions 87
Acknowledgments 88
References 88

4. Teaching Geomorphology in the Field 91

Mary J. Thornbush, Casey D. Allen, Faith A. Fitzpatrick

4.1 Student Learning Styles 93

Miguel Gomez-Heras, Stephen McCabe

4.1.1 Introduction 93
4.1.2 Methodology 95
4.1.3 Discussion of Results 97
4.1.3.1 Student Perceptions of "Novelty Spaces" 97
4.1.4 Educational Implications 99
4.1.5 Conclusions and Recommendations 100

Acknowledgments 102
References 102
Appendix 4.1.1 (Questionnaire)s 103
Appendix 4.1.2 (Results From the Questionnaire) 109

4.2 Fieldwork Going Digital 117
Ian Fuller, Derek France

4.2.1 Introduction 117
4.2.2 A Cook's Tour of Geomorphology: North Island, New Zealand 117
4.2.2.1 The Fieldtrip 118
4.2.2.2 Learning Evaluation and Discussion 120
4.2.3 Field Experiments in Process Geomorphology 122
4.2.3.1 Approach to Teaching and Assessment 122
4.2.3.2 Learning Evaluation and Discussion 123
4.2.4 Web 3.0 Technology 124
4.2.4.1 Mobile Technologies and Fieldwork 124
4.2.4.2 Case Study: Application of Mobile Devices to Residential Fieldwork 125
4.2.5 Recommendations 126
4.2.5.1 Transferable Skills Development for Geomorphologists 126
4.2.5.2 Limitations and Cautions 126
4.2.6 Conclusions 127
Acknowledgements 128
References 128

4.3 Field-Based Learning in Undergraduate Geomorphology Courses 131
Mary J. Thornbush

4.3.1 Introduction 131
4.3.2 Methods 138
4.3.2.1 Introductory Geomorphology 138
4.3.2.2 Environmental Geomorphology 140
4.3.2.3 Field School 144
4.3.3 Results 145
4.3.3.1 Introductory Geomorphology 147
4.3.3.2 Environmental Geomorphology 151
4.3.3.3 Field School 153
4.3.4 Discussion 154
4.3.4.1 Introductory Geomorphology 154
4.3.4.2 Environmental Geomorphology 155
4.3.4.3 Field School 158
4.3.5 Conclusions 160
Acknowledgments 161

Appendix 4.3.1 April 2008 162
Appendix 4.3.2 November 2008 164
Appendix 4.3.3 November 2009 165
References 169

5. Field Methodologies 173
Mary J. Thornbush, Casey D. Allen, Faith A. Fitzpatrick

5.1 Use of Field Experiments in Soil Erosion Research 175
Nikolaus J. Kuhn, Philip Greenwood, Wolfgang Fister

5.1.1 Experiments in Geomorphology 175
5.1.2 Advantages and Disadvantages of Field Experiments in Soil-Erosion Research 178
5.1.2.1 Field Experiments versus Field Monitoring 178
5.1.2.2 Field versus Laboratory Experiments 179
5.1.3 General Considerations for the Preparation of Field Experiments 180
5.1.4 Case Studies: Testing New and Measuring Known Processes 181
5.1.5 Case Study 1: Does the Influence of Wind on Rainfall Alter Erosion Rates on Arable Land? 182
5.1.5.1 Background and Research Question 182
5.1.5.2 Wind-Driven Rain Simulation 182
5.1.5.3 Study Site and Date of Experiments 185
5.1.5.4 Plot Dimensions and Scale Considerations 185
5.1.5.5 Test Sequence and of Sediment Collection 185
5.1.5.6 Supplemental Parameters 186
5.1.5.7 Overview of Results 186
5.1.5.8 Discussion and Conclusions 187
5.1.6 Case Study 2: Runoff and Infiltration Experiments on Terracettes Using Simulated Rainfall 188
5.1.6.1 Introduction 188
5.1.6.2 Expectations of a Field Experiment 189
5.1.6.3 Key Resources and Logistical Constraints 190
5.1.6.4 Study Site 191
5.1.6.5 Rainfall Simulation Experiments 192
5.1.6.6 Antecedent Soil Moisture Conditions 192
5.1.6.7 Surface Hydrology and Volumetric Moisture Content Related to Runoff Generation 193
5.1.6.8 Potential Relevance of Observed Infiltration on Terracettes for Hillslope Hydrology 194
5.1.6.9 Field Experiments: Dealing with Unexpected Findings, or a Case of Serendipity? 195
5.1.7 Conclusions: Field Experiments as a Tool to Explore Surface Processes 195
References 196

5.2 A Geologic Approach to Field Methods in Fluvial Geomorphology **201**

Faith A. Fitzpatrick

5.2.1 Introduction **201**
5.2.2 Prefield Activities **204**
5.2.3 Reconnaissance Trip **208**
5.2.4 Geologic Approach to Fluvial Geomorphology Field Methods **209**
5.2.4.1 Topographic Mapping and Surveying 209
5.2.4.2 Characterizing Fluvial Sedimentary Sequences 214
5.2.4.3 Describing Cores and Exposures in the Field or Laboratory 219
5.2.4.4 Age Determinations of Field Sediment Samples 221
5.2.5 Calculations and Interpreting Field Data **221**
5.2.5.1 Interpreting River Behavior and Floodplain Dynamics 222
5.2.5.2 Historical Floodplain Sedimentation Depths and Volumes and Rates 223
5.2.5.3 Historical Watershed Sediment Budgets 225
5.2.6 Conclusion **225**
References **226**

5.3 Reading the Landscape in Field-Based Fluvial Geomorphology **231**

Gary Brierley, Kirstie Fryirs

5.3.1 Introduction **231**
5.3.2 An Approach to "Reading the Landscape" **234**
5.3.2.1 Step One: Identify Individual Landforms (Geomorphic Units) and the Process-Form Relationships That Determine Their Behavior 236
5.3.2.2 Step Two: Analyze and Interpret the Package and Assemblage of Landforms at the Reach-Scale and How They Adjust Over Time 237
5.3.2.3 Step Three: Explain Controls on the Package and Assemblage of Landforms at the Reach-Scale and How They Adjust Over Time 238
5.3.2.4 Step Four: Integrate Understandings of Geomorphic Relationships at the Catchment Scale 239
5.3.3 Three Fieldwork Exercises to Interpret River Forms, Processes, Patterns, and Evolution **240**
5.3.3.1 Exercise 1: River Reach Analysis 240
5.3.3.2 Exercise 2: Using Longitudinal Profiles as a Teaching and Interpretation Tool: Assessing Downstream Patterns of River Styles and Controls on Their Character and Behavior 245
5.3.3.3 Exercise 3: Scenario Building to Interpret River Evolution and Future Trajectory of Change 249
5.3.4 Discussion and Concluding Comments **252**

Acknowledgements 255
References 255

6. Conclusion 259
Mary J. Thornbush, Casey D. Allen, Faith A. Fitzpatrick

Index 263

List of contributors

Casey D. Allen, Department of Geography and Environmental Sciences, University of Colorado Denver, Colorado, USA

Sam Alpay, Geological Survey of Canada, Natural Resources Canada, Ottawa, Ontario, Canada

Gary Brierley, School of Environment, University of Auckland, Auckland, New Zealand

Mick Day, Department of Geography, University of Wisconsin-Milwaukee, Milwaukee, Wisconsin, USA

Wolfgang Fister, Physical Geography and Environmental Change Research Group, Department of Environmental Sciences, University of Basel, Basel, Switzerland

Faith A. Fitzpatrick, U.S. Geological Survey, Wisconsin Water Science Center, Middleton, Wisconsin, USA

Derek France, Department of Geography and Development Studies, University of Chester, Chester, United Kingdom

Kirstie Fryirs, Department of Environmental Sciences, Macquarie University, North Ryde, New South Wales, Australia

Ian Fuller, Physical Geography Group, Institute of Agriculture and Environment, Massey University, Palmerston North, New Zealand

Miguel Gomez-Heras, CEI Campus Moncloa, UPM-UCM, CSIC: ETS Arquitectura, Universidad Politécnica de Madrid and Instituto de Geociencias (CSIC, UCM), Madrid, Spain and School of Geography, Archaeology and Palaeoecology, Queen's University of Belfast, Northern Ireland, United Kingdom

Philip Greenwood, Physical Geography and Environmental Change Research Group, Department of Environmental Sciences, University of Basel, Basel, Switzerland

Nikolaus J. Kuhn, Physical Geography and Environmental Change Research Group, Department of Environmental Sciences, University of Basel, Basel, Switzerland

Stephen McCabe, School of Geography, Archaeology and Palaeoecology, Queen's University of Belfast, Northern Ireland, United Kingdom; Northern Ireland Environment Link, Belfast, United Kingdom

Roger C. Paulen, Geological Survey of Canada, Natural Resources Canada, Ottawa, Ontario, Canada

Mary J. Thornbush, School of Geography, Earth and Environmental Sciences University of Birmingham, Edgbaston, Birmingham, United Kingdom

Foreword

When Dr. Mary Thornbush first asked me if a book on fieldwork in geomorphology was appropriate for our series on *Developments in Earth Surface Processes*, I leapt at the chance to edit a book about one of my favorite aspects of this discipline. In fact, many students who choose to go into physical geography or geology in the first place are commonly initially drawn to the attraction of working in the out-of-doors, at least for a part of the year, instead of being stuck inside at a desk, perhaps for the rest of their lives. Then, as the years pass and one realizes that regular fieldwork is also a more healthful way of living, the footwork begins to make even more sense in our perpetual pursuits of youthfulness, however futile or seemingly unsuccessful that these may appear. In some cases, in my experience, people who were first drawn toward the pleasures of outside sports, such as mountaineering, rock climbing, or spelunking choose to go on into the study of geomorphology because of their desire to mix pleasure with this discipline. Also, they found that it was a clever way to get paid for doing what they loved anyway.

This book, *Geomorphological Fieldwork*, captures many of the essential aspects and character of the analysis of landforms in their natural habitat. In fact, no matter how much people must utilize the laboratory aspects of the discipline to adequately characterize the research problem, they still need to incorporate fieldwork for a really complete analysis. In fact, if I may be so bold, doing a geomorphological study without any fieldwork, does leave a researcher open to some criticism about perhaps lacking enough "real world" application. In my experience, where a great deal of quantitative computer modeling or remote sensing was to be done in any particular project, for example, the best approach commonly expressed by my co-principle investigators was that they always stressed the essentials of including the fieldwork for greater verisimilitude (Shroder and Bishop, 2003; Copland et al., 2012; Bishop et al., 2014). The numerous chapters presented here show us a number of these interrelated points of view.

Because a number of prior works have reviewed various specific field methodologies from our increasingly sophisticated hardware toolkit for measuring and sampling the plethora of data-acquisition possibilities (Goudie, 1981, 1991; Switzer et al., 2013), this volume focuses more on philosophical and related issues of pedagogy and practicality. Issues about how many students learn best, and the various degrees of thinking about fieldwork versus lectures and laboratories, are described in various fashions. These ways of thinking about fieldwork can be of considerable interest to many; others oriented more toward the technical aspects of their field equipment, or the statistical mechanics of their sampling programs may be less interested. Dr. Thornbush has even pointed out the value of fieldwork for human geographers, many of whom may be considerably less interested in this sort of research experience. Nonetheless, the diversity of points of view expressed in this work may have some broad interest to both physical geographers and geologists, all of whom share a passion for fieldwork.

These chapters have many new points of view that bear reading by anyone who is contemplating fieldwork, especially for those concerned about traveling into foreign lands where different customs prevail. One particular point about fieldwork that has long been a most important point to me after years of working in conservative Islamic countries, has been the need to caution my many excellent female colleagues about taking care to avoid certain male sensibilities in that world, no matter how offensive we may think these are. Commonly, a simple head-scarf (*hijab*) token is enough to keep the prudes and predators at bay. Even so, in some cases unpleasant situations can develop in the presence of illiterate and unsophisticated tribesmen. They may try to take liberties with female field researchers that may lead to attempts to rectify the situation with violence by one's counterpart foreign field leaders, to whom it is a matter of honor to protect their employers. Obviously, it is best to never let such situations develop in the first place; cultural sensitivity is thus a less obvious tool to carry into the field as well.

Another favorite bugaboo to mention here briefly is the need to maintain meticulous record keeping and receipts for field purchases to keep the "bean-counter" accountants at bay. This may entail translating receipts from several arcane languages that the accountants will not understand. Where a soda pop on a hot summer day might be disallowed, for example, oral hydration therapy as medicine that is needed to avoid sunstroke can work miracles instead; it is all in the presentation it would seem.

Other issues mentioned in this volume that periodically affect fieldwork include such difficult problems as obesity or culture shock, the need to discuss such personal issues with students who may have no conception of arduous fieldwork realities, and the problems of diplomatically addressing such with those who can or will take offense. Nonetheless, by considering these many difficult situations, this volume brings attention to issues that must be addressed before naively entering the field; to do otherwise would be to risk unpleasant or even possibly dangerous failure. Field safety of all kinds is another issue that needs a good deal of the useful discussion that this volume provides. In addition, Dr. Thornbush and her coeditors have also given us a number of chapters that deal with fluvial and soils geomorphology, which – because they are such common environments – are quite pragmatic and provided me with a number of details that I did not know beforehand. After many decades of fieldwork all over the world, I am always most pleased to learn new things about my discipline.

REFERENCES

Bishop, M., J. Shroder, Jr., Ghazanfar Ali, A. Bush, Umesh Haritashya, Rakhshan Roohi, Mehmet Akif Sarikaya, and B. Weihs, 2014, Remote sensing of glaciers in Afghanistan and Pakistan, in: Kargel, J., G.J. Leonard, M. Bishop, A. Kääb, B. Raup (Eds.), Global Land Ice Measurements from Space (Springer-Praxis), Chapter 23, ISBN-10: 354079817X | ISBN-13: 978-3540798170.

Copland, L., T., Sylvestre, T., Bishop, M., Shroder, J., Seong, Y., Owen, L., Bush, A., Kamp, U., 2012. Expanded and recently increased glacier surging in the Karakoram. Arctic, Antarctic, and Alpine Research 43(4), 503–516. doi: http://dx.doi.org/10.1657/1938-4246-43.4.503.

Goudie, A.S., 1981, 1991. Geomorphological Techniques, first and second ed. Allen and Unwin, London and New York.

Shroder, J., Bishop, M., 2003. Preface to the special issue: a perspective on computer modeling and field work. Integration of computer modeling and field observations in geomorphology. 31rst Binghamton Geomorphology Symposium, Geomorphology 53, 1–9.

Switzer, A., Kennedy, D., Shroder, J. (Eds.), 2013. Methods in Geomorphology. Treatise on Geomorphology (14). Academic Press, San Diego, CA, (Editor in Chief).

John F. Shroder, Jr.
Editor-in-Chief
Developments in Earth Surface Processes
September 17, 2014

Preface

Despite technological advances in remote sensing technologies over the last couple of decades, fieldwork continues to be a vital component of geomorphological research. Thus, training and learning of geomorphological field methods as part of university undergraduate and graduate curricula are equally important and highly regarded by teachers and students at all levels. This Elsevier volume highlights the important aspects of not only doing fieldwork but getting ready for fieldwork and how to teach geomorphology and fieldwork *in the field*. The contributing authors of this volume originate from around the world from physical geography, geology, and Earth science backgrounds. They contribute toward the development of a field-based education in geomorphology from an international perspective. The volume contains a collection of articles that focus on the rationale for conducting fieldwork (Why Fieldwork?), with direction for doing fieldwork (Getting into the Field) and preparation tips (Preparing for Fieldwork), including general field-safety guidelines (Field Safety: Principles, Practice, and Culture). In addition, the volume importantly incorporates a unique section on pedagogy (Teaching Geomorphology in the Field) that considers learning styles (Student Learning Styles) as well as modern (digital) technology used in teaching geomorphology in the field (Fieldwork Going Digital), and presents student perceptions of conducting geomorphological research as part of undergraduate learning (Field-Based Learning in Undergraduate Geomorphology Courses). The final section in the volume contains a fresh look at field methodologies for soil geomorphology (Use of Field Experiments in Soil Erosion Research) and fluvial geomorphology (A Geologic Approach to Field Methods in Fluvial Geomorphology) while conveying pointers for landscape-scale geomorphology studies (Reading the Landscape in Field-Based Fluvial Geomorphology).

The volume editors are indebted to all those that have contributed directly and indirectly to the preparation of this volume, including the chapter authors, series editor, and the publisher. It is hoped that this volume will continue to inspire fieldwork at all levels and capacities in geomorphology, from the undergraduate learner and postgraduate trainee level to that of the experienced researcher and practitioner.

Chapter 1

Introduction

Mary J. Thornbush*,[1], Casey D. Allen**, Faith A. Fitzpatrick[†]

*School of Geography, Earth and Environmental Sciences University of Birmingham, Edgbaston, Birmingham, United Kingdom; **Department of Geography and Environmental Sciences, University of Colorado Denver, Colorado, USA; [†]U.S. Geological Survey, Wisconsin Water Science Center, Middleton, Wisconsin, USA

[1]Corresponding author: E-mail: m.thornbush@bham.ac.uk

As part of the Developments in Earth Surface Processes Series, Elsevier brings this edited volume on geomorphological fieldwork to the attention of practitioners and teachers of geomorphology, including physical geographers, geologists, and others who are conducting geomorphological fieldwork. The editors are representative of these disciplines within the geosciences, including two physical geographers (from Europe and North America) and a geologist working for the US Geological Survey. The purpose of this edited volume is to present a selection of contributions that address all aspects of conducting fieldwork, particularly how to maximize the benefits of learning geomorphological principles in the field.

This volume expands and complements the contents of a section on field techniques recently published in Elsevier's *Treatise on Geomorphology* (Shroder 2013), a 14-volume set designed for students, instructors, and professionals. The final volume (volume 14 on Methods in Geomorphology) of the set that contains Section 14.2 on the Fundamental Classic and Modern Field Techniques in Geomorphology, including Techniques and Methods for the Field, comprises field surveying; coring and augering; trenching and exposed faces; gravel and boulders; rock sampling; describing soils; and ground-penetrating radar. Although Section 14.2 is geared toward methodological considerations, this volume on geomorphological fieldwork provides a complete view of the pedagogic aspects of fieldwork including why fieldwork is conducted, fieldwork's learning benefits across a variety of levels, as well as how to integrate geomorphological field methods at different spatial and temporal scales. The emphasis on pedagogy in the current edited volume also leads to executing research in traditional geomorphological landscapes (such as karst and fluvial), employing typical media (such as soils), and landscape analysis and study.

Beginning with the rationale for conducting fieldwork (Why Fieldwork?), this volume represents a collection of chapters broken into three overarching sections: (1) Getting into the Field; (2) Teaching Geomorphology in the Field;

Developments in Earth Surface Processes, Vol. 18. http://dx.doi.org/10.1016/B978-0-444-63402-3.00001-7

and (3) Field Methodologies. Each section follows the student and practitioner into the field from the outset of their journey. The first section provides direction in preparation (Preparing for Fieldwork) and safety (Field Safety). Because fieldwork offers experiential learning opportunities for students, the second section incorporates the importance of fieldwork pedagogy, such as how students learn in the field traditionally (Students' Learning Styles) and in the digital age (Fieldwork Going Digital), as well as a chapter centered specifically on undergraduates (Field-based Learning in Undergraduate Geomorphology Courses). The third section focuses on integrated field methodologies and addresses the traditional and commonly examined landscape (Methods in Fluvial Geomorphology), revisits an established topical area (Fieldwork Protocols for Soil Geomorphology), and conveys general strategies for landscape-scale studies (Reading the Landscape in Field-Based Geomorphology). This layout prepares the reader in advance of fieldwork, guiding them through the landscape of fieldwork, causing them to know the purpose of executing fieldwork before they even step into the field and ensuring that they comprehend the approach and preparation so they are ready to make educated decisions as field practitioners.

The importance of fieldwork's pedagogical potential, especially in such a field-based discipline as geomorphology, however, should not be overlooked. Before exploring individual chapters in this volume then, it is worthwhile to outline some background of fieldwork's pedagogical benefits. Field-based learning involves more retention of material and improves initial learning (MacKenzie and White, 1982). The contributions in this volume illustrate why geomorphological fieldwork is more fitting for affective and holistic, rather than reductionist-style, learning (Dalton, 2001, p. 382). Affective learning more closely fits the complexity of spatial and temporal scales in geomorphic theories and paradigms. Davison et al. (2009) consider affective teaching and learning as "active, reciprocal, fully embodied and involved practice" (p. 311). The affective domain has been employed in the engagement of geoscience teaching (encompassing geology, physical geography, meteorology, and oceanography), for instance, van der Hoeven Kraft et al. (2011), who stimulated connections of nonscience majors with Earth (in addition to emotion and motivation) to influence student attitudes (see their adopted model in Fig. 1, p. 72) by evoking emotional attachments. Affective learning acts to engage students more personally so they relate more to the field experience and actually connect more with it (Allen and Lukinbeal, 2011), reinforcing the experience and making it more memorable and engaging.

Through engagement, the fieldworker is able to deeply process information and more easily encode it into lasting memories. British authors Dummer et al. (2008), for instance, have recognized the influence of reflection using fieldwork diaries to encourage deep learning. Critical reflection was spurred by keeping a field course notebook, where students commented on (recorded) and critically analyzed their learning experience during fieldwork relating to tourism (Marvell, 2008). These students became not only reflective, but also more empowered (responsible for learning) presenting at the edge of a volcano at Pompeii in the Mount Vesuvius National Park. Keeping reflective research diaries has similarly helped to engage students through active participation and encouraged them to critically reflect on geographical practice (McGuinness and Simm, 2005). In addition to placements and work experience as well as personal development planning, fieldwork promotes reflection learning, which can be assessed through fieldwork diaries and learning logs as well as group work, research projects, self-assessments, and coursework reviews (Harrison et al., 2003). These authors noted that "[r]eflective learning should empower the student and it encourages deeper investigation into subject matter – promotes self-learning and self-reliance, reinforces and consolidates learning, and promotes learning responsibility and self-improvement" (p. 143).

Approaching fieldwork is influenced by learning predisposition and, for example, experiential learning theory by Kolb (1984) can help physical geographers reach out to human geographers in terms of learning in the field. This theory described four preferred learning styles (divergers, assimilators, convergers, and accommodators) in a four-stage model of learning cycles (active, reflective, abstract, and concrete). Human geographers are most similar to social scientists and are predominantly accommodators, whereas physical geographers as natural scientists are predominantly assimilators. According to Healey and Jenkins (2000, pp. 91, 92), however, "[i]t is important that we establish whether geography students in higher education in the early twenty-first century have a predominant learning style and whether this varies between countries and the stage students are in their studies," Further advocating that "[w]e hope that the next time someone surveys the application of Kolb in geography, there will be a wide range of US and other non-UK examples of its use." Healey et al. (2005), tested over 900 students from 12 different universities in Australia, New Zealand, the United Kingdom, and the United States and discovered that, even though Kolb found that geography as an academic subject scored (similar to zoology and botany) in the Reflective/Abstract orientation (see their Fig. 3, p. 34), and others, such as Nulty and Barret (1996) listed it amongst Concrete-Active Cluster of Accommodation (see their Fig. 4, p. 35), their results indicate (at least for students in entry-level geography courses, in Years 1 and 2) that the majority (69%) are assimilators and convergers and the least accommodators (p. 39). Others, for instance Dunphy and Spellman (2009), have similarly tested learning styles, but in terms of whether it has any impact on the fieldwork experience. Their study of 1191 second- and third-year undergraduates in physical geography at 16 British universities (also tested on Kolb's LSI) showed a majority of respondents to be accommodators and assimilators (representing over a quarter of respondents, p. 24). Importantly, their study revealed that learning styles are not absolute and change throughout an undergraduate degree, as in upper-year courses. Their study focused on British physical geographers, noting that human geographers may respond differently, portraying a range of learner types in geography due to its broad-ranging curriculum. Some published studies comprised of a valid response rate of only 31% from first- and final- (third-) year students at three universities in Brisbane, Australia (Nulty and Barrett, 1996) did not directly address geography, but rather business, computer science, chemistry, and Japanese. Importantly, these authors noted that learning style may not be linked exclusively to discipline, but rather to the way in which a discipline (or subdiscipline) is taught. This is relevant to disciplines like geography, which often consist of human and physical components that are taught differently, making it difficult to conclusively categorize the learning styles of geographers. Studies using a diversity of approaches to learning evaluation have included student questionnaires (perception surveys) to assess quality; pre- versus post-knowledge assessments of student achievement; and standard university-wide (end-of-semester) evaluations to gauge knowledge gains and satisfaction (Rathburn and Weinberg, 2011). Their research conveyed a higher enjoyment of the course by males and that weather is an important consideration in student enjoyment. Other authors have critically addressed gender and sex differences in geographic education, finding a high mean test score for female geography majors (see Hardwick et al., 2000, Table 2, p. 241) over male geography majors, but a better performance of males (and masculine) and among nongeography majors (Hardwick et al., 2000). This pedagogical gender research extends to field-based program as well, with females outperforming their male counterparts, on the basis of concept map scores (Allen, 2011; see Novak and Gowin (1984) for a concept map overview), in a fieldwork setting that included analyzing rock decay (weathering) features. In his study of over 300 undergraduate students, 86% of whom had never taken a college-level geography courses and 46% who had never taken a "lab science" course, Allen (2011) also found that minority students outperformed their nonminority counterparts significantly (a concept map score increase of 23% vs. 11%, respectively). Results such as these speak volumes for including field-based pedagogy in geomorphology courses, especially when fieldwork can increase interest in female and minority students. Past results from student questionnaires speak volumes for including field-based pedagogy in geomorphology courses, especially for students who have not had much of a chance to be outside of an urban environment and need to connect with natural landscapes and experience the spirit of place (Hardwick et al., 2000; Allen, 2011).

It is also stimulating for students to learn about field-related research studies. Undergraduate courses that incorporate some such form of research can be (1) research led; (2) research oriented; (3) research based; or (4) research informed (Griffiths, 2004). At the University of the West England, for instance, Hill and Woodland (2002) found that a research proposal was an appropriate form of course assessment for second-year undergraduates because it granted them opportunities for independent work and enabled them to apply their knowledge to something of interest. Still, as these authors each note, more research is needed to investigate the effectiveness of fieldwork in improving undergraduate learning, as measured through performance. Jenkins (2000, p. 333) recommended that traditional lectures be replaced with methods that mirror the research process. Geographical expeditions, for example, have been considered as a type of experiential learning experience, requiring active participation that provides students with a sense of discovery, challenge, and autonomy (Pawson and Teather, 2002), even though there are considerations of expense and elitism in overseas expeditions (e.g., Nairn et al., 2000). However, others have more recently used unmarked coursework to assess the value of field-based learning in a 6-week filed course held in Thailand (Malam and Grundy-Warr, 2011). They used this approach to foster engagement in students and to encourage deep learning with student encounters of local places and people, enhancing student cultural exposure as well as allowing for critical inquiry. Group-based research projects were employed in the course in order to promote deep (experiential) learning, including, for instance, the use of a photo essay. By incorporating unassessed tasks in their course, these authors were able to reduce performance anxiety in their students and permitted them to take intellectual risks. This approach stimulated personal development, in addition to their learning of the course content and acquiring of skills, such as critical thinking. Jackson et al. (1997) found it difficult to use traditional assessment tools to measure student learning executed in team-based projects. Still, they observed students develop connections in the field with laboratory work, making it worth some excruciating conditions of some fieldwork performed under hot conditions (of temperature over 90°C).

Given the technology breakthroughs of computing power and visualization software, virtual fieldwork can provide some of the same learning benefits of outdoor fieldwork for those that are limited in mobility from physical or financial restrictions. Stainfield et al. (2000) presented the case for using virtual fieldtrips (VFTs) for students who are unable to partake of regular fieldwork, with the learning emphasis on visualization, involvement, and presentation (including C&IT skills, Gratton, 1999). Using desert geomorphology as the subject, Stumpf et al. (2008, p. 387) found that the virtual fieldtrip produced no distinguishable learning difference than "real field trips". Another study engaged students by maintaining a website of field progress, which enhanced their sense of empowerment over the learning experience and gave them a sense of ownership (Hasse and Colvard, 2006), in a way that research-led learning also stimulated. Project management in a capstone course has also elicited in students a sense of ownership (Gomezdelcampo, 2006). Students also gained leadership skills through job titles, such as project manager and coordinator, in a real-world application of knowledge (in a concrete project) with ramifications outside of the classroom, which challenges students and at the same time enables them to acquire skills required in the workforce and in their careers. By integrating a visual technique (namely, digital photography) into an upper-year undergraduate field course in Berlin, it was possible to engage students in urban environments as well as acquire materials for other (class- and web-based) course components (Latham and McCormack, 2007). Field projects have enabled students to employ lecture-based theory as well as research methods in a real-world setting of interviews (Herrick, 2010). Hudak (1999, p. 23) stated that "[fieldwork] is important for groundwater students because it illustrates the relevance of important concepts, enhances comprehension by enabling students to see and do rather than memorize, and breaks up monotonous classroom lectures." Visualization engages the learners and enables them to relate more closely to the work and its real-world application regardless of whether or not it is actually executed in the field. However, the traditional field tradition does dictate that students step into and encounter the outdoors so that they may broaden their experience not only of how the work is done, but also what it is like to be an active (team) member and be in on the action and part of it all where it actually takes place; so, again to develop a real sense of place and acquire a feel for how it operates.

Field courses are also held in the social sciences, including human geography. Arreola (2001) considered this to be the "muddy-boots" variant of fieldwork to the "scuffed-shoe" tradition. May (1999) offered a 10-day residential field class for the past 3 years in Los Angeles and Las Vegas in the areas of social and cultural geography. The course begins in the classroom and incorporates more independent student work that culminates in a project work in Los Angeles and an independent ethnographic exercise in Las Vegas. Empirical research from a human geography perspective illustrates the importance of field-based learning during festivals as well as through the experience of natural hazards, such as floods and fires (Measham, 2007). It has been advocated that "journal writing might add human geography dimensions to physical geography field-trips" (Nairn, 1999, p. 279). Geographers have tested for racism in fieldwork involving undergraduate human geography students, finding that they cannot be expected to "interrogate the ideological underpinnings of direct experience alone and through writing in journals for assessment" (Nairn, 2005, p. 306). Field-based "applied learning" should be encouraged early and often in the undergraduate student curriculum for human geography, as provided by opportunities to survey the field even in (local) urban settings, in applied urban geography research (Walcott, 1999). The University of Georgia has held an urban field study for a field assignment in urban geography in metropolitan Atlanta, Georgia, where students partake of a field study project that covers much ground and encompasses a variety of landscapes, in order to develop observational skills through first-hand knowledge (Martin, 2003). According to Whitlock (2001), human geographers are encouraged to participate in fieldwork, despite any apprehensions, as it is "one of the great rewards of the discipline [of geography], and it counterbalances long, sometimes tedious hours in the lab" (p. 25), at least for physical geographers. It is thought that traditionally applied courses that require practical training, such as in planning, are more likely to incorporate experiential learning in their curriculum that need justification in theory-based courses (Kotval, 2003).

All the chapters in this volume share a common thread: fieldwork has an experiential element that enhances student learning and also connects researchers to real-world phenomena. Fieldwork lends an application to theoretical learning that is both informative and enlightening. It enriches not only student experience and perception during their undergraduate years, but also as they continue into postgraduate study and become trained researchers. This is true for geology as it is for physical geography. Fieldwork has a long, and established, tradition in these disciplines that is continuing in modern times, engaging researchers with landscapes from the local to international scale, from simple day-trips to planned expeditions. This volume takes the reader through the fieldwork experience, from planning and preparing for fieldwork, to actually executing fieldwork through established and more modern approaches, to the priceless experience of being a member of a learning party in the field. The authors wish to share their experiences with you, the reader, and to provide you with what they have already gained (the benefits) from this approach to learning, from the time they went on their first fieldtrip to the more advanced stages of field-based research as academics and practitioners.

REFERENCES

Allen, C.D., 2011. Concept mapping validates fieldwork's capacity to deepen students' cognitive linkages of complex processes. Res. Geogr. Educ. 13, 30–51.

Allen, C.D., Lukinbeal, C., 2011. Practicing physical geography: an actor-network view of physical geography exemplified by the Rock Art Stability Index. Progress Phys. Geogr. 35, 227–248.

Arreola, D.D., 2001. Teaching the borderland. Geogr. Rev. 91, 480–486.

Davidson, J., Huff, L., Bridgen, J., Carolan, A., Chang, A., Ennis, K., Loynes, K., Miller, J., 2009. Doing gender' at Body Worlds: embodying field trips as affective educational experience. J. Geogr. Higher Educ. 33, 303–314.

Dalton, R.T., 2001. What do they bring with them? The fieldwork experiences of undergraduates on entry into higher education. J. Geogr. Higher Educ. 25, 379–393.

Dummer, T.J.B., Cook, I.G., Parker, S.L., Barrett, G.A., Hull, A.P., 2008. Promoting and assessing 'deep learning' in geography fieldwork: an evaluation of reflective field diaries. J. Geogr. Higher Educ. 32, 459–479.

Dunphy, A., Spellman, G., 2009. Geography fieldwork, fieldwork value and learning styles. Int. Res. Geogr. Environ. Educ. 18, 19–28.

Gomezdelcampo, E., 2006. Using student-managed projects to enhance the capstone experience. J. Geosci. Educ. 54, 572–577.

Gratton, J., 1999. Who would have thought it? C&IT skills development via a geography field-course. GeoCal 20, 5–9.

Griffiths, R., 2004. Knowledge production and the research-teaching nexus: the case of the built environment disciplines. Stud. High. Educ. 29, 709–726.

Hardwick, S.W., Bean, L.L., Alexander, K.A., Shelley, F.M., 2000. Gender vs. sex differences: factors affecting performance in geographic education. J. Geogr. 99, 238–244.

Harrison, M., Short, C., Roberts, C., 2003. Reflecting on reflective learning: the case of geography, earth and environmental sciences. J. Geogr. Higher Educ. 27, 133–152.

Hasse, J., Colvard, C., 2006. Inverse distance learning: digitally enhancing a geography field-course. J. Geogr. 105, 167–174.
Healey, M., Jenkins, A., 2000. Kolb's experiential learning theory and its application in geography in higher education. J. Geogr. 99, 185–195.
Healey, M., Kneale, P., Bradbeer, J., 2005. Learning styles among geography undergraduates: an international comparison. Area 37.1, 30–42.
Herrick, C., 2010. Lost in the field: ensuring student learning in the 'threatened' geography fieldtrip. Area 42.1, 108–116.
Hill, J., Woodland, W., 2002. An evaluation of foreign fieldwork in promoting deep learning: a preliminary investigation. AEHE 27, 539–555.
Hudak, P.F., 1999. Groundwater field station for geosciences students. J. Geogr. 98, 23–28.
Jackson, N.L., Cerrato, M.L., Elliott, N., 1997. Geography and fieldwork at the secondary school level: an investigation of anthropogenic litter on an estuarine shoreline. J. Geogr. 96, 301–306.
Jenkins, A., 2000. The relationship between teaching and research: where does geography stand and deliver? J. Geogr. Higher Educ. 24, 325–351.
Kolb, D.A., 1984. Experiential Learning: Experience as the Source of Learning and Development. Prentice-Hall, Englewood Cliffs, New Jersey.
Kotval, Z., 2003. Teaching experiential learning in the urban planning curriculum. J. Geogr. Higher Educ. 27, 297–308.
Latham, A., McCormack, D.P., 2007. Digital photography and web-based assignments in an urban field course: snapshots from Berlin. J. Geogr. Higher Educ. 31, 241–256.
MacKenzie, A.A., White, R.T., 1982. Fieldwork in geography and long-term memory structures. Am. Educ. Res. J. 19, 623–632.
Malam, L., Grundy-Warr, C., 2011. Liberating learning: thinking beyond 'the grade' in field-based approaches to teaching. NZG 67, 213–221.
Martin, D.G., 2003. Observing metropolitan Atlanta, Georgia: using an urban field study to enhance student experiences and instructor knowledge in urban geography. J. Geogr. 102, 35–41.
Marvell, A., 2008. Student-led presentations *in situ*: the challenges to presenting on the edge of a volcano. J. Geogr. Higher Educ. 32, 321–335.
May, J., 1999. Developing fieldwork in social and cultural geography: illustrations from a residential field class in Los Angeles and Las Vegas. J. Geogr. Higher Educ. 23, 207–225.
McGuinness, M., Simm, D., 2005. Going global? Long-haul fieldwork in undergraduate geography. J. Geogr. Higher Educ. 29, 241–253.
Measham, T.G., 2007. Primal landscapes: insights for education from empirical research on ways of learning about environments. Int. Res. Geogr. Environ. Educ. 16, 339–350.
Nairn, K., 1999. Embodied fieldwork. J. Geogr. 98, 272–282.
Nairn, K., 2005. The problems of utilizing 'direct experience' in geography education. J. Geogr. Higher Educ. 29, 293–309.
Nairn, K., Higgitt, D., Vanneste, D., 2000. International perspectives on fieldcourses. J. Geogr. Higher Educ. 24, 246–254.
Novak, J.D., Gowin, D.B., 1984. Learning How to Learn. Cambridge University Press, New York.
Nulty, D.D., Barrett, M.A., 1996. Transitions in students' learning styles. Stud. High. Educ. 21, 333–345.
Pawson, E., Teather, E.K., 2002. 'Geographical Expeditions': assessing the benefits of a student-driven fieldwork method. J. Geogr. Higher Educ. 26, 275–289.
Rathburn, S.L., Weinberg, A.E., 2011. Undergraduate student satisfaction and achievement at the GetWET Observatory: a fluid learning experience at Colorado State University. J. Geosci. Educ. 59, 47–55.

Shroder, J.F., 2013. Treatise on Geomorphology. Elsevier, Oxford, UK.
Stainfield, J., Fisher, P., Ford, B., Solem, M., 2000. International virtual field trips: a new direction? J. Geogr. Higher Educ. 24, 255–262.
Stumpf, II, R.J., Douglass, J., Dorn, R.I., 2008. Learning desert geomorphology virtually versus in the field. J. Geogr. Higher Educ. 32, 387–399.
van der Hoeven Kraft, K.J., Srogi, L., Husman, J., Semken, S., Fuhrman, M., 2011. Engaging students to learn through the affective domain: a new framework for teaching in the geosciences. J. Geosci. Educ. 59, 71–84.
Walcott, S.M., 1999. Fieldwork in an urban setting: structuring a human geography learning exercise. J. Geogr. 98, 221–228.
Whitlock, C., 2001. Doing fieldwork in the mud. Geogr. Rev. 91, 19–25.

Chapter 2

Why Fieldwork?

Casey D. Allen[1]
Department of Geography and Environmental Sciences, University of Colorado Denver, Colorado, USA
[1]Corresponding author: E-mail: casey.allen@ucdenver.edu

2.1 INTRODUCTION

Observation and imagination are two binding forces behind fieldwork and geomorphology. From its beginnings as a discipline, geomorphology has seemingly always been intertwined with the exploratory sciences and fieldwork (Nielsen, 2012). Though some may suggest the relationship superficial, it is in conducting geomorphological fieldwork that observation and theory become "symbiotic" (Rhoads and Thorn, 1996, p. 51). Tying in imagination to the symbiosis, Inkpen and Wilson (2013, p. 133) noted, "...the role of imagination in deciphering reality...is one that should be acknowledged and studied." During the formative years of geomorphology, nearly all discoveries, every encounter with a landscape, was "unfamiliar" as Inkpen and Wilson (2013, p. 133) remind the scientist, and interpreting those landscapes into pertinent information required strong creativity and imagination. Baker and Twidale (1991) similarly noted that when the discipline was fledgling, new geomorphologists used their creativity to interpret the sometimes bizarre landscapes they encountered. Yet when *in* the field, they remind geomorphologists, observation remains a powerful tool, even if the ability is now a characteristic often lacking in geomorphology. But that ability alone, no matter how keen and developed, requires interpretation, which in turn often requires dedicated imagination to generate a useful explanation, and *that* may need to change depending on a field person's contexts (Inkpen and Wilson, 2013).

In each field encounter, the geomorphologist becomes a scientific witness, akin to their "gentlemen" counterparts in seventeenth century England (Shapin, 2010, p. 73). This important science component a few hundred years ago – witness of an experiment by a credible and reliable source – still remains embedded in science today, though such "gentlemen" are now called referees, editors, and (anonymous) peer-reviewers, irrespective of whether they have been in the field or not. Likewise, as Hall (2006) (see also Section 3 in Dorn et al. (2013)) has observed so astutely, those incorporating fieldwork should also strive for being objective, and resist hyper-focusing, lest they remain in

Developments in Earth Surface Processes, Vol. 18. http://dx.doi.org/10.1016/B978-0-444-63402-3.00002-9

the same paradigm when all evidence might point to the contrary. Further, as modeling gains more presence within the discipline, fieldwork often remains relegated as a side note, or as Twidale (1996, p. 373) suggested, "…an excuse for a holiday…" Those conducting research "in the field" however, and whose research contains strong field components, know that field-based data gathering, analysis, explanation, and interpretation often go well beyond modeling. Ground truthing can also help in rectifying mismatched spatial and temporal datasets, since some components of the natural world cannot be modeled. Fieldwork then remains necessary to inform those geomorphologic models (Mol and Viles, 2012; Inkpen et al., 2012; Bowker et al., 2008; Jennings and Huber, 2003; Wilson and Burrough, 1999; Convey, 1994; Baker, 1987; Box, 1981; Johnson et al., 2005), forcing researchers to think outside personal models and biases when gathering new data.

Certainly fieldwork in geomorphology has ancillary benefits such as enhancing teaching practices – what budding geomorphologist does *not* like to explore outside? (Dorn et al., 2013; Allen, 2011a; Kent et al., 1997). However, it also has the power to strengthen subfield and interdisciplinary relationships (Allen, 2011b; Scott et al., 2006; Warburton and Higgitt, 1997) by verifying and enhancing form-process linkages and connections (Church, 2013; Allen and Lukinbeal, 2011; Viles et al., 2008). More important for the discipline of geomorphology, however, fieldwork can aid in challenging long-held theories based on casual and/or nonfield observations, or those entrenched ideas merely taken for granted (Hall et al., 2012; Hall and Thorn, 2011; Bracken and Wainwright, 2006; Schaffer, 1997; Bretz, 1923; Wegener, 1920), and help overcome these perceived weaknesses (lack of objectivity or a hyperfocus, for example).

Using several examples of geomorphologic field-based inquiry related to the aforementioned topics, this chapter elaborates on why fieldwork remains an important facet of the discipline. But this chapter also serves as a plea of sorts to geomorphologists, regardless of subfield specialty: continue including fieldwork in research and teaching (for those that already do), or begin incorporating fieldwork into research programs and add it to pedagogy (for those who do not).

2.2 WITHER FIELDWORK? THE NECESSITY OF "GROUND TRUTHING" AND THE IMPORTANCE OF PLACE

Although increasingly finer resolution technology and access to remotely-sensed data becomes easier, the perceived need for fieldwork (in terms of ground truthing) decreases. Why endure potentially arduous environments, long stays away from home, and get muddy, experience frostbite or hypothermia, or encounter dehydration and sunburn, when gathering data remotely is so easily achieved? Technologies that allow for such landform-process interactions as 3-D renderings of mass-wasting events, real-time mapping of earthquakes, LIDAR, and time-lapse photography of coastal retreat are exciting. Even advanced laboratory techniques such as scanning electron microscopy, backscatter microscopes, and varnish microlaminations allow us to see the landscape in nanometer size, peering into a whole new universe of understanding, sometimes helping us understand past environments and perhaps prepare for the future. Yet despite these advancements in landform processes and cross-scalar underpinnings, popular sentiment regarding fieldwork, especially by those who *do* it, centers on anecdotal evidence that while fieldwork is critical to geomorphology, it commonly gets rejected or neglected in favor of technological and laboratory-heavy disciplinary expansions. What these often magical-seeming hi-tech marvels overlook, however, is that each of them began with some type of fieldwork and, to be further validated as a model and gain stronger credibility, verification *on the ground* becomes necessary. And that is one component of geomorphological fieldwork's great importance: ground truthing.

Without on-the-ground verification, mistakes in data may occur. For example, mismatched imagery may occur in a final product (Fig. 2.1), incorrect distances might be measured due to imperceptible-from-the-image topography, thalweg depth is difficult to measure, paleoflood imbrications may be overlooked, damage from mass-wasting events might be miscalculated or dismissed, and

FIGURE 2.1 An image from Google Maps displaying the northern rain forest in Grenada, West Indies. Misaligned imagery, denoted by rectangles, suggests some kind of rigid topographic barrier. This is not the case, as the topography is uniformly mountainous in the region. Similar satellite and aerial misalignments occur frequently, but more so with commercially available (and more affordable) imagery. Misrepresentations like these, however, demonstrate the need for ground truthing (and fieldwork), especially when using these technologies to identify potential study sites or, more importantly, analyze a region.

the list continues. In each instance, potential errors could be thwarted and/or corrected by ground truthing. The question that when the ground truthing occurs remains irrelevant, as long as it happens – sometimes ground truthing after the fact, perhaps for model/data verification might be the best course of action for a research plan; sometimes it is planned in advance. In any case, the importance of ground truthing cannot be understated. It *needs* to transpire.

Instigators of modern geomorphology like W.M. Davis and G.K. Gilbert understood this concept precisely, and certainly considered ground truthing as representing the discipline's foundation with "the field," *their* place, often playing the role of laboratory. Not in the traditional, controlled, and sterile sense of course, but as a type of "open system" where it is possible to "bodily enter the field of enquiry" (Baker, 2004, p. 136–137), regardless of how chaotic it may be. Irrespective of rigidity or lack thereof, these sentiments ring true for many of today's geomorphologists. But beyond that, and similar to what Richards (2011) has suggested, because geomorphology's practice – whether modeled or not – occurs in a particular place, the field then represents where the geomorphologist acts and performs to create some appearance of control as in the conventional laboratory. This is why the distinctiveness of place selection remains vitally important for geomorphologists. Whether a site is chosen for aesthetics, convenience, historical, epistemological, or any other reason, it should offer up some larger truth(s), even if the actors or location changes as new understandings and developments occur while solving a problem (Richards, 2011).

Regardless of action or location, people remain part of the landscape, entwined with it, performing with/in it (Allen, 2011b). Whether their enthusiasm stems from work or play, when people engage with landscape, they invariably become a part of it (Ingold, 1993). Goudie (2002, p. 245) has reminded geomorphologists that one of the discipline's most alluring traits are the "…visually appealing... great landscapes," and that early explorer geomorphologists were often enchanted with the landscapes they saw to the point of producing poetic prose in their field journals. What time duration geomorphologist takes to produce poetry while in the field nowadays? Goudie (2002, p. 246) continued, citing the desires of "ordinary people" who endeavor to explore awe-inspiring landscapes by highlighting the millions of annual visitors to the vast National Parks of the US Southwest. The lure of such radical landscapes is what drew then-geomorphologist Yi-Fu Tuan to the grand pediments of Arizona where he completed his dissertation. The geomorphological lens remains apparent in Tuan's later humanistic endeavors, demonstrating the broad reach geomorphology can have because of a personal connection to a landscape, further illustrating the importance of place in the discipline of geomorphology (Tuan, 2004, 2001a, 2001b, 1979, 1974, 1968). Regardless of what positivistic science may suggest, people and landscape *are* connected. Separating them often leads to biased outlooks of the landscape under study. They are consistently drawn to each other, and particularly, those "great landscapes" as Goudie has noted, even if for solely aesthetic or personal reasons. Yet that may be enough to spark an interest in geomorphology (see Dorn et al.

(2013), specifically Section 2.1). Geomorphology represents an extraordinary medium for unraveling the mysteries of a landscape lain down by millennia of processes. Perhaps, given its importance on understanding the science of scenery, geomorphology *is* the landscape: that place to which people become endeared, a place they can enjoy, explore, discover, and ultimately become linked to in space.

In the end, those entrenched in and practicing fieldwork, perhaps using it as a primary data source, may feel ignored. In truth though, geomorphological fieldwork remains as important today as it has been in the past, and it will continue to be just as important in the future (Church, 2013). The key task to keep fieldwork prevalent in geomorphology rests in getting people excited about fieldwork and then sustaining that excitement. As (lack of) time and the strains on pocketbooks increase, keeping fieldwork at the forefront of the discipline through carefully selected place-based inquiry (for whatever reason) could be one way to keep younger geomorphologists involved in fieldwork while simultaneously highlighting fieldwork as the vanguard of the discipline (and keeping it at the forefront of funding mechanisms' and entities' minds). Students enjoy learning outside the four walls of formal education, and involving them in ground truthing is one reliable way to "hook" them into geomorphology (Dorn et al., 2013; Day, 2012; Allen, 2011a). Most budding geomorphologists relish in the chance to conduct fieldwork, regardless of how straightforward (or tedious) the tasks may be, and irrespective of whether they consider themselves a geomorphologist (Dorn et al., 2013). They may do so, in part, because of an innate attraction to the landscape, their connection in space to that place, brought on by imagination and observation of the connections generated and/or practiced with/in the landscape.

2.3 TEACHING AND LEARNING THROUGH FIELDWORK

From a learning standpoint, fieldwork occurs when, according to Lonergan and Andresen (1988, p. 64), "...supervised learning can take place via first-hand experience, outside…the classroom setting." But this is perhaps easier stated than done. Indeed, various scholars often relegate fieldwork to the "look–see" tour/field-trip or suggest that the same learning can occur just as well in structured classroom settings (Gold et al., 1991). These types of studies however – mostly compiled at the turn of the twentieth century and mostly from the geography–geomorphology side of the discipline – seem to forget the discipline's very foundation. Though some may be quick to push fieldwork in geomorphology aside, all of them learned and gained at least some their knowledge from direct fieldwork and first-hand experiences, even if they are slow to admit it. So, how exactly does fieldwork in geomorphology influence teaching and learning? Why would that influence be important?

First, Gold (1991, p. 3.3) noted six overarching facets related to fieldwork's value: "Developing observation skills...Facilitating experiential learning… Encouraging students to take responsibility for their own learning...Developing analytical skills...Experiencing'real' research...Developing a respect for the environment..." Similarly, Kent et al. (1997) list "observational fieldwork," "participatory fieldwork," and "learner–practitioner and participant observation" as fieldwork-activity categories. Based on these two meta-analyses, observational awareness and first-hand experience emerge as the overarching foundation for geomorphological fieldwork when it comes to pedagogical value, and represent two aspects not as readily achievable in a classroom setting, no matter what type of technology or modeling is available. It is, in fact, these two seemingly all-encompassing categories that tend to drive an instructor's desire to incorporate fieldwork into their pedagogy.

Second, students participating in fieldwork, or conducting it under guidance, learn valuable, transferable skills. These can range from the practical (e.g., Brunton compass, field mapping, data-logger programming, field journaling, etc.) to the theoretical where a student may use multiple working hypotheses to delve deeper into the form–process connections. Recently, growing evidence supports the notion that fieldwork in geomorphology (and its specialties) deepens students' capacities to understand complex concepts through networking in the landscape (Allen and Lukinbeal, 2011). From such fundamental concepts as rock decay (Dorn et al., 2013; Allen, 2011a) and soil science (Mellor, 1991), to more advanced concepts in coastal geomorphology (Ellis and Rindfleisch, 2006), glacial geomorphology (Lewis, 2007), and even aesthetics (Dixon et al., 2013), students consistently learn concepts better when conducting fieldwork than when sitting in a classroom and/or laboratory setting. Some even incorporate technology from the classroom into field settings (Roering et al., 2013; Thorndycraft et al., 2009; Fletcher et al., 2007). In each instance, students also learn about themselves, their likes and dislikes, and often discover hidden attributes (good and bad) – all significant aspects to developing a healthy professional behavior.

This overlooked benefit of geomorphology fieldwork rests in individual growth and development, and promoting fledgling enthusiasm is not a necessarily a bad thing. Indeed, once they become comfortable with their (sense of) place, many students embrace being a "nerd" (Dorn et al., 2013, p. 60). Allowing for students to experience life outside their comfort zones – in the field – helps expand their worldview, increasing their ability to gain a big picture perspective. The empowerment that can follow after a student completes a field-based project is noticed almost immediately. The overall experience may not just encourage students to be part of geomorphology as a major field of study (Dorn et al., 2013), but can also affect them on a deeper level, helping them understand, and sometimes helping create, a sense of place (Dixon et al., 2013; Allen, 2012; Kudryavtsev et al., 2011; Allen, 2011b; Semken and Brandt, 2010; Semken et al., 2009; Semken and Freeman, 2008; Semken, 2005; Massey, 1997). Further, fieldwork brings out both positive and negative attributes of those who engage in it. A student (or colleague) who performs well in the classroom may well flounder when tasked with straight-forward hands-on tasks involved with fieldwork. Their personality, demeanor, and health may all be affected adversely once outside the four walls of traditional education. Conversely, an under-performing student in the classroom may thrive in a field setting, discovering their passion, growing and developing into a full-fledged geomorphologist, and someone who will, hopefully, pass on their love of the field to others and contribute strongly to the discipline.

Third, the majority of students like fieldwork. Each instructor who drags students into the field invariably has their own anecdotal evidence to support this statement. Regardless of how many "field trips" or "outdoor laboratories" or "self-selected field problems" are required for a geomorphology course (including specialty courses), students consistently want more. Yet engaging students is fieldwork is not easy. It takes time, dedication, and funding to engage students in meaningful fieldwork – all three of which a department and/or instructor may be lacking – and careful precautions must be taken to ensure safety regulations are met, learning goals kept at the forefront of the experience, and assessments remain entwined throughout the fieldwork. To aid the instructor, many strategies have been outlined in recent years (Evenson, 2013; Day, 2012; Phillips, 2012; Dunphy and Spellman, 2009; Lewis, 2007; Ellis and Rindfleisch, 2006; Hudak, 2003; Fuller et al., 2000; Warburton and Higgitt, 1997), but each has at its core what Kent et al. (1997, p. 325) suggest is perhaps the most important components for fieldwork becoming pedagogy: to "...engage and motivate students." Each study also notes the need for flexibility and adaptability in both the fieldwork and pedagogy.

Finally, engaging students and piquing their curiosity of geomorphology and its facets via fieldwork could increase the discipline's visibility across the spectrum, especially if students understand that what they are doing in the field *is* geomorphological. Of course, the quintessential field trip remains the traditional way to incorporate fieldwork in geomorphology classes and, to be sure, these should not be relegated to the storage closet. But two persistent items may thwart even the local field trip agenda: money and urbanization.

Monetary woes will always be present, especially during recessions and when economies are globally connected. As funding agencies at all levels feel the repercussions of a downward-spiraling economy, less and less becomes available for fieldwork, whether part of a course, or otherwise. When funds dry up, field programs always seem like the first line item to be deleted, so instructors need to be creative. Local field trips, perhaps on the campus, can be just as informing as the more traditional weekender field trip. Entwining fieldwork by connecting separate courses, for example, sharing the field between an introductory landform-process course and an advanced fluvial geomorphology course not only saves money but also increases the learning potential as beginning students become engaged with advanced students. As top-notch instructors know, one of the best ways to learn a subject rests in teaching it, and geomorphological fieldwork, led by an astute instructor, allows for that pedagogy, regardless of the fieldwork's setting.

Before urbanization became a worldwide phenomenon, instructors had little problem getting their students in the field. Rural settings were abundant and near to campuses. Some fortunate institutions still have wonderful rural settings or nearby lands set aside for such experiences that make traditional field excursions more viable. Although urban settings might seem less than ideal, instructors in urban areas and on urban campuses still have plenty of opportunities to explore locally in the field students, because most all landform-process connections are scale *in*dependent (Allen and Lukinbeal, 2011; Phillips, 2001, 1999). This instance represents an opportunity for instructors to be creative when it comes to geomorphological fieldwork. Of course the grand alluvial fans of Death Valley, CA (USA) represent textbook examples of the quintessential form and process. But alluvial fan formation can be seen and studied in real time along a campus gutter after a rainstorm, for example, as students gather around to see particle size distribution's relationship to amount of flow. Pedagogical techniques like these have been used in highly urbanized campuses to great effect (Dorn et al., 2013; Frink, 2009), and sometimes make the important crossover and linkages (Massey, 1999) between physical geography/geomorphology and the cultural and humanistic realms (Allen et al., 2013; Semken et al., 2009; Semken and Freeman, 2008; Daniels, 1992). Imagine the rich possibilities if students learn the power of building connections between seemingly disparate disciplines, as Dixon et al. (2013) have noted. Evidence suggests that students generate and see these networks in the landscape, and that helps them understand complex processes better (Allen and Lukinbeal, 2011; Allen, 2011a). Indeed, embracing geomorphology's ability to integrate into other disciplines – especially through fieldwork – keeps practitioners at the forefront of the discipline.

Although observation and imagination may be the basis for geomorphology fieldwork, it goes hand-in-hand with practicing (*doing*) geomorphology in the networked landscape which, in turn, keeps students *and* instructors connected to and excited about exploring and understanding the landscape's forms and processes. In short, fieldwork (in a geomorphological setting or otherwise) deepens student understanding of form–process connections and helps students grow and develop – find their place – while at the same time, when assessed properly, aiding in stronger pedagogical practices for the instructor.

2.4 FIELDWORK ENHANCES SUBFIELD AND INTERDISCIPLINARY RELATIONSHIPS

While geomorphology becomes more and more reductionist (Hall et al., 2012), there exists at the same time, somewhat paradoxically, a call for more inter-/multi-/transdisciplinary thinking. The idea behind this rests in synthesizing concepts and information from different disciplines (even subdisciplines) to open realms of possibilities never before conceived. This remains an arduous task, but a task that can be made more feasible through fieldwork (Simon et al., 2013). When it comes to geomorphology specifically, connecting with other disciplines may seem disparaging at first – especially given the multiple resistances found throughout the establishment (see Section 2.6). Yet when enhanced by fieldwork, geomorphology and geomorphological concepts lend themselves very well to other disciplines by:

- Promoting interconnectivity in ecology and biology (Gimenez et al., 2013; Scott et al., 2012; Viles et al., 2008; Pérez, 1997; Viles, 1995);
- Connecting climate to geomorphology (Fawcett et al., 2011; Guitierrez, 2005; Powlson, 2005; Kidron and Yair, 1995; Sharon and Margalit, 1990; Pielke, 1990);
- Increasing dynamic understanding of hydrology and (paleo)flooding (Baker et al., 2002; Gillieson et al., 1994; Jarrett, 1991; Baker, 1987; Branson et al., 1981);
- Generally identifying, understanding, and enhancing networks throughout the physical landscape (Thornbush, 2013b, 2010, 2008; Allen, 2011b; Allen and Lukinbeal, 2011; Opdam et al., 2006; Campbell, 2005); and
- Linking geomorphology to culture and place (see Thornbush (2013a), special issue on archaeogeomorphology).

Further, these broad endeavors do not include the many contributions geomorphological fieldwork makes in biome-specific research (arid, alpine, tropical, tundra, etc.), regional studies (Australia, China, Europe, etc.), or a myriad of other topical pursuits (aesthetics, biogeomorphology, fire, soils, etc.) Evidence also points to geomorphologic fieldwork verifying those connections (Church, 2013; Allen, 2011a).

Fieldwork in geomorphology represents a very social undertaking, and rarely is it conducted as a solo endeavor. The professor drags students into the field. Students may, in turn, persuade their counterparts to tag along with them in the field. A research team accompanies each other into the field. Regardless of *how* they get into the field, once there, debates occur, common understandings promulgate, and new research questions and hypotheses get discussed, dissected, and tested or discarded. Alternative techniques and methods are tested and learned. Ultimately, this slew of happenings results in new knowledge being gained. Indeed, the social aspect of fieldwork should not be overlooked. As Richards (2011) has noted, not only does the field offer a place for intellectual acculturation, but increased socialization. It also helps preserve important research values and train colleagues and future disciplinary visionaries – often times without realizing it (Richards, 2011).

Expeditions abound in Earth science disciplines, and geomorphology is no exception. By definition, however, an expedition suggests a one-time short trip, for a particular purpose (Nielsen, 2012). In contrast, fieldwork suggests a returning, whether for replication, validation, or more data gathering. Even Latour (1999, p. 74, original italics) noted that his botanist, geomorphologist, and pedologist colleagues were, while packing up their Amazonian camp, "... also preparing to return," because they had, through mutual realization and interacting in their fieldplace, discovered that they needed an entomologist to help answer their deeper queries. But if fieldwork generates, supports, and enhances the inter-/multi-/transdisciplinary paradigm, then it can never be *only* geomorphological. This, on one hand, Richards (2011) suggests, may undermine the geomorphologist's project(s). On the other hand, he quickly notes, taking away of the strictly geomorphologic project is fine. A shift of topic, technique, method, hypothesis, or even research question and design should not be seen as detrimental, especially where fieldwork is involved. On the contrary, fieldwork is fraught with changes because, once on the ground, the best-laid plans can become (more often than admitted) ineffective. When it comes to fieldwork in geomorphology, then change of focus should be embraced. Flexibility, adaptability, and patience remain the keys in geomorphological fieldwork for, as a construct, fieldwork embodies an ever-changing social character, actively promoting disciplinary exchanges. This trait should be lauded.

2.5 FIELDWORK CHALLENGES ESTABLISHED PARADIGMS

Many examples of new discoveries challenging the established *modus operandi* of the day exist throughout the sciences, and geomorphology is no exception. From the science of geology becoming more accepting of catastrophism to the foundation of rock-decay science being shaken, instances are numerous. This section outlines briefly a few cases where fieldwork specifically, often backed up by models and technology after the fact, takes the lead role in changing a dominant paradigm. Taking examples from global- to hand-sample scales, this section's point rests in the reader gaining awareness that fieldwork not only informs practice (why there is geomorphology, what geomorphology is, and how geomorphology is practiced – the discipline's epistemology, ontology, and methodology, respectively), but can and does shape fact, especially when it comes to landform processes.

2.5.1 Global-Scale: Continental Drift

Perhaps the quintessential example of fieldwork challenging the establishment is theory proposed by Alfred Wegener (1920) – theory of continental drift. His visionary work initially conceived at the beginning of the twentieth century took ridicule for decades, but he never stopped defending his theory nor did he stop gathering evidence to support it. Coming from a meteorology background, and thus lacking geology credentials, pre-eminent geologists of the day had difficulty accepting his seemingly outlandish claims that the continents were once a single landmass. Each time a continental drift detractor questioned his hypothesis, he worked even more diligently to strengthen his position through additional field-based evidence. Wegener spent years in the field, gathering paleontological and geological evidence to support his claims. In his book's fourth edition, he included further supportive evidence by claiming that, geologically speaking, the shallower oceans were younger. Although he died young (age 50) and never saw his hypothesis verified or accepted by the establishment, his fieldwork-based observations and data gathering – his ground truthing – laid the foundation for modern day plate tectonics which, consequently, remains a driving force behind geomorphology's general landform processes of diastrophism (tectonic) and degradation (erosional) processes. Had he lived a few decades longer, perhaps his ability to make giant intellectual leaps by incorporating many facets of Earth science *from field-based observations*, would have allowed him to lead the way in modern-day development of plate tectonics modeling.

2.5.2 Regional-Scale: The Channeled Scablands

After exploring many different landform processes in the field (Chamberlin, 1890) using *Method of Multiple Working Hypotheses*, J. Harlen Bretz came to a stunning conclusion for the time. The resultant article, *The Channeled Scabland of the Columbia Plateau* (Bretz, 1923), received less than enthusiastic responses from the geology establishment of the day. Although he had a few supporters, his catastrophism hypothesis – that the Channeled Scablands were created by a great flood – was ridiculed for more than 40 years by the establishment, in favor of the slower processes brought on by uniformitarianism, such as formed the Grand Canyon (though there is still current debate, for example (Douglass et al., 2009; Meek and Douglass, 2001, 2003). This paradigm, at least in the Channeled Scablands, shifted when a field trip to the Scablands by the International Union for Quaternary Research was suggested as part of their 1965 meeting. Invited by the field trip leader – who *was* one of Bretz's supporters – to accompany them, he declined, being then in his 80s, and not wanting a repeat of the 1927 Geological Society meeting where he was ambushed by detractors such as Richard Foster Flint (Soennichsen, 2008). On the field trip's last day however, Bretz received a telegram from the field trip leader announcing that, "We are all catastrophists!" (Soennichsen, 2008, p. 231). A little less than a decade later, NASA's Landsats 1 and 2 were launched, and satellite imagery of the Scablands provided almost identical outlines of Bretz's field maps, made more than 50 years previously. Remnants of a paleolake (Lake Missoula) have since been found, further corroborating Bretz's initial catastrophism hypothesis. In 1971, *Mariner 9* took imagery of Mars, revealing similar features on the Red Planet. Although vindicated by the establishment in 1979 with the Geological Society of America's highest honor, the Penrose Medal, Bretz is said to have lamented that all his enemies were dead, so he had no one over which to gloat (Soennichsen, 2008). A somewhat-agreeable end to a sad period in the history of geomorphology as a discipline that may have been lessened if the establishment would have taken Bretz's claims more scientifically (objectively?) and conduct a little geomorphological fieldwork for verification.

2.5.3 Local-Scale: Yosemite Valley, CA (USA)

Glaciers and large-area erosional processes acting along joints and recent geological uplifts over the past few million years carved out Yosemite. With basic geomorphologic knowledge and looking out over Yosemite's giant glacial troughs and large *roche moutonées*, that statement seems valid. A proponent of this hypothesis, Muir (1880) observations stood in stark contrast to the more gruff Matthes (1930), and in the end, the more gentle-seeming Muir had more supporters, and his ideas were propelled forward. Even today, with modern technology and modeling techniques, this still remains the scientific consensus (Guyton, 1998; Huber, 1987; Wahrhaftig, 1962). After many years of tedious fieldwork, Schaffer (1997), on the other hand, suggested that Yosemite Valley 30 million years ago would look very similar to what we see today, with glacier-sculpting having only minor influences. Schaffer's work clearly bucks traditional views enough to produce vehement detractors, and perhaps being so controversial that, as a result of his book being published, the University of California, Berkeley's geography department did *not* grant Schaffer his doctorate (Dorn, 2005). As a reviewer of Schaffer's work notes, eschewing this new hypothesis outright because it fails to fit into the current paradigm may be premature as, "Researchers a century from now will find it simple to isolate the poor scholars who simply dismissed the whole book from the good scholars who took the time to analyze implications of solid empirical observations and direct inference" (Dorn, 2005, p. 492).

The "solid empirical observations and direct inference" refer to the thousands of hours of fieldwork Schaffer conducted to demonstrate the validity of his claims (Dorn, 2005). Why must it take "a century" for established bodies of scholars to accept – or even consider, instead of dismissing it in an outright manner – a strong evidence-based argument?

2.5.4 Hand Sample-Scale: Rock Decay Science

At the opposite end of catastrophic hypotheses, the field of weathering science recently underwent a paradigm shift of sorts in regards to an age-old assumption of the freeze-thaw process (Hall and Thorn, 2011), and is currently rebranding itself as "rock decay" science (Dorn et al., 2013; Hall et al., 2012). In regards to freeze-thaw, Hall and Thorn (2011) argued that traditionally, debris shape (angular) was the main mechanism by which to identify freeze-thaw occurrences – the generally accepted premise that alternating freezing and thawing cycles eventually breaks apart rock. Although occurring at two scales, macrogélivation (block-scale) and microgélivation (grain- and flake-size), traditional paradigm says the processes are different (Tricart, 1970). Hall and Thorn (2011, p. 83) note, correctly, that freeze-thaw is such "...an important geomorphic concept that needs to be well-founded and clearly conceptualized or defined." Yet laboratory experimentation, backed up by very close monitoring *in situ* – in the field – demonstrate these two long-held beliefs to be, at best, flawed in their assumptions (Hall and Thorn, 2011; Hall and André, 2001).

When it comes to weathering itself, Hall et al. (2012, p. 1) suggested that, "As process studies become ever more reductionist in nature, so the functionality of the term comes more and more into question." Weathering as traditionally described (mechanical or chemical) is plagued by being a misnomer that, according to Hall et al. (2012), leads people to believe that rock decay processes are climate dependent. This climate-dominated view can be found in secondary schools (Dove, 1997) and even permeates university settings (Allen and Lukinbeal, 2011; Allen, 2008). Yet anyone who has observed weathering in the field, however, understands that climate plays a small role in the decay process, and that even in more extreme climes (e.g., the tropics or polar regions) other factors are at work just as much as the weather and climate. Hall et al. (2012) ascribed to the notion – as do many of his colleagues (Dorn et al., 2013) – that a more appropriate term is "rock decay," since evidence from the field accompanied by studies in the laboratory strongly points to intrinsic rock properties being the driving factor behind decay, not necessarily the "weather" or climate where the rock is found.

Although not as heated of debates as Bretz, these two local-scale instances are nonetheless important to geomorphological fieldwork, as rock decay is the beginning to the changing of the Earth – the very crux of geomorphology. Rock decay remains fundamental to geomorphology, landform-process explorations, soil science, and structural geology, even though it may be glossed over in classes and taken for granted as driving processes in landform development and change. Forthright geomorphological fieldwork remains a necessary component at all scales of the landscape if potential pitfalls and missteps are to be avoided in the discipline (and in science generally).

2.6 WHY *NOT* FIELDWORK?

From its inception, fieldwork has been the lifeblood of geomorphology. The discipline's very foundations are based on fieldwork. Early geomorphologists spent copious amounts of their time in the field. More recently, the Forty-Third Binghamton Geomorphology Symposium (2012) was dedicated to fieldwork. The concepts found in fieldwork praxis remain inherent in geomorphology (Fig. 2.2). If fieldwork is paramount for the discipline of geomorphology – even part of its epistemology – then, why this book? "Why fieldwork?"

Since its beginnings with casual observations to its more grand and general theories often brought on through imaginations, this chapter illustrates fieldwork's importance in geomorphology. Yet there is so much more to it. Fieldwork can increase potential for learning *and* teaching about the Earth, provide verifications of models and GIScience data and results, and inform and enhance overall geomorphic and geomorphologic-related research activities. It can also strengthen disciplinary relationships, and help people find their place in the world at large and perhaps the discipline more specifically. Fieldwork permeates geomorphology. It connects doing and performing with landscape. Escape from its strong grasp remains impractical at best, for the discipline resounds with it. That fieldwork continues to be practiced by many (most?) geomorphologists is a testament to its prominence as a method, ontology, and epistemology.

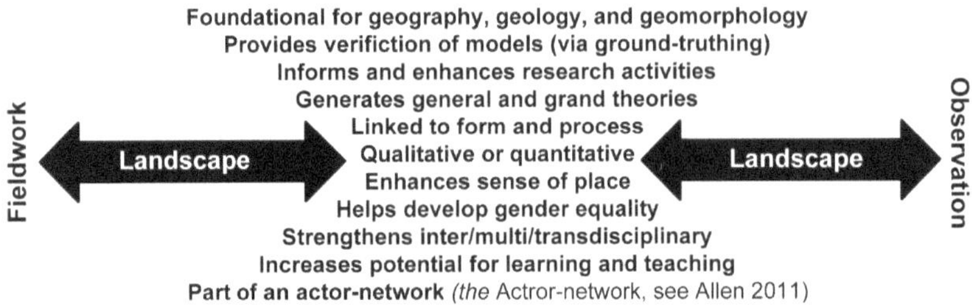

FIGURE 2.2 How geomorphology praxis centers on the interactivity and interconnectivity of fieldwork and observation with/in landscape. This figure is not suggested as being holistic. Statements represent topics outlined in this chapter and are in no particular rank-order. Many more aspects, elements, and topics occur that could be included, and rank-order of them could be argued depending on distinct landscape, fieldwork, or observation parameters.

Still, the sense of tyranny among the majority in science remains at large. No matter the evidence to the contrary, as with other important scientists like Wegner and Bretz, Schaffer remains shunned from the conventional crowd, as it seems that those entrenched in their beliefs refuse to review and study for themselves any contrary evidence, especially when that evidence is field based. Given all the tests and evidence, Hall and his colleagues have certainly experienced their share of tyranny among the established paradigm as well. Why would scientists, presumably with a working knowledge of the scientific method's evidence-based process be worried about paradigm change? Is it from a lack of field experience on their part? A lack of desire to field-test alternative hypotheses? Fear of being "proven" wrong? As Chamberlin (1890) noted more than a century ago, multiple working hypotheses, while also following the scientific method, provide a very strong way to understand science. Even in geomorphology his techniques are continually used, perhaps even more so when in the field. Why then must alternative explanations – especially those backed up by field-based evidence – be discounted in lieu of established ideas? If the history of geomorphology as a discipline demonstrates anything, it is that being entrenched in paradigms and failing to look at other evidence – however skeptical detractors may be and even when *field-based evidence points strongly to the contrary* – leads to disciplinary stagnation.

In the end, geomorphology remains entwined with fieldwork: it cannot escape its own episteme. Fieldwork represents the ultimate way to generate new ideas, create new techniques and methods, challenge cherished beliefs, learn new skills and knowledge, and gain stronger connections with both colleagues and the landscape. Performing solid, useful fieldwork includes open and respectful interaction among fieldmates, regardless of experience or (sub)disciplinary background, and camaraderie experienced in the field remains an attribute unattainable any other way or in any other location. Fieldwork is – or should be, as this chapter argues – geomorphology's *raison d'être*. And perhaps, instead of justifying its existence ("Why Fieldwork?"), geomorphology's guiding mantra should be, "Why *NOT* fieldwork?" Since on-the-ground, in-the-field endeavors have no substitute, and geomorphology remains poised, set, and ready to lead the way in performing fieldwork, why not return to the landscape, put into practice what is known, strive for collaboration with others willingly, engage in new dialogues, come away with new perspectives and knowledge, and pass on lessons learned to subsequent generations, so they can be empowered by and prosper from the most essential facet of geomorphology: fieldwork.

REFERENCES

Allen, C.D. 2008. Using rock art as an alternative science pedagogy. School of Geographical Sciences, Arizona State University: Tempe, AZ.

Allen, C.D., 2011a. Concept mapping validates fieldwork's capacity to deepen students' cognitive linkages of complex processes. Res. Geogr. Educ. 13, 30–51.

Allen, C.D., 2011b. On actor network theory and landscape. Area 43, 274–280.

Allen, C.D., 2012. Los Lavadores: discovering place in the Andes. Geogr. Rev. 102, 364–371.

Allen, C.D., Lukinbeal, C., 2011. Practicing physical geography: an actor-network view of physical geography exemplified by the rock art stability index. Prog. Phys. Geogr. 35, 227–248.

Allen, C.D., Thompson, T.J., Hansen, M.K., 2013. Using music to learn geomorphology: an undergraduate experience. Geogr. Bull. 54, 37–48.

Baker, V.R., 1987. Paleoflood hydrology and extraordinary flood events. J. Hydrol. 96, 79–99.

Baker, V.R., Twidale, C.R., 1991. The reenchantment of geomorphology. Geomorphology 4, 73–100.

Baker, V.R., Webb, R.H., House, P.K., 2002. The scientific and societal value of plaeoflood hydrology. House, P.K., Webb, R.H., Baker, V.R., Levish, D.R. (Eds.), Ancient Floods, Modern Hazards: Principles and Applications of Paleoflood Hydrolofy Water Science and Application, 5. American Geophysical Union, Washington, American Geophysical Union, pp. 1–19.

Baker, V.R., 2004. Fieldwork. Harrison, S., Pile, S., Thrift, N. (Eds.), Patterned Ground, 350. Reaktion Books, London.

Bowker, M.A., Belnap, J., Bala Chaudhary, V., Johnson, N.C., 2008. Revisiting classic water erosion models in drylands: The strong impact of biological soil crusts. Soil Biol. Biochem. 40, 2309–2316.

Box, E.O., 1981. Macroclimate and plant forms: an introduction to productive modeling in phytogeography. W. Junk, The Hague.

Bracken, L.J., Wainwright, J., 2006. Geomorphological equilibrium: myth and metaphor? T. I. Brit. Geogr. 31, 167–178.

Branson, F.A., Gifford, G.F., Renard, K.G., Hadley, R.F., 1981. Rangeland Hydrology. Society for Range Management, Denver, Colorado.

Bretz, J.H., 1923. The channeled scabland of the columbia plateau. J. Geol. 31, 617–649.

Campbell, M.O., 2005. Actor networking, technological planning and conceptions of space: the dynamics of irrigation farming in the coastal savanna of Ghana. Appl. Geogr. 25, 367–381.

Chamberlin, T.C., 1890. The method of multiple working hypotheses. Science (old series) 15, 92–96.

Church, M., 2013. Refocusing geomorphology: field work in four acts. Geomorphology. 200, 184–192.

Convey, P., 1994. Modelling reproductive effort in sub- and maritime Antarctic mosses. Oecologia 100, 45–53.

Daniels, S., 1992. Place and the geographical imagination. Geography 77, 310–322.

Day, T., 2012. Undergraduate teaching and learning in physical geography. Prog. Phys. Geogr. 36, 305–332.

Dixon, D.P., Hawkins, H., Straughan, E.R., 2013. Wonder-full geomorphology: sublime aesthetics and the place of art. Prog. Phys. Geogr. 37, 227–247.

Dorn, R., 2005. The geomorphic evolution of the Yosemite Valley and Sierra Nevada landscapes: solving the riddles in the rocks. Ann. Assoc. Am. Geogr. 95, 491–493.

Dorn, R.I., Gordon, S.J., Allen, C.D., Cerveny, N., Dixon, J.C., Groom, K.M., Hall, K., Harrison, E., Mol, L., Paradise, T.R., Sumner, P., Thompson, T.J., Turkington, A.V., 2013. The role of fieldwork in rock-decay research: case studies from the fringe. Geomorphology 200, 59–74.

Douglass, J., Meek, N., Dorn, R.I., Schmeeckle, M.W., 2009. A criteria-based methodology for determining the mechanism of transverse drainage development, with application to the southwestern United States. Geol. Soc. Am. Bull. 121, 586–598.

Dove, J., 1997. Student ideas about weathering and erosion. Int. J. Sci. Educ. 19, 971–980.

Dunphy, A., Spellman, G., 2009. Geography fieldwork, fieldwork value and learning styles. Int. Res. Geogr. Environ. Educ. 18, 19–28.

Ellis, J.T., Rindfleisch, P.R., 2006. A coastal environment field and laboratory activity for an undergraduate geomorphology course. J. Geogr. 105, 216–224.

Evenson, M.E., 2013. Preparing for fieldwork: students' perceptions of their readiness to provide evidence-based practice. Work J. Prevent. Assess. Rehabil. 44, 297–306.

Fawcett, P.J., Werne, J.P., Anderson, R.S., Heikoop, J.M., Brown, E.T., Berke, M.A., Smith, S.J., Goff, F., Donohoo-Hurley, L., Cisneros-Dozal, L.M., Schouten, S., Sinninghe Damste, J.S., Huang, Y., Toney, J., Fessenden, J., WoldeGabriel, G., Atudorei, V., Geissman, J.W., Allen, C.D., 2011. Extended megadroughts in the southwestern United States during Pleistocene interglacials. Nature 470, 518–521.

Fletcher, S., France, D., Moore, K., Robinson, G., 2007. Practitioner perspectives on the use of technology in fieldwork teaching. J. Geogr. Higher Educ. 31, 319–330.

Frink, D.S., 2009. Demonstrating emergent properties in dynamic systems. J. Coll. Sci. Teach. XXXIX, 68–70.

Fuller, I., Rawlinson, S., Bevan, R., 2000. Evaluation of student learning experiences in physical geography fieldwork: paddling or pedagogy? J. Geogr. Higher Educ. 24, 199–215.

Gillieson, D.S., Cochrane, J.A., Murray, A., 1994. Surface hydrology and soil movement in an arid karst: the Nullarbor Plain Australia. Environ. Geol. 23, 125–133.

Gimenez, O., Abadi, F., Barnagaud, J.Y., Blanc, L., Buoro, M., Cubaynes, S., Desprez, M., Gamelon, M., Guilhaumon, F., Lagrange, P., Madon, B., Marescot, L., Papadatou, E., Papaix, J., Peron, G., Servanty, S., 2013. How can quantitative ecology be attractive to young scientists? Balancing computer/desk work with fieldwork. Anim. Conserv. 16, 134–136.

Gold, J.R., 1991. In: Gold, J.R., Jenkins, A., Lee, R., Monk, J., Riley, J., Sheppard, I., Unwin, D. (Eds.), Fieldwork. Teaching Geography in Higher Education: A Manual of Good Practice. Blackwell, Oxford, pp. 21–35.

Gold, J.R., Jenkins, A., Lee, R., Monk, J.R., Riley, J., Shepherd, I.D.H., Unwin, D.J., 1991. Teaching Geography in Higher Education: A Manual of Good Practice. Blackwell, Oxford.

Goudie, A.S., 2002. Aesthetics and relevance in geomorphological outreach. Geomorphology 47, 245–249.

Guitierrez, M., 2005. Climatic Geomorphology. Elsevier, Amsterdam.

Guyton, B., 1998. Glaciers of California: Modern Glaciers, Ice Age Glaciers, the Origin of Yosemite Valley, and a Glacier Tour in the Sierra Nevada. University of California Press, Berkeley and Los Angeles, CA.

Hall, K., 2006. Perceptions of rock weathering in cold regions: a discussion on space and time attributes of scale. Geomorphologie: Relief, Processus, Environnement 3, 187–196.

Hall, K, André, M. -F., 2001. New insights into rock weathering from high-frequency rock temperature data: an Antarctic study of weathering by thermal stress. Geomorphology 41, 23–35.
Hall, K., Thorn, C., 2011. The historical legacy of spatial scales in freeze–thaw weathering: misrepresentation and resulting misdirection. Geomorphology 130, 83–90.
Hall, K., Thorn, C., Sumner, P., 2012. On the persistence of'weathering'. Geomorphology, 149–150, 1–10.
Huber, K.N., 1987. The geologic story of Yosemite National Park. US Geol. Surv. Bull., 1595.
Hudak, P.F., 2003. Campus field exercises for introductory geoscience courses. J. Geogr. 102, 220–225.
Ingold, T., 1993. The temporality of landscape. World Archeol. 25, 152–174.
Inkpen, R., Viles, H., Moses, C., Baily, B., 2012. Modelling the impact of changing atmospheric pollution levels on Limestone erosion rates in Central London, 1980–2010. Atmos. Environ. 61, 476–481.
Inkpen, R.J., Wilson, G., 2013. Science, Philosophy, and Physical Geography. Routledge, Abingdon.
Jarrett, R.D., 1991. Paleohydrology and its Value in Analyzing Floods and Droughts. U.S. Geol. Surv., 106–116.
Jennings, S.A., Huber, T.P., 2003. Campus-based geographic learning: a field oriented teaching scenario. J. Geogr. 102, 185–192.
Johnson, D.L., Domier, J.E.J., Johnson, D.N., 2005. Reflections on the nature of soils and its biomantle. Ann. Assoc. Am. Geogr. 95, 11–31.
Kent, M., Gilbertson, D.D., Hunt, C.O., 1997. Fieldwork in geography teaching: a critical review of the literature and approaches. J. Geogr. Higher Educ. 21, 313–332.
Kidron, G.J., Yair, A., 1995. Factors affecting runoff over encrusted dune slopes, Nizzana, Western Negev, Israel. In International Conference on Geomorphic Response of Mediterranean and Arid Areas to Climate Change. Israel.
Kudryavtsev, A., Stedman, R.C., Krasny, M.E., 2011. Sense of place in environmental education. Environ. Educ. Res. 18, 229–250.
Latour, B., 1999. Pandora's Hope: Essays on the Reality of Science Studies. Harvard University Press, Cambridge, MA.
Lewis, A.R., 2007. Teaching glacial geomorphology in the field: linking fieldwork to global-scale scientific questions. 2007 GSA Denver Annual Meeting, 28–31.
Lonergan, N., Andresen, L.W., 1988. Field-based education: some theoretical considerations. HERD, 7.
Massey, D., 1997. Reading human geography. In: Barnes, T., Gregory, D. (Eds.), A Global Sense of Place. Arnold, London, 315-323.
Massey, D., 1999. Negotiating disciplinary boundaries. Curr. Sociol. 47, 5–12.
Matthes, F.E., 1930. Geologic history of the Yosemite Valley. US Geol. Surv. Prof. Paper, 160.
Meek, N., Douglass, J., 2001. Colorado River origin and evolution. In: Young, R.A., Spamer, E.E. (Eds.), Lake overflow: An alternative hypothesis for Grand Canyon incision and development of the Colorado River. Grand Canyon Association, Grand Canyon National Park, AZ, pp. 199–205.
Meek, N., Douglass, J., 2003. Colorado River origin and evolution. Lake-Overflow: An alternative hypothesis to explain Grand Canyon incision and development of the Colorado RiverGrand Canyon Association, Grand Canyon National Park, AZ, 188–204.
Mellor, A., 1991. Experiential learning through integrated project work: an example from soil science. J. Geogr. Higher Educ. 15, 135–147.
Mol, L., Viles, H.A., 2012. The role of rock surface hardness and internal moisture in tafoni development in sandstone. Earth Surf. Proc. Land 37, 301–314.
Muir, J., 1880. Ancient glaciers of the Sierra. Californian 2, 550–557.

Nielsen, K.H., 2012. Scientists and Scholars in the Field: Studies in the History of Fieldwork and Expeditions. Aarhus University Press, Aarhus.
Opdam, P., Steingrover, E., van, S., Rooij, 2006. Ecological networks: a spatial concept for multi-actor planning of sustainable landscapes. Landscape Urban Plan 75, 322–332.
Pérez, F.L., 1997. Geoecology of erratic globular lichens of *Catapyrenium lachneum* in a high Andean Paramo. Flora, 192.
Phillips, J.D., 1999. Earth Surface Systems: Complexity, Order, and Scale. Blackwell, Oxford.
Phillips, J.D., 2001. Methodology, scale and the field of dreams. Ann. Assoc. Am. Geogr. 91, 754–760.
Phillips, R., 2012. Curiosity and fieldwork. Geography 97, 78–85.
Pielke, R.A., 1990. Influence of landscape structure on local and regional climate. Landscape Ecol. 4, 133–155.
Powlson, D., 2005. Will soil amplify climate change? Nature, 204–205.
Rhoads, B.L., Thorn, C.E., 1996. The Scientific Nature of Geomorphology. John Wiley & Sons, Chichester.
Richards, K., 2011. In: Agnew, J., Livingstone, D. (Eds.), The Field. Handbook of Geographical Knowledge. Sage, London, pp. 53–63.
Roering, J.J., Mackey, B.H., Marshall, J.A., Sweeney, K., Deligne, N. I., Booth, A.M., Handwerger, A.L., Cerovski-Darriau, C., 2013 'You are HERE': Connecting the dots with airborne lidar for geomorphic fieldwork. Geomorphology, 200, 172–183. doi:10.10106/j.geomorph.2013.04.009.
Schaffer, J.P., 1997. The Geomorphic Evolution of the Yosemite Valley and Sierra Nevada Landscapes: Solving the Riddles in the Rocks. Wilderness Press, Berkely.
Scott, I., Fuller, I., Gaskin, S., 2006. Life without fieldwork: some lecturers' perceptions of geography and environmental science fieldwork. J. Geogr. Higher Educ. 30, 161–171.
Scott, G.W., Goulder, R., Wheeler, P., Scott, L.J., Tobin, M.L., Marsham, S., 2012. The value of fieldwork in life and environmental sciences in the context of higher education: a case study in learning about biodiversity. J. Sci. Educ. Technol. 21, 11–21.
Semken, S., 2005. Sense of place and place-based introductory geoscience teaching for American, Indian and Alaska native undergraduates. JGE 53, 149–157.
Semken, S., Brandt, E., 2010. Cultural studies and environmentalism. In: Tippins, D.J., Mueller, M.P., Eijck, M., Adams, J.D. (Eds.), Implications of Sense of Place and Place-Based Education for Ecological Integrity and Cultural Sustainability in Diverse Places. Springer, Netherlands, pp. 287–302.
Semken, S., Freeman, C.B., 2008. Sense of place in the practice and assessment of place-based science teaching. Sci. Educ. 92, 1042–1057.
Semken, S., Freeman, C.B., Watts, N.B., Neakrase, J.J., Dial, R.E., Baker, D.R., 2009. Factors that influence sense of place as a learning outcome and assessment measure of place-based geoscience teaching. Electron. J. Sci. Educ. 13, 136–159.
Shapin, S., 2010. Never Pure: Historical Studies of Science as if it was Produced by People with Bodies, Situated in Time, Space, Culture, and Society, and Struggling for Credibility and Authority. Johns Hopkins University Press, Baltimore, MD.
Sharon, D., Margalit, A., 1990. Local climatic studies on dunes at the Nitzana research station. Arid Ecosystem Research Center.
Simon, G.L., Wee, B.S.-C., Chin, A., Tindale, A.D., Guth, D., Mason, H., Allen, C.D., Barbour, J.M., 2013. Synthesis for the interdisciplinary environmental sciences: integrating systems approaches and service learning. J. Coll. Sci. Teach. 42, 42–49.
Soennichsen, J., 2008. Bretz's Flood: The Remarkable Story of a Rebel Geologist and the World's Greatest Flood. Sasquatch Books, Seattle, WA.
Thornbush, M.J., 2008. On postcards used to track environmental history. Environ. Hist. 13, 360–365.

Thornbush, M.J., 2010. Photographic surveys of building exteriors in central Oxford, UK. Int. J. Archit. Herit. 4, 351–369.
Thornbush, M.J., 2013a. Introduction: archaeogeomorphology (geomorphology, culture, and place). Appl. Geogr. 42, 155–156.
Thornbush, M.J., 2013b. Photogeomorphological studies of Oxford stone: a review. Land. Anal. 22, 111–116.
Thorndycraft, V.R., Thompson, D., Tomlinson, E., 2009. Google earth, virtual fieldwork and quantitative methods in physical geography. Planet, 22, 48–51.
Tricart, J., 1970. Geomorphology of Cold Regions. McMillan, London.
Tuan, Y.-F., 1968. The Hydrologic Cycle and the Wisdom of God. University of Toronto, Toronto.
Tuan, Y.-F., 1974. Topophilia: A Study of Environmental Perception, Attitudes, and Values. Prentice-Hall, Englewood Cliffs, NJ.
Tuan, Y.-F., 1979. Landscapes of Fear. New York, NY, Pantheon Books.
Tuan, Y.-F., 2001a. Life as a field trip. Geogr. Rev. 91, 41–45.
Tuan, Y.-F., 2001b. Space and Place. University of Minnesota Press, Minneapolis.
Tuan, Y.-F., 2004. Place, Art, and Self. University of Virginia Press, Santa Fe, NM.
Twidale, C.R., 1996. The scientific nature of geomorphology. In: Rhoads, B.L., Thorn, C.E. (Eds.), Derivation and Innovation in Improper Geology, aka Geomorphology. John Wiley & Sons, pp. 361–380.
Viles, H.A., 1995. Ecological perspectives on rock surface weathering: towards a conceptual model. Geomorphology 13, 21–35.
Viles, H.A., Naylor, L.A., Carter, N.E.A., Chaput, D., 2008. Biogeomorphological disturbance regimes: progress in linking ecological and geomorphological systems. Earth Surf. Proc. Land. 33, 1419–1435.
Wahrhaftig, C., 1962. Geomorphology of the Yosemite Valley region. Bulletin-California, Division of Mines and Geology, California, 33-46.
Warburton, J., Higgitt, M., 1997. Improving the preparation for fieldwork with 'IT': two examples from physical geography. J. Geogr. Higher Educ. 21, 333–347.
Wegener, A., 1920. Die Entstehung der Kontinente und Ozeane (German Edition). University of Michigan Library, Ann Arbor, MI.
Wilson, J.P., Burrough, P.A., 1999. Dynamic modeling, geostatistics, and fuzzy classification: new sneakers for a new geography? Ann. Assoc. Am. Geogr. 89, 736–746.

Chapter 3

Getting into the Field

Mary J. Thornbush*,[1], Casey D. Allen**, Faith A. Fitzpatrick†
*School of Geography, Earth and Environmental Sciences University of Birmingham, Edgbaston, Birmingham, United Kingdom; **Department of Geography and Environmental Sciences, University of Colorado Denver, Colorado, USA; †U.S. Geological Survey, Wisconsin Water Science Center, Middleton, Wisconsin, USA
[1]Corresponding author: E-mail: m.thornbush@bham.ac.uk

Organizing a fieldwork foray represents no small challenge. The importance of fieldwork notwithstanding, preparing for it while a seemingly tedious task should not be overlooked – this section focuses on this topic. Proper preparation before heading into the field remains paramount for successfully completing a field-based project, as Day outlines in Chapter 3.1, "Preparing for Fieldwork." A seemingly simple component can quickly turn into a nightmare, if left unchecked. Although the tasks involved may appear mundane and monotonous, that does not negate their importance. Utilizing examples from his many years of fieldwork spanning four continents – plus various locations in the tropics – Day reminds us of forethought's importance when it comes to the field. Setting goals; assessing strengths and weaknesses of personnel, institutions, equipment, and money; and asking: "What if?" are key driving forces behind this chapter, as are the significance of both spatial and temporal scales. This chapter leads the reader from straightforward issues, such as site access, emergency planning, and transportation, to complex issues involving fiscal constraints, personnel selection and management, and selecting a field team. Although Day touches briefly on select issues of field safety, Chapter 3.2, "Field Safety: Principles, Practice, and Culture," by Alpay and Paulen breaks down its components into manageable pieces. When it comes to safety in the field, they remind the reader that there is no substitution for critical thinking, especially because fieldwork occurs in unquestionably diverse settings. Taking a systematic and hierarchical approach, Alpay and Paulen prompt field practitioners to identify potential risks and hazards beforehand, so mitigation can preemptively occur whenever possible and, should a reactive situation arise, plans are already in place to deal with it plainly and decisively, negating any potentially harmful second-guessing. "Safety first," as the saying goes, should remain at the forefront of anyone's mind who expects to conduct fieldwork. Being safe in the field begins with proper preparation and, taken together, the following two chapters give the field practitioner a solid grounding from which to enter the field in a reliable, planned out, and managed approach to safety.

Developments in Earth Surface Processes, Vol. 18. http://dx.doi.org/10.1016/B978-0-444-63402-3.00003-0

Chapter 3.1

Preparing for Fieldwork

Mick Day[1]
Department of Geography, University of Wisconsin-Milwaukee, Milwaukee, Wisconsin, USA
[1]Corresponding author: E-mail: mickday@uwm.edu

3.1.1 INTRODUCTION

Professionally productive and personally rewarding fieldwork in geomorphology, as in the gamut of other disciplines, necessitates keen forethought and careful preparation, without which unique academic opportunities may remain unrealized, and long-term, personal well-being may be compromised. Although the focus in fieldwork is often on "being there" and "making sense of it," neither of these can prosper without adequate preparation, which is all too easily overlooked. Put another way, it is crucial to recognize that fieldwork occurs not in a void, nor as a distinct experience, but rather takes place within a broader philosophical context, and as but part of an extended educational process.

Field observation is a major branch of geomorphology (Chorley et al., 1964; Chorley, 1966; Keylock, 2003; Trudgill, 2003; Gregory and Goudie, 2011; Church, 2013), and a wide range of techniques and protocols have been developed (Dackombe and Gardiner, 1987; Goudie, 1990; Slaymaker, 1991; Switzer and Kennedy, 2013). A full consideration of exactly what constitutes fieldwork in geomorphology is provided in this volume and elsewhere (Trudgill, 2003; Church, 2013). Relevant too are broader considerations of geomorphological methodologies (Sherman, 1999) and transdisciplinary considerations of the nature of fieldwork (Hobbs and Wright, 2006). In general, fieldwork occurs away from the home or place of employment, necessitating travel and adjustment to another location, and it involves assessment, often measurement of entire or partial landforms, landform assemblages and/or Earth surface processes. As such, it involves a broad array of potential techniques, a wide range of possible research locations and scales, and variable temporal frameworks. Much geomorphological fieldwork occurs literally "in the field," but fieldwork *sensu lato* may also encompass travel to acquire information from remote depositories or other researchers. Increasingly, fieldwork in geomorphology also involves consideration of related aspects of the atmosphere, hydrosphere, and biosphere, and recognition of human constructs and/or actions.

Methodological and ethical dimensions of fieldwork have been discussed widely in the social sciences, including geography (Latour, 1999; Driver, 2000;

Developments in Earth Surface Processes, Vol. 18. http://dx.doi.org/10.1016/B978-0-444-63402-3.00004-2

Gerber and Chuan, 2000; Clifford and Valentine, 2003; Hay, 2010; Phillips and Johns, 2012), but parallel considerations in the natural sciences have until recently been muted and sporadic (Kuklick and Kohler, 1996; Sherman, 1999; Trudgill, 2003; Church, 2013). Nevertheless, it is clear that fieldwork is both a complex and contested human enterprise (Georges and Jones, 1980) in which "...the representational, the spatial and the practical are closely entwined" and whose phenomenology needs to consider "...the materiality of the field, the contingency of encounters within it, and the embodied practices of fieldworkers themselves" (Driver, 2000, p. 268).

Fieldwork in geomorphology enjoys an enormous potential range of activities, venues, and timeframes, ranging from casual landscape observation to precise measurement of geomorphic processes, from local studies to exotic international expeditions, and from one-time, day-long excursions to career-spanning repetitive monitoring. Similarly, the potential range and scale of geomorphic foci are enormous, from Antarctic mountains to tropical caves, from discrete stream reaches and channel banks to continental-scale drainage basins, and from individual weathering features to massive volcanoes.

Although fieldwork in geomorphology plays an important instructional role prior to and during undergraduate studies, for example, in teaching students how to "read the landscape" or experience the sense of place, how to use basic equipment, how to negotiate gender and other differences, and how to conduct themselves in the field while gathering information and respecting the environment (Nast, 1994; Kent et al., 1997; Maguire, 1998; Bracken and Mawdsley, 2004; McGuinness and Simm, 2005), it often assumes critical significance during graduate studies and in subsequent professional life. Despite this, undergraduate fieldwork in geography, geology, and other disciplines is increasingly curtailed as a result of concerns about "health and safety," personal and institutional legal liabilities, and costs (Kent et al., 1997; Cann and Mayor, 2004), and subsequent experience is generally acquired through personal "trial and error," at best with the formal or informal guidance of academic supervisors or peers.

Pedagogic considerations notwithstanding, the primary purpose or role of fieldwork in geomorphology is as a component of the academic or otherwise professional research process, primarily as a means of acquiring "real world" information/data to answer questions or test hypotheses that form the core of graduate theses or dissertations, professional research projects, or consultancy. This information may be qualitative or quantitative, commonly both, and it may be required at a wide range of spatial and/or temporal scales, depending on the precise scope and nature of the inquiry (Millar, 2013). Similarly, the information required may be about components of or complete landforms, or assemblages thereof, and/or it may be about geomorphic processes in the short or long term. In addition, it may be necessary to acquire relevant information about atmospheric, biological, and human interactions with geomorphic processes and forms (Trudgill, 2003; Gregory and Goudie, 2011).

All of this takes physical and mental effort, time, and financial resources, and it is in the optimization of these expenditures that preparation for fieldwork takes on particular significance. In an idealized cyber world of invincibility and limitless resources, one could afford multiple failures, costly mistakes, and unlimited repetition, but in the current geomorphologists' world opportunities, resources, and human well-being cannot be so squandered.

Above all, fieldwork in geomorphology, as in other disciplines, requires sensible reconciliation of philosophical, academic, or professional goals with practical, logistic considerations. Even the most laudable, innovative, comprehensive, and paradigm-changing field research program will inevitably suffer from problems and setbacks, and it is often prudent, at least during one's early career or in the formative stage of a project, to think seriously about the scale of inquiry, and consider just what one can reasonably expect to accomplish given the probable constraints.

Although practical considerations weigh heavily upon all fieldwork programs, in geomorphology or otherwise, it is important to always keep in mind professional objectives, to adhere to rigorous academic standards, and to behave professionally, unless this conflicts with considerations of personal safety, or other overriding concerns. Professional geomorphological organizations, such as the International Association of Geomorphologists and the British Society for Geomorphology provide sound reflection on decisions of this nature, as do resources such as the Association of American Geographers statement on ethics (AAG, 2009).

Although induction, intuition, and other methodologies have their places in scientific research, most geomorphic fieldwork is formulated and constrained by the precepts of the hypothetico-deductive method, which rests upon broad paradigms, the existing body of theory, and the testing of single or multiple working hypotheses, or statements of statistical probability (Rogerson, 2001; Medawar, 2009). Identifying tenable hypotheses, and devising appropriate testing procedures, with the aim of rejecting null hypotheses, is thus a critical precursor to most geomorphic fieldwork, and has received due scrutiny elsewhere (Rogerson, 2001; Haining, 2003). Fieldwork may also be employed in obtaining field verification or "ground truth" of information derived from digital and other satellite data.

3.1.2 CONSULTING THE EXISTING INFORMATION

A key element in preparing for fieldwork is to first become thoroughly acquainted with the existing information and relevant literature, widely defined, as they relate to the academic topic, the methodology, and the study area. Typically this involves the acquisition and perusal of historical and contemporary scholarly literature, including books, journal articles, and professional reports, plus maps (both hard copy and digital), aerial photographs, satellite imagery, and any other sources that seem germane. Specific scholarly information may be acquired through traditional personal and institutional library resources, or via database searches, using online library resources such as WorldCat, or via tools such as Google Scholar (Clifford and Valentine, 2003). Digital maps, digital elevation models, air photos, and/or high resolution satellite images may be downloaded from numerous government, institutional, or commercial websites, and may be taken readily into the field on personal electronic devices. Information technology has much to offer in this regard (Warburton and Higgit, 1997).

Personal contacts may also be valuable, particularly communication with others who work on similar topics, employ the same methodology, or who are familiar with the field area. Supervisors and colleagues are often significant in this context. "Background," or informal literature, such as travel guides and various online and social media resources may also prove valuable in alerting the fieldworker to potential problems and opportunities. No potentially useful resources should be overlooked, but diligence and persistence may be required.

3.1.3 TIME

One critical consideration in fieldwork is its timing: outset, duration, and completion, each of them is equally important. By definition, preparation takes place before fieldwork commences, but it is no less important throughout the following stages. Preparation is, of course, a time-dependent process, in that one prepares both individually and collectively, and implicitly or explicitly for various episodes and periods of time, ranging from momentary and hourly activities, through daily routines, to professional programs spanning weeks, months, or years. As such, preparation is actually continuous and ongoing, with no real beginning or end, and it is constantly evolving, and in flux. Therefore, preparing for fieldwork in geomorphology involves nested, multidimensional, and nondiscrete sets of processes, partly explicit, public, and deliberate, but also partially intuitive, unspoken, and intensely personal. The, 20-point preparatory checklist (Table 3.1.1) should be regarded as a flexible framework, to be manipulated as appropriate.

The focus here is on fieldwork preparation at the temporal scale of days to months, rather than hours or years, but it is important to recognize that shorter and longer time frames are also involved. Just as fieldwork is but a component of a broader objective, so preparation cannot be considered in isolation or as a time-discrete process, rather it must be recognized as continuous, fluid, and holistic.

Thorough preparation may delay the start of fieldwork, but it can more than compensate for this by reducing its duration, and hastening its completion, as well as bolstering academic rigor during fieldwork, and providing the necessary information for meaningful analysis afterwards. This, then, is a key consideration: not to rush into fieldwork without being adequately prepared both academically and logistically. To do otherwise invites failure on multiple fronts.

Of course, this invocation is less critical where fieldwork is local, allowing for numerous repeat performances and where, by design, it is exploratory or

TABLE 3.1.1 Abbreviated Preparatory Checklist

1. Attain comprehensive understanding of the fieldwork topic and context. Consult all pertinent resources.
2. Insure that the proposed fieldwork is academically rigorous, and appropriately scaled.
3. Ascertain what data and information are critical, and what is optional.
4. Consider how the data will be used and analyzed.
5. Develop a realistic fieldwork timetable. Anticipate delays.
6. Assess practical feasibility and risks. Develop "backup" plans. Anticipate "worst-case" scenarios.
7. Insure that institutional and/or personal support is in place.
8. Attain complete familiarity with field equipment and techniques. Anticipate problems.
9. Consider security – personal, equipment, and data.
10. Consider potential impacts.
11. Assess personnel needs, availability, and potential management issues.
12. Consider health, safety, and fitness issues.
13. Develop a realistic budget; insure that adequate funds are available and accessible.
14. Arrange all necessary transportation. Arrange vehicles, flights, passports, and visas.
15. Arrange appropriate accommodation.
16. Consider provision of food, water, and medications.
17. Prepare for field physical conditions – weather, disease, hazards, vaccinations, and first-aid.
18. Prepare for field human characteristics – culture, customs, language, appearance, distractions, and safety.
19. Obtain necessary permissions and documentation.
20. Survive, enjoy, learn, recover, disseminate, and move on.

incremental in scope or sophistication. In contrast, if the fieldwork is a one-time, "make or break" opportunity, involving long-distance travel, considerable expense, and the participation and goodwill of others, particularly within a timeframe of their prescription, then it is imperative to take full advantage of the opportunity, and maximize the probability of success while working within the available time limits.

In some instances, then, for example, when participating in an international, interdisciplinary expedition whose start and finish are constrained by climatic, political, or other logistic considerations, the individual geomorphologist may have no control over the timing of fieldwork. This may be quite acceptable to experienced individuals, but it can represent a major challenge to those in the early stages of their careers, although it should not necessarily dissuade them from participating, perhaps in the opportunity of a lifetime.

Both students and faculty members may also be temporally constrained by the instructional calendars of their academic institutions, whereby fieldwork is effectively impossible during terms, semesters, or other teaching periods, and is thus restricted to vacations or breaks. Institutions and individuals, of course, vary in their flexibility in this regard, but the matter warrants consideration. What is to be done, for example, if the only time period available for graduate student fieldwork at a potential location coincides with extreme conditions, such as the Antarctic or alpine winter, or tropical monsoon, or hurricane seasons? Should a potential research topic be abandoned in favor of one more logistically probable to succeed, or is it better to reschedule the fieldwork, perhaps risking financial repercussions, or additional time to graduation? Thus, temporal considerations must enter into decisions about research topics involving fieldwork, both for the success of the research, and for the well-being of the researcher.

As important as it is to ensure thorough preparation before starting fieldwork, it is equally important to avoid undue delay. Experienced individuals can often prepare themselves quickly, and supervisors have an important role both in preparing students, and assessing their readiness. As throughout the research process, this may involve compromises. For example, before fieldwork starts it may be vital that a student has mastery of a field technique, whereas it may not be critical to have read a significant, but perhaps tangential scholarly paper, or fully understand the statistical analysis to be implemented later on. Yes, ideally the student should "be on top of" all facets of the research, but sometimes unfortunate priorities have to be established.

The time available for the completion of fieldwork is inevitably limited, and that time often seems to pass disconcertingly swiftly. Preparation must involve a realistic assessment of how long a fieldwork project will take to complete, at least to an acceptable threshold, and/or must recognize explicitly just what can be accomplished in the time available. Again, these decisions intertwine both academic and practical considerations, and they are not always easy, but they cannot and should not be avoided. It is surely a cardinal sin for a supervisor to dispatch a student into the field to undertake a task that simply cannot be accomplished in the time available!

Even when temporal needs or limits have been assessed realistically, it must be recognized that things may go wrong in the field, potentially derailing even the best-laid plans. Although preparation involves acknowledging and trying to avoid potential problems, environmental conditions and human attributes dictate that it is best to plan for difficulties, and to build temporal redundancy into fieldwork plans. Despite our best efforts, individuals contract illnesses, equipment malfunctions, unexpected floods occur, thieves strike, and political coups unfold.

Practical constraints aside, fieldwork preparation also needs to take into account the spatial and temporal issues of sampling and representativeness, both of which merit consideration beyond that which is appropriate here (Richards, 1996). For example, are repeated, short-term process measurements appropriate or adequate, or is there a need for continuous, long-term monitoring? Is either or are both feasible, given distance to fieldwork locations? What is the significance of a single extreme episode, and how does fieldwork accommodate considerations of periodicity and duration (Thornes and Brunsden, 1977; Cullingford and Davidson, 1980; Millar, 2013)?

Preparing to start fieldwork is an obvious necessity, but it is also important to prepare for its termination to make sure that goals have been accomplished to the extent possible before time expires or, if time is not the arbiter, to recognize when sufficient information has been acquired, and the point of diminishing returns has been reached. In the case of numerical data, this may be determined by the minimal data needs of statistical analysis; in other cases it is more difficult to define, and will be determined by a more subjective assessment of accomplishments and completeness.

3.1.4 RECONCILING ACADEMIC AND LOGISTIC CONSIDERATIONS

Ultimately, perhaps the most difficult aspect of fieldwork is to reconcile academic or professional goals with logistic constraints. What is academically desirable may not be practically possible given contemporary technology and personnel limitations. Or, it may be possible, but only at unreasonable risk to field researchers themselves. Until late in the twentieth century, for example, it was essentially impossible to conduct primary research in water-filled (phreatic) caves, but improvements in SCUBA equipment and safety protocols now render such research increasingly possible, although potentially at a high cost in human lives (NSS, 2014). Similarly, on-site field research on active extrusive volcanism, processes on near-vertical slopes, or extreme river floods will always remain constrained by considerations of human safety, although the use of remotely sensed imagery, continuous optical recorders, data loggers, and other remote devices minimizes personal risk and shifts the potential burden of loss from human life to expensive, sophisticated equipment.

Balancing personal philosophies, professional goals, and practical limitations lies at the heart of successful preparation for fieldwork, yet it remains difficult to teach, and it is underestimated and underrepresented in the scholarly literature. Most geomorphologists initially acquire the relevant knowledge through informal advice, rather than formal instruction, and this is honed subsequently by personal experience. As Sherman (1999, p. 688) has noted, "Every geomorphologist applies a singular methodology to a particular circumstance....a characteristic methodology that represents an integration of personal experience, perception, and ability."

3.1.5 INSTITUTIONAL AND PERSONAL SUPPORT

Although fieldwork is often considered as an individual enterprise, its success is largely dependent on a network of supporting actors, structures, and institutions, both professional and personal. Paraphrasing John Donne (1624/1959), "No man is an island," and the field researcher is ultimately both constrained and empowered by his or her own support systems, whose components' involvement may be tangible, tangential or imperceptible, emotional, financial, or logistic.

Sophisticated professional fieldwork in geomorphology is increasingly dependent on institutional support from universities, government departments, or other agencies that can provide the necessary equipment, the finances, and the logistic foundation upon which rests any field endeavor beyond the rudimentary. Basic equipment, such as compasses, clinometers, measuring tapes, sample bags, and shovels may still be within the personal budgets of individual researchers, as may digital cameras, laptop computers, and other electronics, but the precise instruments needed for detailed chemical and mechanical measurements and analyses are increasingly expensive enough to require institutional purchase, and perhaps to mandate interdepartmental or interinstitutional cooperative use. Likewise, fieldwork budgets may require substantial, long-term underwritten financial commitments, involving multiple and evolving personnel, and numerous budgetary cycles, necessitating professional economic expertise. Beyond that, universities and other home institutions can provide necessary introductions, and documentation of a researcher's authenticity, and may have reciprocal arrangements for accommodation and other facilities with other institutions in the field region. Provision of appropriate insurance for personnel and equipment is often appropriate, and in emergencies, home institutions may be able to arrange medical treatment, provide legal representation, or assist with repatriation.

Professional support may be an obvious requirement, but personal support from friends and family may prove even more critical. Acknowledgements in theses and dissertations testify to the various roles that individuals play in assisting research efforts, and the emotional stability provided by personal relationships represents a foundation without which few individuals can flourish. How many researchers have not enlisted friends or family in some aspect of fieldwork: as driver, photographic scale, or the person "at the other end" of the tape measure? In exigency, institutions may prove cumbrous and hesitant, whereas individuals may be agile and indefatigable. Those preparing for fieldwork should enlist both sets of allies.

3.1.6 FIELD TECHNIQUES AND EQUIPMENT

Fieldwork in geomorphology potentially involves a wide range of field equipment and techniques (Dackombe and Gardiner, 1987; Goudie, 1990). In preparation, one needs to determine what is really necessary and what is optional, how their employment possibly complements or prohibits other fieldwork activities, and how they affect personnel requirements. For example, if a slope study is concerned primarily with processes, how important is it to measure slope form accurately, given that the latter is time consuming and that cross-slope variations may have negligible influence on downslope movements? In another sense, how much time and effort should be devoted to context, and how much to the specific focus?

Both pros and cons can apply to tightly focused, and broader fieldwork objectives. The former potentially risks providing an accurate assessment of something that may be of only local significance, whereas the latter may miss important details. Dilemmas such as this must be resolved before fieldwork begins, informed by due considerations of context, scale, and objectives. Similarly, single-field techniques and/or reliance on one piece of field equipment may yield large volumes of invaluable and specific data, but may leave the field researcher open to questions of context, and vulnerable to failure of the crucial equipment.

Dependence on field equipment and understanding its dependability are critical issues to be addressed in advance, particularly before departing for distant sites. Field equipment varies widely in its reliability and durability, manufacturers' claims notwithstanding, and more sophisticated, expensive equipment is not necessarily better in this context than basic, inexpensive tools. Tropical heat, humidity, and microorganisms continue to wreak havoc with electronic field equipment, and a functioning instrument is always more valuable than an inoperable one. Rigorous prefield equipment testing under simulated field conditions is always sensible, and backup equipment, if available, is strongly recommended, although it is important to ascertain whether measurements made by different pieces of equipment are consistent. Familiarity with field equipment, and the ability to remedy problems, is highly desirable.

Electric power supplies can be problematic in remote locations, where mains supply is unavailable. Even with electric mains supplies, the compatibility of equipment with different voltage supplies must be assessed, and adapters and surge protectors employed as necessary. Off-the-grid electricity generation at field sites using small wind turbines or solar panels has limitations but also advantages. Batteries, particularly rechargeable AA or AAA sizes, seem increasingly reliable and durable, but they are not universally appropriate. Recharging may not be easy, although carrying ready-charged spares is not a great burden.

Field equipment security may also be an issue. Cameras, computers, and GPS receivers are obvious targets for thieves (and repressive regimes), and equipment may be taken, even if its use is unclear. A Schmidt test hammer (Day and Goudie, 1977) is of little use to anyone other than a materials specialist, but that may not preclude its theft. Unattended HOBOS and other data loggers are particular targets, for animals as well as people, and uncounted limestone weight-loss tablets (Trudgill, 1977) have been disturbed by unknown agencies, or simply lost by researchers who cannot locate them months or years after emplacement.

Georeferencing via GPS, and digital mapping are increasingly important aspects of fieldwork in geomorphology. Georeferencing has benefits and limitations (Cornelius et al., 2006), but it is now integral to data acquisition. As indicated by Seijmonsbergen (2012, p.1) "...modern geomorphological maps are digital collections of geomorphological information layers consisting of georeferenced vector, raster, and tabular data which are stored in a digital environment such as a GIS geodatabase, and are easily visualized as, for example, "birds" eye' views, as animated 3-D displays, on virtual globes, or stored as GeoPDF maps in which georeferenced attribute information can be easily exchanged over the internet." This is not in dispute, but there are limitations, for example, in rugged, densely forested landscapes, where GPS receivers have limited functionality.

Field data security also warrants consideration in advance. In the past, useful field data has been lost in dropped notebooks, erased by immersion in water, or confiscated by overzealous officials, and digital data is no less at risk in this context than its predecessors. Temporary field data storage and field-to-home data transfer both represent potentially weak links in the overall data security system. Digital files and data should be downloaded or saved frequently on laptops, tablets, or other devices, perhaps to external hard drives, or should be uploaded to "cloud" storage immediately. Field sheets should ideally be scanned and stored electronically as pdfs or images, and data may be distributed or integrated using tools such as DataPad. Data archival also warrants consideration, particularly in the contexts of replication, data sharing, and long-term storage stability (Lane and Chandler, 2003).

Field techniques and equipment may also have potentially adverse impacts, both on personnel, and on the environment, and it is important to recognize and minimize those risks. For example, research in caves, beneath glaciers, or on active volcanoes is always accompanied by some risk, and even the most seemingly benign locations may have hidden dangers. Disturbing slopes or potentially unstable materials during testing or sampling may produce dangerous material failures, and the use of any heavy machinery in the field always carries some risk. Dyes used for underground water tracing should be handled with care, although modern versions contain generally environmentally safer chemicals than in the past, and are best used under permit by experienced personnel, with such assistance as is needed, to avoid cross-contamination or causing public anxiety (Bednar, 2005; Smart and Laidlaw, 2010). Turning local water supplies bright green or purple is guaranteed to cause concern under the best of circumstances, but can be avoided with due prior consideration of discharges, dilution factors, potential travel times, and detection thresholds.

Even seemingly minor acts may have cumulative adverse effects, as evidenced by the masses of rock detritus at many well-known fossil sites, and unnecessary digging of soil pits or excessive "cleaning" of sediment sections should be discouraged. "Humans as geological agents" (Ehlen et al., 2005) applies to geomorphologists too!

3.1.7 PREPARING FOR PROBLEMS

Even the best-laid field research plans are at the mercy of developments beyond the researcher's control, and it is always sensible to prepare for problems by employing backup or fallback plans, alternative strategies, and "safety nets."

Some rationale exists behind the assertion that any data are better than no data, although this is not a universal truth, and the well-prepared fieldworker should always be asking the question "What if?," and considering what to do under any predictable set of adverse circumstances. Along these lines, it might be prudent to develop, at least in outline, plans to be implemented under the "worst case scenarios," such as equipment failure or loss, natural disaster, personal injury, or civil unrest. Undue pessimism may not be healthy, but undue optimism may prove fatal, and a realistic appraisal of academic and logistic probabilities is always wise.

"Thinking on one's feet," and adapting to unforeseen circumstances in the field is commendable, but a modicum of preparation can go a long way towards making such adjustments easier than might otherwise be the case. Prefieldwork formulation of alternative hypotheses or questions, prior acquisition of backup equipment, identification of alternate fieldwork sites, and/or consideration of improvised data-gathering techniques may prove unnecessary, but sometimes prove well worth the time investment.

Both in preparing for and during the commission of fieldwork, patience, flexibility, and adaptability are enormous logistic advantages, if not necessarily universally recognized academic ideals. A reserve plan to substitute qualitative information for quantitative data may not be optimal, but it is surely better than returning from fieldwork empty handed. Unanticipated qualitative information may well lead to fresh insights, and to new research agendas. Qualitative approaches to geomorphology have a long history, much of which is illustrious; some of which illustrates their pitfalls (Chorley et al., 1964 et seq.; Chorley, 1966), but qualitative information is currently an underutilized tool in the geomorphologist's cabinet.

3.1.8 PERSONNEL CONSIDERATIONS

Fieldwork in geomorphology potentially involves highly variable numbers of participants, with requirements, limitations, and advisability depending on objectives, techniques, locations, and available expertise. Much traditional fieldwork has been carried out by individuals, but small teams convey many advantages, and large, multipersonnel, interdisciplinary forces, for example, on international scientific expeditions provide unique opportunities, as does the recruitment of "citizen scientists." Although individual researchers may be better focused and more dedicated than others, and are often able to operate "under the radar," there are many situations in which an individual is simply unable to accomplish fieldwork tasks, and there are substantive bases for adages such as "two heads are better than one," and "safety in numbers."

Small teams, perhaps 4–10 in number, have much to recommend them. They are often capable of accomplishing routine and repetitive field assignments quickly, and without attracting unwanted attention, and they provide opportunities for exchange of ideas, and for students to learn techniques and more from experienced researchers. Pairs of individuals may work well together, particularly if they are equally committed, and familiar with each other's abilities, but a field team of one or two "principle investigators," together with a pair of (post) graduate students and half a dozen willing undergraduates can generally total more than the sum of its components, providing exponential academic and logistic benefits.

Of course, with increasing numbers and variable expertise come ancillary concerns. Mentoring students is commendable, but it may compromise research agendas, particularly if time is short, and if conditions are extreme. Variable levels of participant expertise require careful supervision to assure data consistency and compatibility, and larger teams inevitably bring greater logistic challenges. More is not always better!

In addition to the academic participants themselves, fieldwork, particularly in distant locales may require nonacademic local-field assistants and guides, the recruitment and selection of whom may prove pivotal. The contributions of these individuals are sometimes underappreciated, but they are commonly crucial to fieldwork success. Paid workers add a different, sometimes complementary, sometimes challenging dimension to fieldwork, but they may be essential where extensive manual labor is necessary. Local guides and interpreters may play a similarly critical role in facilitating research, locating sites, avoiding unnecessary problems, solving problems, and conserving resources. Their financial and other requirements may be unpredictable, but knowledgeable field guides, eccentricities included, are not only invaluable in a confusing physical landscape, such as Jamaica's Cockpit Country, or where there is civil unrest, as in parts of the Philippines, but they may also be lifesavers.

The potential need for personnel beyond the principal investigator requires myriad considerations of personnel interactions and management, most of which are beyond the scope of this review. Predicting how individuals will perform in the field, and how they will interact under field conditions is inherently difficult, but it is easier when participants are familiar with each other before fieldwork, and/or when they have worked together previously. "Personal chemistry" should not be discounted, group dynamics often assume considerable importance, and potentially disruptive individuals should be anticipated, identified, and treated as seems appropriate. Personalities, and issues such as age, gender, sexuality, and religious beliefs may well enter into these considerations, and resolution of problems may be time consuming, and requisite of delicate interpersonal skills. Rarely are geomorphologists trained in personnel management, counseling, or related fields, but fieldwork success may hinge upon individuals' personal interactive skills, which should not be overlooked.

Personnel considerations raise important issues of leadership, organization, advising, and morale – again largely beyond the remit of this discussion. Good leadership may be defined in numerous ways: survival of all concerned may be the ultimate arbiter, but it is not necessarily always foremost in fieldworkers' minds. Individual personality and experience, allied with others' perceptions are always in play, as in the parallel context of professional academic advising (Solem et al., 2009). Leadership styles are infinitely variable, and no "one size fits all" eists; some individuals "hold light reins," whereas others are more direct and authoritative; some individuals appreciate clear, firm direction, whereas others prefer greater flexibility and freedom of choice. All this needs to be anticipated and negotiated at the personal and group level.

Academics may regard comparisons of fieldwork with military, business, or other organizational endeavors as distasteful and/or inappropriate, but there may be useful parallels. Just as effective leaders earn, rather than demand respect, and reputations may or may not be warranted, it is commonly helpful to have a clear chain of command, with clearly identified individual responsibilities under various, perhaps changing conditions. Individuals bring various skills to fieldwork, and although the benefits of flexibility are never to be underestimated, primary roles and responsibilities need clear advance definition.

Lest these considerations seem mechanistic and prescriptive, essential humanity, individual ethics, and personal empathy are also vital ingredients. Fieldwork in geomorphology is a human endeavor that may be important to those concerned, but it is rarely of immediate, critical societal significance. Professional success must be weighed against humanitarian costs, no field program is worth death or serious injury, and maintenance of dignity and morale may be of greater consequence than the accomplishment of any academic goal.

The preceding paragraphs also draw attention to the importance of physical and emotional fitness and health as contributors to fieldwork endeavors. Fieldwork demands are highly variable, but appropriate preparation is nonetheless important. At best, a lack of physical fitness may contribute only to temporary personal discomfort; under other circumstances it may compromise the research program, or place additional responsibilities on other participants; at worst, it may result in serious injury, incapacitation, long-term disability, even death. Obesity is likely to be problematic, as is a lack of physical strength, although much is context specific. Accidents can and do happen, but appropriate physical preparation can offset many eventualities.

Personal health considerations are equally important; individuals with a wide range of medical conditions, such as asthma, allergies, high blood pressure, diabetes, or heart disease should assess realistically whether and how these might affect them during any given fieldwork program. They may not preclude participation, they may not impede performance, but they certainly warrant consideration. Beyond personal consideration, and with due respect for personal privacy or confidentiality, such health concerns also should be brought to the attention of other fieldwork participants, particularly supervisors or primary investigators. Doing so ensures that others will understand the circumstances, and be prepared to assist should difficulties arise.

Emotional well-being and good mental health are also important in preparing for fieldwork. Diagnosed mental illnesses carry the same provisos as other medical conditions, but participants also need to be aware of less obvious emotional conditions, such as undiagnosed depression or anxiety, and to not underestimate the potential effects of events such as bereavement, separation, or other domestic upheavals. These may not manifest in obvious or significant ways, but they may result in distraction and loss of focus, which can compromise field agendas. By contrast, although there may be a sense of adventure, change, and escape from routine, homesickness and loneliness may afflict those traveling to distant fieldwork sites for extended periods.

Citizen science, in which nonprofessional volunteers gather large volumes of field data over wider spatial ranges and greater time spans than would be otherwise possible, is increasingly employed in disciplines such as phenology and ornithology (Schwartz, 2003; Bonney et al., 2009,), but it has yet to be utilized widely in geomorphology. Its potential benefits and limitations have been discussed elsewhere (Wiggins and Crowston, 2011), and clearly there are trade-offs involved in its adoption. Although the acquisition of large amounts of data has advantages, the utilization of such programs calls for extended consideration of the personnel and other issues raised previously, particularly regarding operator (in)consistency, and the clarity of technical protocols. Crowdsourcing, volunteered geographic information, metadata analysis, and other data-mining techniques may also have potential roles contributory to geomorphic fieldwork, although their application in this context has to date been limited (Sui et al., 2013).

3.1.9 BUDGETS, EXPENSES, FUNDING, AND AUTHORIZATION

Fieldwork involves inevitable financial expenses, and provision for these outlays, be they modest or major is another critical element of the preparatory process. Some expenses may be met from personal resources, but fieldwork is increasingly dependent upon the acquisition of funds through internal institutional budgets, or via external granting agencies. The preparation of grant proposals has been discussed elsewhere (Solem et al., 2009; Gatrell et al., 2012), and the details will not be pursued further here. Nonetheless, it is important to recognize that successful grant acquisition depends on preparation of a well thought-out, carefully prepared, and clearly articulated proposal, such that the grant application process often goes hand-in-hand with, and complements the overall program of preparation for fieldwork. In both, among other things, it is imperative to identify clear, concise objectives, delineate realistic timeframes, specify techniques and personnel, anticipate outcomes, and justify budgets.

Funding proposals typically need to be concise and focused, rather than identifying all the ancillary possibilities that fieldwork invites, but their preparation can be instructive, particularly when reviewers provide objective and constructive feedback, which can lead to increasing precision and clarity. Obtaining the funds may be the primary objective, but the insights gained from the review process may be equally helpful. Granting agencies increasingly encourage and favor interdisciplinarity, local institutional participation, student involvement, and ongoing pedagogic commitments, so these may need to be incorporated preemptively into funded fieldwork programs.

Academic institutions increasingly require that field and other research be preapproved by internal institutional review boards (IRBs), or similar bodies. This is particularly the case when proposed research involves human or other animal subjects, particularly if these constitute "vulnerable" or "sensitive" communities (DHHS, 2014). Because of its physical landscape focus, most geomorphological fieldwork falls beyond the concerns of such committees, except where direct human–environmental interactions are central, where adverse environmental impacts may occur, or where fieldwork is proposed in authoritarian states (Koch, 2013), or regions of political or military instability, where researchers themselves may be at unacceptable risk. Professional ethics and institutional transparency are commonly identified drivers of these programs, but institutional and personal legal liabilities are also implicated.

Management of research funding, including that used during fieldwork, is a contentious issue, with organizations increasingly reluctant to cede control to individuals whom, it is feared, may misappropriate funds, or handle them in unprofessional ways. Institutional budget officers may insist on making financial arrangements, including paying in advance for transportation and accommodation. As with many other things, this has its pros and cons: it lessens the administrative burden on researchers, but it also decreases their decision-making capabilities, and reduces flexibility. In some circumstances using personal funds (particularly if such expenses are tax deductible) and circumventing "official" channels may be worthwhile. Institutionally affiliated travel agencies, for example, may not always provide the cheapest airfares or the most convenient routing, but they may be the sole option if one is dependent on funding from or through the institution.

3.1.10 RECONCILING ACADEMIC AND LOGISTIC PREPARATION

The preceding sections illustrate the importance of recognizing explicitly that preparing for fieldwork in geomorphology necessitates accommodation of both academic goals, and practical constraints. Although individual endeavors are time and location specific, it is only with an appropriate reconciliation of sometimes conflicting priorities and realities that a successful outcome is probable.

One approach to this dilemma is to develop an "ideal" academic fieldwork program initially, allowing for the accomplishment of all the geomorphic goals, and assuming that everything "falls into place" under ideal circumstances. Then, recognizing that this scenario represents optimal and improbable conditions, a series of "What if?" queries can be employed to assess realistically the individual, and ultimately the collective probabilities of the programmatic components proceeding according to plan. One advantage of this approach is that it places priority on the fieldwork goals, as is appropriate with practical considerations, then assuming secondary consideration. Almost inevitably, however, potential problems can be identified that suggest possible modifications, and alternatives. Simple, single-technique plans often present fewer possibilities for foreseeable problems, but any such development could derail the project completely. By contrast, complex multiphase proposals raise multiple prospects for things going awry, but at the same time failure of individual components of a multifaceted program will still allow for the accomplishment of some goals, leading to better overall outcomes.

An alternative, reverse approach, and sometimes the only possibility, is to assess realistically the practical considerations, then address what can possibly be accomplished given the known constraints, and the available resources. This is a less desirable approach, since it relegates professional goals to secondary consideration, but under duress, and in the face of evolving exigencies it may prove the only way of rescuing an otherwise impossible situation.

Each fieldwork project is in some respect unique, and approaches and solutions will need to be tailored to these circumstances. Retention of flexibility, and maintenance of options will always be advantageous, as will the ability to recognize and seize opportunities. No amount of preparation can prepare the fieldworker for every possible eventuality, but considering as many academic and practical options as possible will always be worthwhile.

3.1.11 TRANSPORTATION

Even if fieldwork is local and short term, transportation of personnel and equipment is likely to require organization, expedition, and completion, involving expenditure of time, finances, and other resources. Distant, international, and/or long-term fieldwork involves even greater transportation considerations, which can be severely limiting.

Distance apart, myriad other practical, social, and ethical factors enter into decision-making about appropriate fieldwork transportation. Once on-site, much can be accomplished by walking, or by other environmentally friendly means, such as cycling, on horseback, or via canoe or kayak, assuming these are available. Swimming may be an option, but water and fieldwork equipment are not always compatible. Sailing to islands may be an option, and hot-air balloons or ultra-light aircraft provide often-illuminating landscape views, although their utility is otherwise limited. Donkeys, mules, llamas, camels, and other animals may also be employed, but livestock requires food, water, and tending, which may prove more effort than is worthwhile, although this depends on the circumstances. In remote areas these may be the only on-site options, and they are certainly not to be discounted, particularly where local, slow, low-technology options are appropriate.

Commonly, fieldwork transportation involves motorized vehicles: personal or rented cars, vans, trucks, or motorcycles; powerboats, buses, trains, or aircraft. Public transportation undoubtedly has many merits, but transport of field equipment via bus or train is always likely to prove awkward, and public transportation may not provide adequate flexibility of destination or timing. Domestic or international air travel involves its own suites of pleasures and discomforts; it may be the only viable option for overseas fieldwork, but it comes at a cost, both financial and otherwise. In the context of increasing public concern about global climate change, for example, is the "carbon footprint" of long-distance air travel for fieldwork justifiable? Does the fieldwork bring about sufficient advances in the understanding of nature, or contribute to human well-being enough to override such concerns? Perceptions will differ, and personal decisions are to be made.

The use of personal or rented vehicles for geomorphic fieldwork is widespread and frequent, understandably so given the ubiquitous availability of such means, the inherent flexibility available, and the relatively low costs incurred, particularly with shared expenses. By such means limited numbers of personnel can be moved to fieldwork sites quickly and conveniently, as can all but the bulkiest of fieldwork equipment and supplies. Dual-purpose vehicles, such as pickup trucks, can readily transport both personnel and equipment to and around fieldwork locations, and they may be of even greater utility if they have off-road, four-wheel-drive capabilities. Although they may not be suitable for all purposes, Toyota pickups, Land rovers, and Jeeps all have their adherents, and not without good reason.

Fieldwork transportation may be a relatively simple, straightforward, personal, or group consideration, or it may involve difficult decisions about costs, itineraries, timetables, and participants; but in most cases it is a fundamental matter that should not be ignored or downplayed. Getting there, getting around, and getting back are critical ingredients of successful fieldwork in geomorphology, and they merit appropriate preparation.

3.1.12 ACCOMMODATION

Likewise, accommodation in the field is a central concern of preparation, particularly if the fieldwork duration exceeds a single day by design, or by necessity. The potential range of accommodation options is extremely wide, depending on location, timing, numbers of personnel, financial resources, personal preferences, weather, and other physical and societal considerations. Overnight accommodation in a vehicle, hammock, sleeping bag, or tent may well be adequate and appropriate, but such options, although inexpensive, are not necessarily conducive to successful longer-term fieldwork, especially under less than ideal conditions.

Although there is much to commend self-sufficiency during fieldwork (and beyond), broader cultural and societal considerations may also come into play. Sleeping in a vehicle or in the open may actually be illegal in certain locations, and may arouse suspicion or unwanted attention elsewhere. A hammock and/or sleeping bag may be appropriate in a tropical or temperate forest or in a cave, but not elsewhere. Camping and backpacking provide useful analogies for fieldwork preparation in general, since they encourage preparation for challenging conditions (Curtis, 2005). They may well be appropriate under a wide range of circumstances, but they are not everyone's preferences, and they call for appropriate equipment that not all fieldwork participants may possess. Under certain circumstances, they may be the only viable options, and they certainly build camaraderie, but they are not universally appropriate.

Caving, climbing, hiking, and other recreational, academic or social organizations often have "huts" and other similar premises available for the field use of members and/or guests, and these can be ideal for geomorphic fieldwork, in part because they are designed and/or accustomed to the demands and impacts of people who spend extended periods in wet, muddy, "field" conditions. Of course, locations of such facilities may not accord with ideal fieldwork sites, but selecting a site with such facilities surely has merits. Charges are usually reasonable, facilities are adequate, although perhaps Spartan, provisioning is flexible, and space for personnel and equipment is required, all under the aegis of organizations that understand field needs and imperatives.

Universities, schools, government, and other organizations may also provide appropriate accommodation in potential fieldwork sites, under the umbrella of research stations, field stations, field laboratories, and other facilities. Similarly, individual professionals or their friends and/or relatives in potential fieldwork locations may be willing to provide accommodation and understanding to visiting scientists in similar fields. Such options are always worth exploring, not only for their cost savings but also for their capacity to foster professional relationships and personal friendships, and for their potential to integrate field research into local knowledge and decision-making.

The preceding options all have considerable merits, but sometimes accommodation in hotels, guesthouses, or similar commercial facilities is necessary, if only on a temporary basis. Costs are usually greater, but the convenience and the available facilities may make this worthwhile. Fieldwork is demanding, both physically and mentally, and a period of "rest and recuperation" can contribute substantially to a prolonged fieldwork campaign. Commercial premises are commonly amenable to the needs of fieldwork personnel, although they generally appreciate due warning of possible impacts, and they may be intolerant of what they perceive as abuse of laundry and other facilities.

The broad issue of an individual's or team's contribution to the local community and/or its economy is also a consideration. Although locals may have no objection to field researchers "doing their thing," they are likely to be more appreciative and/or supportive if they perceive a tangible benefit to the community, perhaps in the form of environmental problem-solving, or a contribution to the local economy. Spending time and money in local establishments, such as locally owned hotels, striking up conversation, and explaining professional objectives often go a long way towards local acceptance and support, although they are not always without risk. As in all fieldwork, common sense, judgment, and discretion are vital. Some individuals will always regard "outsiders" with suspicion, where others may welcome variety and fresh insights. Local contacts may serve as mutually beneficial intermediaries, professional arrogance never results in meaningful advances, humility usually works well, and honest exchanges of considered opinions, particularly if tempered by mutual interests, may lead to unforeseen opportunities and long-term relationships.

3.1.13 PROVISIONS

Transportation and accommodation aside, anything beyond very brief-duration fieldwork requires acquisition of sustenance – food and beverages – and other supplies as may become necessary, such as toiletries, medications, batteries, and other items. These may all be brought in to the fieldwork site from outside, and adequate preparation suggests that at least basic or provisional needs should be so met, but extended fieldwork will inevitably require resupply from local sources, whose capabilities and capacities should be assessed in advance.

Food and beverages represent few fundamental supply problems anywhere, although what is available locally may differ from that to which the researcher is accustomed. Making demands on local supplies of food and water where these are already stressed during drought or famine is unconscionable, but otherwise provisioning provides opportunities for considerable mutual material and cultural exchange and benefit, and is not to be underestimated. It is advisable, if not necessary, to adapt to the local diet that reflects the realities of local supply, and to adopt meal schedules, perhaps flexible ones that both accommodate fieldwork agendas and fit with the routines of local providers. Unaccustomed diets may prove difficult for those with specific dietary requirements or food allergies, and these issues need to be negotiated in advance, or remedied subsequently. "Specialized" items may not be available in the field, or they may be prohibitively expensive, so they are best brought in initially. Any items that are not used can perhaps be donated to appropriate local individuals or organizations, although care should be taken not to encourage unrealistic future expectations. Adequate and additional supplies of personal medications and toiletries should always be carried into the field.

Personal lifestyles and preferences, even vices, warrant due consideration. For example, alcohol is widely available internationally, but its quality is highly variable and its consumption in some regions carries cultural stigmas, and it may be prohibited. Elsewhere it is widely accepted, and even its abuse may be condoned or encouraged. The pros and cons of drinking or otherwise engaging with locals should be weighed carefully; often it is unavoidable if local acceptance is to be gained, but it may not necessarily be in the best interests of fieldwork, at least in the short term. As with alcohol, drug laws and their enforcement vary widely, even within countries, states, and districts, and caution, if not abstinence, is always advisable. Local tolerances regarding gambling may vary enormously, and sexuality is a subject to be broached with caution in the field. LGBT (lesbian, gay, bisexual, transgender) individuals may be embraced by some individuals and/or communities, but others may be unwelcoming or hostile. Wherever possible, try to know before you go.

3.1.14 CONSIDERING THE PHYSICAL ENVIRONMENT

Adapting to and working with, rather than against the physical environment is a prerequisite for successful fieldwork in geomorphology. Combating nature is neither sensible nor laudable, and geomorphologists should know better. Professional goals are admirable, but recognizing and adapting to physical constraints, impacts, and realities is a sign of professional and personal maturity, and is to be valued and appreciated.

Acclimatization, both literally and metaphorically, is an essential component of fieldwork, and its importance should not be underestimated. Age, place of birth and upbringing, physical fitness, and other variables play roles, but no fieldworker should ignore the potential physical impacts of undertaking fieldwork at unaccustomed altitudes or latitudes. Beyond interfering with fieldwork plans, altitude sickness is a potentially serious medical condition, as are hypothermia and heat exhaustion (CDCP, 2014a, 2014b, 2014c). Other environmental impacts on personnel health and vitality may be less dramatic, but nonetheless they should be anticipated and prepared for or, better, avoided.

Climate should be on every geomorphologist's preparatory fieldwork checklist, but seasonality and shorter-term weather phenomena may evade due advance recognition among the considerations of other factors. Daily maximum temperatures may be relatively constant in tropical locales, but wet/dry season hydrologic regimes and water supplies may differ dramatically, and unpredicted dry season rains may disrupt travel more than expected wet season deluges. "Unseasonable" warmth or cold has the potential to interfere with fieldwork plans, although not always adversely, and the effects of altitude and aridity on daily temperature regimes, particularly at night, should not be forgotten. Hurricanes and typhoons can kill, so can tornadoes, floods, lightning, and winter storms, but even short-term damp, wind and exposure under "benign" conditions can threaten the unprepared, and curtail fieldwork plans.

Provision of safe and adequate drinking water is problematic in some field areas, particularly in arid regions and those where surface and/or groundwater is contaminated. It may be necessary to bring in potable water, but this is not always feasible, especially for extended field periods, and/or where numerous personnel are involved. Water is also bulky and heavy; 1 gallon/4 L bottles or smaller are convenient, but many of them may be necessary; 5 gallon/20 L containers are good for extended use, but they weigh in excess of 40 pounds, roughly 20 kg, when full, and may be difficult for some personnel to maneuver. Local water treatment in the field may be necessary, but can be time consuming. Single or combined approaches, such as boiling, filtration, activated charcoal adsorption, chemical disinfection, ultraviolet purification, and solar disinfection or distillation may need to be employed; these can eliminate or minimize microorganisms, but will not remove chemical toxins (Curtis, 2005).

Geomorphological fieldwork also involves physical challenges related to the landscape itself, particularly, for example, in mountains, on slopes and in rivers and caves. These challenges should not be underestimated, but should be assessed realistically and holistically in light of participants' physical fitness and experience, and with due regard to the wider range of environmental considerations. Safety considerations will be paramount, and repeated risk assessment may be necessary. Those who work in the field in, on, or by sizeable bodies of water should probably be able to swim! Even seemingly straightforward tasks, such as digging pits and measuring hill slopes, require preparation and physical effort, and are not without risk. Falls, fractures, and abrasions are common fieldwork hazards, as in other aspects of life. The analogy with gardening is an apt one, in that both involve similar efforts, risks, and rewards.

Some landscapes may appear more challenging and threatening than others, but in reality there are risks in all settings, and adequately prepared field researchers can adapt to virtually any set of conditions. Remoteness and isolation pose particular challenges, but with sufficient preparation such locales should be no more inappropriate than local, familiar terrain. Tropical karst, with its climate, biota and ruggedness, represents a daunting challenge to some individuals, whereas others thrive under the conditions. People have their own preferences and tolerances, and it is wise to take these into account. Providing and taking opportunities is commendable, but field personnel should take care before putting themselves and others in situations where they are ill at ease and unlikely to perform satisfactorily.

Biological considerations may also be significant, depending on fieldwork locations and participants' overall health. Fieldworkers should ensure that routine vaccinations are current, particularly with respect to widespread and soil- and water-borne diseases, such as tetanus and hepatitis A. Influenza may be a greater than normal risk for certain individuals, who should plan immunization accordingly. International travel may require proof of vaccination against diseases such as yellow fever for immigration clearance, and vaccinations against regional risks such as typhoid and cholera are highly advisable. Other risks may be less obvious and more specific to individuals, sites, and circumstances; rabies is a threat to those interacting with wildlife and/or domestic animals, and hepatitis B and C may be risks given accidents, contaminated blood supplies, and/or individual lifestyle choices. Antimalarial prophylaxis is important in tropical regions particularly, and there may be other diseases, such as dengue fever and Ebola, against which there are currently no vaccines. Fieldworkers should consult appropriate authorities, such as the World Health Authority, or the Centers for Disease Control and Prevention, prior to international travel, and should heed all relevant advice.

Animals pose threats to fieldwork personnel, some more obvious than others. Large, mobile carnivores and poisonous reptiles are clear risks, but peccaries, army ants, domestic animals, and cattle may be just as dangerous in certain circumstances, and are easily underestimated. Beyond risks of bites and trampling, infection may also be a possibility, and due caution should be exercised. Insects, such as mosquitoes and ticks may or may not pose dangers, but they are often irritants sufficient to deter and disrupt. Appropriate standards of personal hygiene should be observed under all circumstances, understanding that field conditions impose definite constraints.

The preceding advice pertains to a wide array of potential field hazards, and the possibility of accidents should always be acknowledged, with appropriate preparation. Personnel safety is a fundamental consideration, and fieldworkers should be as prepared as possible to deal with all probable accidents and emergency situations. A first aid kit, the sophistication of which will be circumstance-specific, should be carried by any fieldwork party, as should cell/mobile phones or other means of contacting emergency services. Field personnel with medical and/or first aid training are highly desirable, and all participants would benefit from basic knowledge of CPR and other emergency procedures. In general, emergency planning may initially appear to be an unnecessary concern, but experience suggests that it is really a wise precaution.

Numbers inevitably enter into safety considerations. For example, it is generally accepted that caving teams, whether involved in research or otherwise, should have a surface contact who is aware of their whereabouts and plan, and should include a minimum of four individuals so that, in the event of personal injury, one person can remain with the victim while the other two summon assistance, nobody being left alone. Likewise, experienced cave researchers always carry three separate, independent sources of light: a main source, a backup, and an emergency source. Such precautions, although context specific, provide good advice for fieldwork preparation.

Beyond health and safety, general personnel well-being and comfort also merit consideration. Temporary discomfort may be inevitable during fieldwork, but fatigue and irritation impede performance, and may compromise data reliability. During extended fieldwork programs, periods of rest and recuperation may be appropriate, and should be planned and implemented accordingly. Workloads and responsibilities should be made as explicit as possible in advance, so that personnel are cognizant of expectations.

Geomorphologists will have their own considered opinions about the natural world, the importance of wilderness, and the concept of the Anthropocene, but surely no geomorphologist can be professionally unaware of the scope and intensity of human impacts on the Earth, its landforms, and geomorphic processes. As a karst geomorphologist, one sees this particularly in groundwater extraction and contamination, diminished spring discharges, water supply problems, damage to cave formations, accelerated soil erosion, and "rocky desertification" (Wang et al., 2004), and there are clear parallels in other branches of the discipline.

What are our professional ethical and practical obligations in this respect? To what extent should we prepare to speak out about human impacts on the environment (Goudie, 2006)? What are our own impacts during fieldwork, and can we justify them? To what extent are the forms and processes that we measure truly natural, and what is the magnitude of the interference that our professional investigations produce? Do we accept that the emplacement of limestone weight-loss tablets (Trudgill, 1977) causes sufficient disturbance to negate the meaning of any such data acquired by the methodology? If not, can the methodology be justified or its impacts minimized?

3.1.15 CONSIDERING THE HUMAN ENVIRONMENT

Whatever the relationship between geomorphic fieldwork and the natural environment, it is always probable that many of the more immediate considerations and frustrations will be those associated with human constructs, cultures, relationships, and fallibilities, all of which have significant local, regional, and international dimensions. Ultimately, a geomorphologist preparing for fieldwork must, above all, be honest with him/herself about personal character, ethics, integrity, objectives, capabilities, and responsibilities, because these individual introspections will eventually weigh most heavily in later considerations about accomplishment and satisfaction. To be happy with oneself and one's research, fieldwork or no, is worth more than any professional recognition or accolades, however merited and gratifying they may be.

Personal safety, and physical and emotional well-being are vital to the completion of fieldwork, and one should always be attuned to possible dangers and unwanted attention that may emerge from the "... range of unexpected events that can happen while undertaking research" (Diprose et al., 2013, p. 296). Female researchers may or may not be at greater risk than males in this context, particularly in terms of unwanted sexual advances, but all fieldworkers are potential targets, and need to be "...emotionally prepared and supported to cope with unanticipated interactions" (Diprose et al., 2013, p. 296).

Language may be an important consideration in the context of international fieldwork, and even regional dialects may require some attention. Even though English is the *lingua franca* of science, international fieldwork may require the preparatory reading of scholarly articles written in various foreign languages; similarly the interpretation of maps with foreign language legends and/or symbols. The logistics of an international fieldwork campaign will also necessitate familiarity with, and perhaps mastery of the local language in order to organize transportation, accommodation, and provisions. At least one member, preferably several of the field team, should understand and/or speak the relevant language, and local personnel with bilingual and translational abilities are often invaluable. Understanding and being able to communicate in the local language is not only advantageous logistically, but it may also be critical in emergency situations.

Negotiating cultural, religious, and other differences may also pose challenges and opportunities before and during fieldwork. Just as it is necessary to understand and appreciate the physical field environment in advance, so a sound comprehension of the human context is also imperative. Much can be achieved in the field given adequate flexibility and adaptability, but adequate preparation makes this easier. Sacred spaces and culturally sensitive topics should be identified and respected. Local customs and traditions should be recognized similarly. The calendar and timing of local holidays and/or festivals should be noted, because these can interfere with fieldwork programs, although they may also provide considerable opportunity for rest, reflection, and/or social engagement.

Broader funding and financial considerations aside, adapting to local currencies may be straightforward, or not, depending on location and timing. Foreign cash may be welcome, acceptable, or unacceptable, depending on the currency and where one is trying to use it. The US dollar remains widely acceptable globally, and Euros provide similar flexibility in some locations, but currencies with more restricted circulation are generally of less utility abroad. In many parts of the world cash remains the only acceptable field currency, an inevitability that can be daunting, but to which the field scientist should become accustomed. Currency exchange can most safely be negotiated officially, through banks or other venues, with payment of official conversion charges. Unofficial exchange may have the attraction of greater convenience, and/or more advantageous exchange rates but it carries increased risks. Gray- and black-market economies are ubiquitous and widely tolerated, but engaging in them calls for a broad consideration of the potential costs and benefits. Acquiring sufficient fieldwork currency is only one facet of the operation; converting any surplus to another international currency may prove even more difficult, and/or expensive.

International money transfer operations, online banking, and the inexorable diffusion of cash dispensing ATMs render international fieldwork much easier than used to be the case, but they still have their limitations. Local power supplies and internet connections may be unpredictable, ATMs malfunction, and bank holidays interfere with plans. Travelers' checks, once a low-risk mainstay of international fieldwork transactions, may still prove useful for official transactions at banks and hotels, although they are increasingly less used, and even viewed with suspicion in some circumstances. Alternative currencies, even bartering or exchange of goods or services may have their role to play, and should not be ignored.

Attitudes, behavior, appearances, and dress (codes) all merit preparatory consideration because they affect local engagement/acceptance, and influence the potential outcomes of fieldwork plans. Individual personality notwithstanding, deference rather than aggression usually pays dividends, and respect and politeness always accomplish more than condescension and intemperance. Much recommends maintaining a low profile, although this may not always be possible, and it is often advantageous to "blend in" with local appearances and behaviors to indicate and foster acceptance. Being prepared to remove footwear before entering structures, to sit cross-legged on the floor, use one hand rather than the other for specific tasks, or avoid certain offensive hand gestures can make a big difference.

Maintaining appropriate attitudes and behaviors requires self-awareness and practice, as do attention to appearance and dress. Comfortable fieldwork clothing is important, but so may be maintaining the appearance of professionalism. Fieldwork inevitably leads to a state of rumpled accomplishment, and this is usually not incompatible with local standards, or at odds with residents' perceptions of a seasoned budget traveler or a diligent worker, rather than a rich, idle tourist. Adherence to local dress codes is always sensible because inappropriate appearance, even before any behavioral confirmation, can cause offense, and suggests intolerance. Although seasoned and expendable clothing, such as that used for gardening or camping is generally appropriate, fieldwork is often interrupted by invitations to more formal occasions, and having clothes appropriate for these is also a worthwhile consideration, rather than declining invitations, and appearing impolite.

Potential fieldwork sites may be on public or private land, and issues of accessibility and permission will vary accordingly. Gaining access to a fieldwork site clearly is of paramount importance, and this should be negotiated in advance wherever possible. Official or unofficial permission may be necessary; in either instance it is imperative to develop and maintain good working relationships with landowners, land managers, and/or residents. Access to public land is usually not problematic, but gaining permission to carry out field research there may require considerable bureaucratic negotiation. Fieldwork on privately owned land poses other problems, not the least of which may be confirming ownership, and contacting the owner. Individual landowners' understanding of geomorphology, and their inclination to allow research on their property is highly variable and often unpredictable. Persuasion may be time consuming, but trespassing may lead to greater difficulties. Verbal agreements are often sufficient, but some landowners may require written contracts, and signed liability releases, depending on their individual preferences and the prevailing local legal climate. Universities and other organizations are increasingly concerned about legal liabilities, and they may not allow individual researchers to negotiate separate contracts or agreements that involve legal matters. Field researchers will need to deal with these issues on a case-by-case basis, and to make decisions as necessary, sometimes while in the field. Taking chances may reap dividends, but it may also result in serious problems.

The need for fieldwork authorization or permission may extend well beyond individual fieldwork site(s), to encompass local, regional, or national entities. In some places, tribal consent may be necessary, and it is not unusual for mayoral, elder, police, or even military permission to be required. This may be a major problem in authoritarian or totalitarian states (Koch, 2013). Negotiating these matters is time consuming, and often frustrating, but often it cannot be avoided, and it is always a wise precaution in order to avoid confrontation, and subsequent difficulties.

The question of international travel/research/work visas is not always simple. It is often easiest, if not entirely honest, to travel on a visitor or tourist visa, or to state the purpose of a visit as "holiday/vacation," but this can be awkward, if not worse, if one is subsequently questioned by customs officers about scientific equipment, or if one is later detained by local authorities for undertaking fieldwork that arouses concerns or suspicions. By comparison, obtaining an official research permit, if not actually a work permit, can be expensive and time consuming, but it may well prove worthwhile for the official sanction and the peace of mind that it brings. Decisions of this nature will always be case specific, and will hinge upon individual fieldworkers' needs, plans, experiences, and comfort levels. Unnecessary caution may be detrimental to timely fieldwork preparation, but haste and incaution may lead to unanticipated consequences.

Similarly, taking rock, sediment, soil, water, or other fieldwork samples, particularly shipping any such samples internationally, may prove challenging. Rock samples are often bulky and heavy, but they usually cause only minor customs and freight concerns, both outgoing and incoming, unless they are perceived as valuable, or for commercial business purposes. Shipping rock samples by sea is slow but relatively inexpensive; air freight is quicker but more expensive; carrying them in personal luggage may or may not be feasible and cheaper. Which method to employ depends on sample sizes, weights, and considerations of cost and time constraints. Sediment samples too rarely prove problematic, especially if they are relatively small, and apparently inorganic. In both cases, official permits may be required, particularly if rock samples might be interpreted as cultural artifacts, or if sediment samples could be mistaken for narcotics or explosives.

By contrast, shipment of soil samples, particularly internationally, can be much more problematic, in most cases understandably so because of the public and official perceptions of the risks of transmission of diseases, pathogens, or plant and animal pests. Rocks and sediments, being "inorganic," are usually regarded as lower risks, whether or not this realistic, and importing soil samples is almost always more troublesome than exporting them, whether or not that is rational. Soil import permits can usually be obtained by institutions, laboratories, or other entities that can demonstrate that they have the appropriate facilities to safely analyze, handle, and dispose of foreign soil samples, and this usually facilitates import in small, sealed containers. Bulk samples, particularly any with obvious plant or animal constituents, arouse greater concern.

Much water sample analysis needs to be conducted within short time frames in the field, but international shipment of such samples for laboratory analysis can also be challenging, particularly if the samples require refrigeration, or more specialized treatment. Where on-site or local analysis is not feasible, international shipping permits may again be necessary, and appropriate protocols should be observed. Increasingly, water is regarded as a commodity, and its shipping raises questions about potential commercial extraction. Current international airline travel regulations make personal transportation of water samples virtually impossible.

As with the physical environment, much in the human setting presents potential distractions from a fieldwork program. The extent to which one resists or embraces these influences will vary widely, depending on the individual. On the one hand, fieldwork should not be allowed to be derailed by distractions, but on the other hand, accommodation of opportunities that could have benefits: professional, immediate, and tangible, or otherwise, should not be precluded. Being aware of these potential distractions beforehand will not erase them, but it will make their management easier.

It is important also to consider any potential human impacts and ramifications of any proposed geomorphic fieldwork. What impacts are possible or probable? Will they be positive or negative? Do they constrain or preclude the proposed work? Will benefits accrue only to the researcher, or will there be improvements in local living conditions? Are there ramifications for infrastructure, or agriculture, water supply, or slope stability? In a similar vein, what are the field geomorphologist's obligations to local institutions and professionals? With whom should the results of fieldwork be shared? Is the international researcher involved, knowingly or otherwise, in a form of academic intellectual colonialism, or abiding by a professional code of conduct (Atkinson, 2005)? These are practical, professional, and personal ethical considerations, again largely beyond the scope of this review, but they are nonetheless integral to fieldwork preparation, and they may render any proposed research, even that which has a seemingly straightforward Earth-science agenda, susceptible to the attention of institutional review boards, or similar entities.

Returning to an earlier theme, it is important to treat fieldwork, the subsequent analysis of field data, and the writing up and dissemination of the results as integral and holistic, rather than independent. Field data are useless if they are not used, and made available for public and professional scrutiny, and ideally these components should mesh seamlessly. No reason exists for not integrating data analysis and writing into the fieldwork program. Formulating a report/publication outline before starting fieldwork provides a framework that can largely be fleshed out during fieldwork, facilitating identification of problems, and needs. Similarly, writing as much of a report as is possible while still in the field serves to integrate the process and to provide real-time assessment of progress toward professional objectives.

3.1.16 CONCLUSION

Temporary setbacks and difficulties notwithstanding, fieldwork is for many the most enjoyable and rewarding aspect of geomorphology, and the pleasure it brings renders any amount of preparation worthwhile. Colloquial adages, such as "a stitch in time saves nine," and "an ounce of prevention is worth a pound of cure," are not without practical substance, and they provide convenient metaphors for fieldwork preparation, likewise the scouting admonition to "be prepared."

Although it is important to accomplish professional and academic goals in geomorphology, it is imperative to survive the fieldwork experience, and it is also important to enjoy and learn from it. Fieldwork in geomorphology may be dangerous, but it is more often challenging; it is rarely simple. Either way, it calls for physical and mental preparedness, and a determination to make the best of situations and opportunities. Professional objectives notwithstanding, fieldwork should be enjoyed, and savored, like life itself. The information acquired may have short-term professional benefits, but it will also have longer-term positive personal implications. Sharing the experience and the outcomes offers numerous rewards, tangible and otherwise.

Geomorphological fieldwork is a central component of the discipline, meriting the attention that is bestowed on it in this volume, and elsewhere, but each fieldwork endeavor, each piece of geomorphic research is but a small part of a much larger professional, personal, and societal picture – perhaps important, but not the be-all and end-all, and not necessitating obsession. Individuals' approaches to fieldwork will inevitably vary; for example, some researchers favor protracted, ongoing, closely related field research with multiple phases that build upon their predecessors, while others prefer more discrete and variable topics. A place exists for both. Eventually, however, one needs to recognize when it is time to complete a project or a program, and move on to new opportunities and challenges. In fieldwork, as in life, all good things must come to an end, but that end is often made a good deal easier for all concerned by forethought and preparation.

ACKNOWLEDGEMENTS

This contribution draws on a local through international spectrum of short- and long-term fieldwork experience, primarily in the geomorphology of carbonate karst terrains, particularly those in developing areas in the humid tropics of Central America, the Caribbean and insular Southeast Asia, and also in Europe, North America, China, and Australasia. Much of this fieldwork has been carried out individually, or with small teams of colleagues and/or students, but some of it has occurred within the contexts of large multidisciplinary expeditions, and broad programs of international cooperation. In large part, the focus has been on the morphometry and dynamics of karst landforms, such as sinkholes/dolines, dry valleys, and caves, and on processes such as carbonate dissolution, slope failure, and surface subsidence. More recent fieldwork has turned attention to human interactions with karst landscapes, including land uses, and issues of "development," conservation, and designation and maintenance of protected areas. I am deeply grateful to those colleagues, students, friends, family members, and acquaintances, too numerous to list individually, who have accompanied me in the field, or assisted with my fieldwork in any way. I could not have succeeded without them. I have invested a lot of my own resources in fieldwork, but it was worth every penny; for the rest I acknowledge all those institutions, organizations, and individuals who have provided financial assistance. I thank Mary Thornbush, Casey Allen, and Faith Fitzpatrick for their editorial input; omissions, errors, or eccentricities herein are my fault, not theirs.

REFERENCES

Association of American Geographers, 2009. Statement on professional ethics. http://www.aag.org/cs/about_aag/governance/statement_of_professional_ethics.

Atkinson, D. (Ed.), 2005. Cultural Geography: A Critical Dictionary of Key Concepts. I. B. Tauris, London and New York.

Bednar, D.M., 2005. Groundwater Dye Tracing in Karst. Wiley, New York.

Bonney, R., Cooper, D.B., Dickinson, J., Kelling, S., Phillips, T., Rosenberg, K.V., Shirk, J., 2009. Citizen science: a developing tool for expanding science knowledge and scientific literacy. BioScience 59, 977–984.

Bracken, L., Mawdsley, E., 2004. "Muddy glee": rounding out the picture of women and physical geography fieldwork. Area 36, 280–286.

Cann, J., Mayor, H., 2004. Liability in the field. Geoscientist 14 (11), 16–17.

Centers for Disease Control and Prevention, 2014a. Altitude illness. http://wwwnc.cdc.gov/travel/yellowbook/2014/chapter-2-the-pre-travel-consultation/altitude-illness.

Centers for Disease Control and Prevention, 2014b. Winter weather. http://emergency.cdc.gov/disasters/winter/.

Centers for Disease Control and Prevention, 2014c. Heat stress. http://www.cdc.gov/niosh/topics/heatstress/.

Chorley, R.J., 1966. In: Dury, G.H. (Ed.), The application of statistical methods to geomorphology. Essays in Geomorphology, Heinemann, London, pp. 275–388.

Chorley, R.J., Dunn, A.J., Beckinsale, R.P., Burt, T.P., Brunsden, D., Cox, N.J., Goudie, A.S., 1964. The History of the Study of Landforms; or, the Development of Geomorphology4Wiley and New York.

Church, M., 2013. Refocusing geomorphology: fieldwork in four acts. Geomorphology 200, 184–192.

Clifford, N.J., Valentine, G., 2003. Key Methods in Geography. SAGE, London and Thousand Oaks.

Cornelius, S.C., Sear, D.A., Carver, S.J., Heywood, D.I., 2006. GPS, GIS and geomorphological field work. Earth Surf. Proc. Land. 19 (9), 777–787.

Cullingford, R.A., Davidson, D.A., 1980. Timescales in Geomorphology. Wiley, New York.

Curtis, R., 2005. The Backpacker's Field Manual. Three Rivers Press, New York.

Dackombe, R.V., Gardiner, V., 1987. Geomorphological Field Manual. Allen and Unwin, London.

Day, M.J., Goudie, A.S., 1977. Field assessment of rock hardness using the Schmidt Test Hammer. Brit. Geomor. Res. Group Tech. Bull. 18, 19–29.

Department of Health and Human Services, 2014. Institutional review boards. http://www.hhs.gov/ohrp/assurances/irb/.

Diprose, G., Thomas, A.C., Rushton, R., 2013. Desiring more: complicating understandings of sexuality in research processes. Area 45 (3), 292–298.

Donne, J., 1959. Devotions Upon Emergent Occasions, Together with Death's Duel. Ann Arbor: University of Michigan Press, Toronto, Canada, originally published 1624, London.

Driver, F., 2000. Field-work in geography. Trans. Inst. Br. Geogr. NS 25, 267–268.

Ehlen, J., Haneberg, W.C., Larson, R.A., 2005. Humans as Geological Agents. Geological Society of America, Boulder.

Gatrell, J.D., Bierly, G., Jensen, R.R., 2012. Research Design and Proposal Writing in Spatial Science. Kluwer, Dordrecht.
Georges, R.A., Jones, M.O., 1980. People Studying People: The Human Element in Fieldwork. University of California Press, Berkeley.
Gerber, R., Chuan, G.K., 2000. Fieldwork in Geography: Reflections, Perspectives and Actions. Kluwer, Dordrecht.
Goudie, A.S. (Ed.), 1990. Geomorphological Techniques. Unwin Hyman, London.
Goudie, A.S., 2006. The Human Impact on the Natural Environment: Past, Present and Future. Blackwell, Oxford.
Gregory, K.J., Goudie, A.S., 2011. The SAGE Handbook of Geomorphology. SAGE, Los Angeles.
Haining, R.P., 2003. Spatial Data Analysis Theory and Practice. Cambridge University Press, Cambridge and New York.
Hay, I. (Ed.), 2010. Qualitative Research Methods in Human Geography. Oxford University Press, Oxford.
Hobbs, D., Wright, R., 2006. The SAGE Handbook of Fieldwork. SAGE, London and Thousand Oaks.
Kent, M., Gilbertson, D.D., Hunt, C.O., 1997. Fieldwork in geography teaching: a critical review of the literature and approaches. J. Geogr. Higher Educ. 21 (3), 313–332.
Keylock, C., 2003. The natural science of geomorphology? In: Trudgill, S., Roy, A. (Eds.), Contemporary Meanings in Physical Geography: From What to Why?. Routledge, London, pp. 87–102.
Koch, N., 2013. Introduction – field methods in "closed contexts": undertaking research in authoritarian states and places. Area 45 (4), 390–395.
Kuklick, H., Kohler, R., 1996. Science in the Field. Cornell University Press, Ithaca.
Lane, S.N., Chandler, J.H., 2003. Editorial: the generation of high quality topographic data for hydrology and geomorphology: new data sources, new applications and new problems. Earth Surf. Proc. Land. 28 (3), 229–230.
Latour, B., 1999. Circulating Reference: Sampling the Soil in the Amazon Forest. In: Latour, B. (Ed.), Pandora's Hope: Essays on the Reality of Science Studies. Harvard University Press, Cambridge, pp. 24–79.
Maguire, S., 1998. Gender differences in attitudes to undergraduate fieldwork. Area 30, 207–214.
McGuinness, M., Simm, D., 2005. Going global? Long-haul fieldwork in undergraduate geography. J. Geogr. Higher Educ. 29 (2), 241–253.
Medawar, P.B., 2009. Induction and Intuition in Scientific Thought. Routledge, London and New York.
Millar, S.W.S., 2013. Spatial and Temporal Scales in Geomorphology. Shroder, J., Orme, A.R., Sack, D. (Eds.), Treatise on Geomorphology, The Foundations of Geomorphology, Vol1. Academic Press, San Diego, pp. 130–145.
Nast, H.J., 1994. Women in the field – opening remarks. Prof. Geogr. 46, 54–66.
National Speleological Society, 2014. American caving accidents. http://www.caves.org/pub/aca/.
Phillips, R., Johns, J., 2012. Fieldwork for Human Geography. SAGE, London.
Richards, K., 1996. Samples and Cases: Generalisation and Explanation in Geomorphology. In: Rhoads, B.L., Thorn, C.E. (Eds.), The Scientific Nature of Geomorphology. Wiley, New York, pp. 171–190.
Rogerson, P., 2001. Statistical Methods for Geography. SAGE, London.
Sherman, D.J., 1999. Methodology in geomorphology: traditions and hypocrisy. Ann. Assoc. Am. Geogr. 89 (4), 687–696.
Schwartz, M.D., 2003. Phenology: an integrative environmental science. Kluwer, Dordrecht.

Seijmonsbergen, A.C., 2012. Current trends in geomorphological mapping. Geophysical Research Abstracts 14, EGU2012-6114.

Slaymaker, O., 1991. Field Experiments and Measurement Programs in Geomorphology. A.A. Balkema, Rotterdam.

Smart, P.L., Laidlaw, I.M.S., 2010. An evaluation of some fluorescent dyes for water tracing. Water Resour. Res. 13 (1), 15–33.

Solem, M., Foote, K., Monk, J., 2009. Aspiring Academics: A Resource Book for Graduate Students and Early Career Faculty. Pearson/Prentice Hall. Upper Saddle River, NJ.

Sui, D.Z., Elwood, S., Goodchild, M.F., 2013. Crowdsourcing Geographic Knowledge. Springer, Dordrecht and New York.

Switzer, A.D., Kennedy, D.M. (Eds.), 2013. Methods in Geomorphology. Treatise on Geomorphology, Vol. 14. Academic Press, San Diego.

Thornes, J.B., Brunsden, D., 1977. Geomorphology and Time. Wiley, New York.

Trudgill, S.T., 1977. Problems in the estimation of short-term variations in limestone erosion processes. Earth Surf. Proc. 2 (2–3), 251–256.

Trudgill, S.T., 2003. Meaning, Knowledge, Constructs and Fieldwork in Physical Geography. In: Trudgill, S., Roy, A. (Eds.), Contemporary Meanings in Physical Geography: From What to Why?. Routledge, London, pp. 25–46.

Wang, S.J., Liu, Q.M., Zhang, D.F., 2004. Karst rocky desertification in southwestern China: geomorphology, landuse, impact and rehabilitation. Land Degrad. Dev. 15 (2), 115–121.

Warburton, J., Higgit, M., 1997. Improving the preparation for fieldwork with IT: two examples from physical geography. J. Geogr. Higher Educ. 21 (3), 333–347.

Wiggins, A., Crowston, K., 2011. From Conservation to Crowdsourcing: A Typology of Citizen Science. Proceedings of the 2011 44th Hawaii International Conference on System Sciences. IEEE Computer Society, Washington DC, pp. 1–10.

Chapter 3.2

Field Safety: Principles, Practice, and Culture

Sam Alpay[1], Roger C. Paulen
Geological Survey of Canada, Natural Resources Canada, Ottawa, Ontario, Canada
[1]Corresponding author: E-mail: Sam.Alpay@NRCan-RNCan.gc.ca

3.2.1 INTRODUCTION

As evident throughout this volume, awareness of safety issues permeates all facets of fieldwork, from planning to preparation, execution, and return to headquarters. The purpose of this chapter is to reinforce the concept that field-safety principles and practice are indispensable elements of professionalism in geology and geomorphology and can cultivate a culture of safety. This is meant to exceed a level of safety practice beyond adherence to guidelines, directives, and regulations that one can concede are necessary decrees which have escalated in number and accountability over the last two decades. The intention is for both individuals and scientific organizations to adopt a habit of identifying and mitigating risks, which requires conscious forethought and decision-making. Diversity of fieldwork in the natural sciences presents multiple challenges for field-safety planning. However, the key principles and approach are applicable regardless of terrain, weather, transportation, equipment, instruments, communications, wildlife, partnerships, jurisdictions, and legislation. Although the practices will necessarily vary according to the type of field activity and risks, the approach remains universal.

3.2.2 PRINCIPLES

At the crux of field safety practices are basic principles. Not only do they provide the basis for creating field safety practices, but also an opportunity to develop a profound awareness that each individual must take personal responsibility for safety, regardless of organizational governance structures and hierarchies. Ideally, the approach of logically prioritizing field safety issues and addressing them promptly with reasoned risk management strategies will lead to a direct correlation between principles and practice.

Developments in Earth Surface Processes, Vol. 18. http://dx.doi.org/10.1016/B978-0-444-63402-3.00005-4

3.2.2.1 Preservation of Life, Health, Property, and the Environment

The paramount goal of field safety is the preservation of life, health, property, and the environment. Although this basic precept is obvious, the increasing web of requirements, including labor codes, collective agreements, directives, policies, procedures, regulations, guidelines, controls, standards, and processes, can confound the true intent. Satisfying the mandatory requirements, which can be dispatched mechanically as a matter of course in the name of due diligence, could inadvertently thwart the critical-thinking process needed to meet the primary objective of preservation of life, health, property, and the environment. It takes forethought and integrity, on the part of the individual and the organization, to return to the fundamental purpose of field safety and consciously identify a simple, practical course of action that directly aligns with that objective and simultaneously meets legislated requirements.

3.2.2.2 Culture and Conscience

Within a positive culture of safety, the onus of responsibility lies with everyone. The term, safety culture, was coined by the International Nuclear Safety Advisory Group (INSAG) after the nuclear accident at Chernobyl in 1986. The lack of safety culture was postulated as a causal factor in the disaster (Pidgeon, 1998 and references therein;Edwards et al., 2013). The definition of safety culture continues to be debated (e.g., Cox and Flin, 1998; Chowdry et al., 2007; Antonsen, 2009; Edwards et al., 2013), although an early definition captures several key elements: "the set of beliefs, norms, attitudes, roles, social and technical practices concerned with minimizing the exposure of employees, management, customers and members of the public to conditions considered dangerous or injurious" (Turner et al., 1989; Cox and Flin, 1998). Since then, safety culture has become part of the vernacular of occupational health and safety (OHS) and remains a focus of active research (e.g., Chowdry et al., 2007; Antonsen, 2009; Edwards et al., 2013). The key elements of a safety culture include individual and organizational or legislative components. Although controls, such as policies, processes, and directives, can support increased safety and are relatively straightforward to impose, beliefs and attitudes of employers and employees are less tangible. However, they can be influenced by raising the consciousness of safety issues and individual engagement in critical thinking about safety practices over time (e.g., Reason, 1998).

Within a proactive culture of safety, employers, supervisors, and managers use risk-management tools at their disposal and develop tools, or customize existing tools, for specific applications. They provide relevant, adequate training to employees to mitigate risk, are responsive to employee initiatives and whistle-blowers, and ideally, model a culture of safety in their practices, individually and collectively. Students and field personnel are engaged in field-safety initiatives, both in response to opportunities given by employers and proactively by sharing practices or raising issues for improvement. A major contribution of field parties is to identify risks and hazards on the basis of their expertise, operational experience, and practical knowledge of fieldwork. Incorporating safety into the full spectrum of fieldwork pervades

FIGURE 3.2.1 A blunt, yet effective message placed on an active logging road, northern Vancouver Island near Port Hardy, British Columbia, Canada (*photo courtesy of P. Bobrowsky, Geological Survey of Canada; August 1995*).

the natural-science community and represents a constructive shift in consciousness. It is among one of the key considerations in optimizing fieldwork: scientific objectives, risk, time, cost, productivity, and logistics. Assessing and addressing risk, whether foreseen or unexpected, is a fundamental act of conscience (Fig. 3.2.1).

3.2.2.3 Approach

Being mindful that "one size does not fit all", a preliminary scan of anticipated fieldwork categories can quickly prioritize actions for preparedness and decision-making. For example, given that fieldwork can be conducted in urban, suburban, rural, wilderness, or remote settings, the degree of first aid training requirements will depend on a number of factors, including proximity of field sites to emergency medical care and the nature of hazard associated with the field tasks. Not all strategies are as evident as for the level of first-aid training required, but the exercise of identifying hazards and risk mitigation is pertinent to the array of challenges field parties face. The approach must be flexible enough to accommodate a range of field activities, locations, and working conditions. Note that standardization is not necessarily warranted, particularly if it impedes the thought processes required for field safety in practice and imposes liability planning in place of safety planning. Furthermore, standardization can lead to forced compliance with norms that were developed without full understanding or knowledge of a specific field situation and may actually pose increased inherent risk to safety. Ideally, a variety of supporting mechanisms encourages critical thinking for field safety and can form a practice for sound field operations.

3.2.3 PRACTICE

The number and type of practices vary across organizations, jurisdictions, and as a function of the risks posed by fieldwork. Here, examples of practices are presented, mostly from Canada, which point to a range of options to identify hazards and mitigate risk.

3.2.3.1 Standards

In a truly open, safety culture, standards for field safety are dynamic and depend on a feedback loop for continual renewal and improvement as knowledge advances and experience is gained. Sadly, much advancement in field safety is gleaned in hindsight from field incidents, accidents, illnesses, injuries, and fatalities. Further progress in safety culture research developed after major disasters, such as the nuclear accident at Chernobyl, Three Mile Island in the USA, the Union Carbide toxic gas leak in Bhopal, the Exxon Valdez Alaskan oil spill, and the NASA Challenger space shuttle tragedy, to name a few (e.g., Cox and Flin, 1998; Pidgeon, 1998 and references therein; Reason, 1998).

3.2.3.1.1 Organizational

Most major science organizations develop field-safety guidelines that are tailored to their field activities. As dictated by the nature of fieldwork and in fulfillment of regulatory and legislative requirements, field-safety guides and manuals are customized for different organizations and can provide a substantive resource for frequently asked questions (e.g., Earth Sciences Sector, 2012; Yukon Geological Survey, 2012; United States Geological Survey, 2014). Some field guides offer detailed protocols for specific field situations, whereas others are more general and defer to advanced training for specific fieldwork situations. Field-safety guides can also serve to interpret regulatory requirements for the field operations of their organizations. In some cases, institutional guidelines exceed regulatory requirements and must be clarified explicitly to avoid misinterpretation.

3.2.3.1.2 Regulatory

Legal requirements were formulated with the primary principle of workplace safety in mind. Anyone involved in fieldwork, as a field-party member, manager, supervisor, equipment supplier, communications provider, administrative support, logistical coordinator, or safety trainer, is advised to become familiar with these as part of their responsibilities. For example, in Canada, regulatory provisions in the Canada Labour Code Part II (Parliament of Canada, 2000), Bill C-45 of the Criminal Code of Canada (Parliament of Canada, 2003), Canadian Occupational Health and Safety Regulations (Government of Canada, 2014), and collective agreements must be satisfied. Parallel legislation exists worldwide under different jurisdictions (e.g., United States Department of Labor, 1970; Parliament of the United Kingdom, 1974; Health and Safety Executive, 2013). Compliance with international regulations is also essential, for example for the transportation of dangerous goods (TDG; Department for Transport, 2012; United Nations Economic Commission for Europe, 2013a; United States Department of Transportation, 2014). Some directives for safety training (e.g., Workplace Hazardous Materials Information System (WHMIS)) are also specified for relevant fieldwork and are subject to periodic renewal (e.g., WHMIS to the Globally Harmonized System (GHS) of classification and labeling of chemicals for international standardization; United Nations Economic Commission for Europe, 2013b). Additional support can help employers and employees interpret regulatory requirements for the particulars of their fieldwork activities and environments (discussed in Sections 3.2.3.2.5 and 3.2.4).

3.2.3.1.3 Benchmarking

For some field activities, no clear legislated, manufacturers', or organizational standards are set. In these instances, organizational practices can be developed by consulting with or benchmarking against external practices for similar activities in government, industry, academia, the military, and in other jurisdictions. Benchmarking can also serve to verify and compare practices with different organizations to explore risk mitigation strategies and develop standards. Organizations that have developed standards for field safety are encouraged to make them available for others to consult.

3.2.3.1.4 Training

Training standards are available among training providers for some types of training and not for others. Even when training standards are outlined by training providers, it is up to the science organization to both define standards and set the minimum requirements of training in detail to mitigate risks in the field. If we consider a prior example of first-aid training, it is reasonable that a field crew working in an urban setting in lower-risk activities and close proximity to emergency medical services would receive a different level of training than a crew that is engaged in higher-risk field operations in remote areas of the Canadian high Arctic. It is vital that, in the absence of training standards, scientific organizations consciously set the required standard for their operations rather than taking a convenient training product "off the shelf" that may satisfy the letter of the law, but might not meet their practical needs. Furthermore, individual field-crew members should be trained to a level of personal confidence in their preparedness for the type of field risks they face. If they are not satisfied with the level of training, it is up to them to communicate what is needed to become more prepared. It is, then, the employer's responsibility to respond appropriately. Further consideration of training practices is given in Section 3.2.3.3.

3.2.3.2 Processes

Processes are the risk management tools used to support field safety, formal or informal, proactive or reactive, and individual or collective. The rationale behind field-safety processes is to trigger forethought in field planning. In addition, it is vital to use processes as part of an iterative feedback loop to review field-safety protocols periodically, make recommendations for improvement, and implement them effectively and efficiently.

3.2.3.2.1 Risk Management and Task Hazard Analysis

Task-hazard analysis is the prime tool used in field safety and risk management planning which consists of two priorities: (1) identifying hazards associated with field activities, transportation, equipment, instrumentation, personnel, and working environments, and (2) mitigation strategies to eliminate or reduce risks. Initially, field parties identify hazards of each of their field activities, including everyday tasks which, under field conditions, increase risk. For example, a seismic survey in a downtown environment may need to be conducted off peak hours and warrant appropriate temporary road signs or police escorts to caution motorists and minimize the risk of traffic accidents (Fig. 3.2.2; Pugin et al., 2013). Judicious implementation of risk assessment would include systematic itemization of planned field tasks and associated hazards. Each risk must be accompanied by a practical mitigation strategy. Although the examples in Table 3.2.1 are generalized, to be effective, a task-hazard analysis must be thought through comprehensively and be specific in detail.

FIGURE 3.2.2 Seismic survey during low traffic hours with police escort in a downtown urban setting on Rideau Street, Ottawa, Canada (*photo courtesy of A. J-M Pugin, Geological Survey of Canada; May 2011*).

TABLE 3.2.1 Examples of Task Hazards and Mitigation Strategies

Example of task hazard	Proposed mitigation
Presence of predatory wildlife	Identify the geographic range of foreseeable wildlife encounters that intersect field sites (e.g., polar bear, cougar); train for predator defense strategies and deterrents (lethal and non-lethal); train for and bring necessary equipment (e.g., bear spray, electrified fences, firearms); engage wildlife monitors, if necessary
Challenging terrain (e.g., high altitude)	Evaluate the terrain in advance of fieldwork; obtain relevant training and practice (e.g., in mountaineering and first aid); confirm physical fitness to participate in fieldwork
Availability of potable drinking water	Verify potable water supply in advance (e.g., with named local authorities); bring contingency drinking water supply where feasible; prepare portable water purification systems, tablets and/or ability to boil water
Exposure to insects, spiders, or other vectors of disease (e.g., Lyme disease, cellulitus, hantavirus, rabies)	Learn the geographic range of carriers of disease; find out how to recognize the carriers; avoid or repel carriers (e.g., using insect repellent, sterilization, appropriate clothing, and personal protective equipment); understand exposure routes and symptoms; learn appropriate response in case of exposure; vaccinate when applicable
Contact with poisonous plants (e.g., poison ivy, poison oak, poison sumac, giant hogweed)	Check the geographic distributions of poisonous plants; find out how to recognize them in the field so they can be avoided; learn treatment methods if contact is made; become conversant with preventive measures specific to the risk (e.g., clothing, personal protective equipment)

A myriad of hazards exist that can each be paired with preventive measures and mitigation strategies. Among other general examples of potential hazards are: bush travel; use of firearms for wildlife defense; chainsaw operation; trenching and drilling; all terrain vehicle (ATV) deployment; snow-machine operation; boating; shipboard research; helicopter or aircraft transportation; traffic in regions undergoing active logging or energy development; work in confined spaces; working alone; unstable slopes (avalanche); UV radiation (sunlight); heat stress; lightning strikes; exposures to flu and disease; physical overexertion; chemical handling; exposure to radioactivity; forest fires; natural disasters; winter field conditions (e.g., wind chill, hypothermia, frost nip, frostbite); challenging driving conditions; overlap with hunting seasons; unfriendly landowners; working from ice cover; manual lifting of heavy loads; hitching a trailer; personal security (e.g., in regions of social unrest); and food poisoning (Fig. 3.2.3).

FIGURE 3.2.3 Examples of fieldwork in the Canadian Arctic, boreal forest, and Rocky Mountains, respectively, which show various risk elements. (A) Foot traverse in poor weather conditions on the Cumberland Peninsula, Baffin Island, Canada (*photo courtesy of M. Sanborn-Barrie, Geological Survey of Canada; July 2009*). (B) Large black bear showing a focused, aggressive stare, in northern Alberta, Canada (*photo courtesy of R. Paulen, Alberta Geological Survey; August 2007*). (C) Mobilizing a field party camp by helicopter, Treaty Glacier area, British Columbia, Canada (*photo courtesy of B. Anderson, Geological Survey of Canada; July 2004*).

Field personnel also must assess risk for themselves. For example, a known allergy would necessitate additional personal planning and mitigation (e.g., epinephrine). Individuals within field parties should actively take responsibility for their personal health and task-hazard analysis when planning fieldwork (Section 3.2.3.4.1).

Many organizations use a listing of potential hazards during fieldwork for effective planning (e.g., University of Northern British Columbia, 2014; University of Texas at Austin, 2010; Geological Association of Canada, 2011). These tools are meant to stimulate, rather than replace, forethought by field crews to identify foreseeable risk, prepare for it, and for the unexpected.

3.2.3.2.2 Emergency Planning

Well before departure, planning for an emergency forces field parties to: (1) seek appropriate emergency communication devices; (2) identify the relevant emergency response systems available for their field locations and type of emergency; and (3) ensure that practical emergency information can be accessed rapidly by authorities. It is vital both to choose and to test communication devices for the field location and type of work (e.g., waterproof; viable in remote locations; distress signal with coordinates; two-way communication if warranted; power supply and spare). Identifying the appropriate emergency response systems for each field location is essential (e.g., police; ambulance; fire; hazardous materials and items (Hazmat); search and rescue; emergency medical evacuation; poison control). It is vital to identify the most appropriate emergency responder, and not necessarily the closest, depending on the type of emergency (e.g., nearest hyperbaric chamber for decompression sickness in SCUBA divers rather than nearest hospital). Emergency response systems should not be assumed for a given region; they must be confirmed and authenticated (e.g., by local authorities, jurisdictions, or expediters). The quickest route to receiving an effective emergency response is to send the correct information to the right party without, or with minimal, intermediaries. Therefore, field-party information (location, number in party, type of distress) should be made readily accessible to emergency response providers. Typically, one field-party member initiates an emergency call, yet others in the field crew may be separated from them, at risk, and in need of assistance. Emergency responders need rapid, reliable information from the field crew or headquarters about which personnel are in the field, their activities, and location. As obvious as this may seem, it is not always verified in practice.

3.2.3.2.3 Reporting

A safety reporting system must assemble, examine, and disseminate information from incidents and near misses, in addition to regular proactive checks of safety practices (Reason, 1998). Incidents, accidents, illnesses, injuries, and fatalities are reported through a variety of formal mechanisms which foster review and revision of field-safety practices (e.g., in Canada, Workplace Health and Safety Committees (Human Resources and Skills Development Canada, 2009a); Hazardous Occurrence Investigation Recording and Reporting (Human Resources and Skills Development Canada, 2010); Workplace Safety and Insurance Act in Ontario (Ontario Ministry of Labour, 1990)). Informal avenues of reporting also can cultivate greater learning and sharing of practices. A field-season review of safety (organization-wide or by working groups), daily safety planning during fieldwork, regular fieldwork debriefings within field parties, and field-safety advisory groups are examples of creating a forum for discussion among colleagues. Some field parties extend their discussions by fabricating mock incident scenarios and brainstorming appropriate actions as a team. It can be as important as formal mishap reporting to discuss a near-miss situation that occurred, or to use colleagues as sounding boards to evaluate an action, proactive or reactive. These practices allow working groups and organizations to think through options for field safety so that improved decision-making can take place in the future. Reporting, whether formal or informal, is a means to identify the root cause of a deviation from safety planning and make recommendations that will inevitably lead to better problem solving and preparedness.

3.2.3.2.4 Controls

Controls allow for accountability. They are either regulatory standards or organizational requirements that can be legally binding. It is imperative that these minimum requirements be enforced consistently without exception. Controls are particularly useful mechanisms if employers (supervisors, managers) or employees (professors, students, field parties) are less proactive about safety planning and have not yet joined the positive safety culture of their organization and colleagues. Controls, such as field itineraries, trip plans (e.g., Canadian Red Cross Society, 2013), checklists, work agreements, waivers, emergency planning documents, stepwise or conditional fieldwork approvals, and standard operating procedures, are explicit contracts between employer and employee to document risk analysis and mitigation strategies. However, it can be a sign of complacency to resort to field-safety controls exclusively. If controls are the only recourse to communicate field-safety practices, it is a warning signal that a healthy, operational culture of safety has not been established.

3.2.3.2.5 Interpretation

Legislated requirements or guidelines imposed by organizations are generally open to interpretation in how they are applied in practice. Field parties and managers are not necessarily aware of the implications of regulations on their proposed work. If they exist or can be created, supplementary resources, such as field-safety manuals, checklists, tutorials, workshops, meetings, and answers to frequently asked questions, can provide clarification as warranted. Better still, accredited OHS expertise can be consulted to provide practical interpretations of field-safety requirements that are tailored to the needs of field parties and supervisors (Section 3.2.3.2.4).

Resources for interpretation of field-safety requirements can support both employers and employees. For example, employers in positions to review field-work proposals would benefit from a flow chart or decision tree of the criteria that must be satisfied to proceed with approvals (e.g., appropriate training; valid certifications; current health assessments; documented itinerary; and logical emergency response planning), particularly for those who are new to management of field safety. Minimum training requirements for the organization's field activities should be unequivocally defined in a citable format for both employers and employees. For example, if a general class driving license is sufficient for field personnel to use vehicles operated by the organization, then it is stated as a minimum requirement. However, if the organization has access to other field vehicles under a different license class that requires advanced training and certification, then a general class license cannot be assumed as adequate. The minimum training requirements (e.g., for each class of vehicle) must be made readily available and communicated explicitly for employers and employees to take appropriate action.

3.2.3.2.6 Delegation

Ultimately, responsibility for field safety cannot be delegated. However, the tasks can be shared to alleviate the burden on individuals. In larger field camps, the scientific leader or the logistical coordinator often carries the onus of field-safety tasks, in addition to prioritizing scientific goals, arranging travel, securing permits, ordering supplies, and calibrating field instruments among numerous field preparations. Sensitivity to workloads must be taken into consideration. In larger field parties, it could be effective, within the field crew, to appoint a field-safety officer who would carry out safety tasks and report to the field-party leader. Likewise, consulting with relevant OHS expertise is invaluable to expedite the effectiveness and efficiency of field-safety planning. By sharing tasks for field-safety preparation, more field-crew members and employees are engaged in the process, which inherently generates more discussion and consideration of field-safety issues. That, in itself, is an improvement to the culture of safety in the workplace. However, delegation of responsibility is not an option; fundamentally, responsibility belongs to all.

3.2.3.3 Training

The adage that "one size does not fit all" also holds for training. Training requirements (e.g., content; scope; delivery; standard; specialization) must be specified by the needs of the end-user and is a direct outcome of effective task-hazard analysis (Section 3.2.3.2.1).

3.2.3.3.1 Training Needs Analysis

In practice, training is a key element of risk reduction and can run the gamut of field operations and work environments. A training needs analysis is warranted

TABLE 3.2.2 Components of Training Needs Analysis for Fieldwork

Component	Purpose
Task-hazard analysis (Table 3.2.1)	Evaluate level of skill and training required to mitigate risks of each hazard associated with field tasks, transportation methods, equipment use, deployment of instrumentation, and field environments
Content and delivery	Confirm that training content is directly pertinent to task hazards, level of skill required, and exposures to risk and that it is consistent with legislated safety requirements; verify appropriateness of training delivery method to maximize learning (e.g., practical hands-on component in the field, classroom instruction, on-line)
Trainer/ trainee	Identify who requires the training (i.e., employees and others who have access to the fieldworkplace, equipment suppliers, OHS personnel) and to what degree of competence; identify qualified training provider(s)
Rationale	Justify training as a mitigation strategy against exposures to risk during fieldwork
Alternate risk mitigation	Assess whether or not training is the most effective or the only mitigation strategy to reduce risk; explore alternate strategies
Cost benefit	Consider the investment in training against its effectiveness in mitigating risk; this would include systematic evaluation of training by field-party members and optimizing economies of scale
Structure	Identify minimum core training requirements that are consistent across the organization to ensure critical coverage and optimal balance in a training plan; support development of common basic skill levels within and among field parties before adding specialized or advanced training for variations in task hazards and environments of fieldwork
Review	Evaluate effectiveness of previous training for risk reduction of fieldwork; make recommendations; implement improvements

with leadership from employers, OHS expertise, and front-line, field-party crews. A sound training needs analysis consists of several factors, summarized in Table 3.2.2.

Additional consultation can prove useful during the training-needs analysis. Union representatives, Workplace Health and Safety Committees, external organizations for benchmarking, equipment suppliers, field-safety training providers, and collaborating organizations can provide relevant advice.

3.2.3.3.2 Training Standards

The specific content of training should be determined by the task hazard and training needs analyses and not by the availability of training courses and the lowest bidder. Most field-safety training is not standardized. However, given that employers are legally responsible to set the training standard, they can establish training specifications as required criteria in the statement of work that must be satisfied for a service provider to be engaged. Training specifications, including scope, depth, and restrictions, must both address the practical needs of realistic field situations and communicate expectations of the training to potential providers. Field personnel should be included in the training needs analysis to describe realistic field conditions that necessitate practical training. For example, ATV training on a flat grassy field may qualify a field-party member to operate an ATV, but does not provide an adequate simulation of field terrains.

If not already set by legislation, a minimum common training standard for core field activities can also be set by prioritizing training needs. Ideally, flexible training modules should be incorporated into the courses to match the degree of risk different field teams confront. For example, basic first aid and cardio-pulmonary resuscitation (CPR) may be appropriate for lower-risk activities in urban settings. However, advanced wilderness first aid and survival skills may be necessary for fieldwork in more remote locations. Once minimum training standards are set, optional training modules can be added for advanced or specialized needs according to the task hazards of field duties, transportation, equipment, instrumentation, and environments. It follows that training should also be taken in a logical order, so that the minimum training requirement is met first on a given subject, followed by more advanced training.

Standards for the validity periods of certain types of training or certifications are either legislated or instituted by individual organizations. For example, the Canadian Firearms Safety Course must be taken every 5 years and before applying for a Possession and Acquisition License (Royal Canadian Mounted Police, 2014). However, organizations can add a requisite of practical hands-on training to sharpen firearms skills, perhaps annually, particularly if the end-user is less familiar with handling firearms for predator defense (Earth Sciences Sector, 2014).

3.2.3.3.3 Training Coordination

If pragmatic, centralized training coordination within organizations optimizes a number of factors. Centralized coordination can ensure consistency of minimum training standards delivered to field parties across the organization. Training providers who meet the necessary criteria can be identified on behalf of the entire organization and communicated. Coordinated training can also avoid booking sessions when field parties are heavily occupied with field planning and scientific preparations. Scheduling must be flexible enough throughout the year to accommodate recently hired employees and students, who are often engaged just before departure, and may not have the required level of training yet. If booked in advance, field teams can also take the training together and become familiar with each other's skills and comfort levels; then, once in the field, responsibilities for field activities can be assigned accordingly to optimize pairings and strengths. Training that is centralized also increases economies of scale by booking training for larger groups of field parties simultaneously.

Coordination also includes creating a repository for training records to track employee qualifications to participate in fieldwork (i.e., types of certifications and validity periods). A database can be centralized corporately or created for working groups, depending on organizational structure. Organized and current training records are practical for both employees, by notifying them when renewals are forthcoming, and employers, by confirming that field-safety training requirements have been fulfilled before fieldwork is approved.

3.2.3.3.4 Training Delivery and Expertise Recognition

Effective training simulates field conditions accurately and provides practical hands-on experience. If feasible, live instruction, rather than on-line training, supports dialogue and problem solving among colleagues and experts. Training providers can be approached to provide modular training so that the minimum training standard is achieved in the first module; additional or advanced training can be delivered in optional subsequent modules that build on the first. For example, universal principles exist for nonlethal predator defense that can be delivered in the first module. The training may end there for field personnel who work in locations where firearms are prohibited (e.g., national parks). Additional modules can include firearms training, licenses, and practical hands-on simulations of lethal predator defense. This modular system also makes training more cost effective so that employers are not funding and employees are not spending time in sessions that are not directly relevant to their fieldwork.

Peer-to-peer training takes advantage of the experience of colleagues and allows them to transfer their skills and knowledge to others. This type of training is particularly useful for operation of specialized or custom field instruments for which no formal training exists. It can provide opportunities for novice field-party members to gain more hands-on experience from resident experts (e.g., boating, snow machine, ATV). Delivery of peer-to-peer training can include live demonstrations, practice sessions, presentations, seminars, and workshops that can be recorded to reach a wider user group.

Although ideal, on-site training in the field environment is not generally possible by a formal training provider. However, experiential training (e.g., on equipment use and instrument handling in a field-like setting) can be demonstrated by an instructor (or peer mentor) and practiced with time dedicated to sharpening skills when field results are not at stake. Field-party members need to feel proficient and to practice safe operations without the added pressure of the field environment and schedule.

On rare occasion during fieldwork, field parties can be alerted to specialized training or local knowledge and expertise on site which can decrease risks to personnel. Enough flexibility must exist in the field schedule and budget to accommodate this type of serendipitous opportunity to mitigate risk under the authority of the field-party leader. Examples of unexpected local risks may include presence of certain types of predatory wildlife, identification of local poisonous plants, changes to navigable waterways, or unexpected hazards along a traverse route.

Some field personnel are so well versed in a given field-safety topic that they are qualified to instruct a course themselves. Although periodic renewals are necessary, legally or as determined by an organization, it is not a good investment to force expert employees to repeat a full course. Refresher courses with reduced attendance time and lower tuition fees are one option to recognize both the need for renewal and the degree of competence already in hand. Another option is to provide employees who have achieved a minimum standard of training an opportunity to advance to a higher level of training, to the point of becoming an in-house expert or instructor. A third option is a test of competence or measure of equivalent experience which can be documented.

Holding a current certificate and having passed a training course do not necessarily provide sufficient evidence of practical competence. Enough authority must be given to field-party leaders to assess competence before or during fieldwork. In some cases, assessments may be repeated from one day to another as status of competence can change during fieldwork (e.g., due to illness, fatigue, or psychological distress).

3.2.3.3.5 *Training Evaluation*

To complete the loop of the training needs analysis, which requires periodic review, participants must objectively evaluate the training they receive. Some training providers use evaluation forms for their purposes, however, the science organization must have its own mechanism to give and receive feedback on its investments in training. Training evaluation forms typically ask the participants to assess their understanding of the course materials, relevance of the training to fieldwork activities and environments, expertise of the trainer, training materials, and overall satisfaction. This is usually done at the end of a formal training course and before fieldwork begins. However, the organization must develop an objective instrument to assess how the training affected field operations. Without this component of the training needs analysis, the emerging safety culture, training, and their effectiveness in the reality of fieldwork are left unchecked. Although this can be achieved through formal questionnaires and surveys of field parties, it can also be collated from review meetings and interviews after the field season (e.g., Carroll, 1998; O'Toole, 2002).

3.2.3.3.6 *Examples of Training*

Training for field safety varies according to field operations and task-hazard analyses. Examples of field-safety training include: first aid and CPR (e.g., standard or basic, wilderness, advanced wilderness, wilderness survival); operation of emergency communication devices; transportation in or to the field (e.g., operation of ATV, snow machine, boat; marine safety, helicopter safety, trailer safety, defensive driving); equipment (e.g., chain saw operation, small motor repair, deployment of specialized field instruments); predator defense (e.g., wildlife, including black bears, grizzly bears, polar bears, cougar, moose; lethal and nonlethal defense, such as pepper spray and firearms); chemical handling; WHMIS or GHS; fire extinguisher safety; personal protective equipment; transportation of dangerous goods (by road, rail, air; radioactive goods); field conditions (e.g., working from ice cover, mountaineering); and awareness of occupational health and safety for employers and employees.

3.2.3.4 Field Parties

Fieldwork relies on the competence, experience, preparation, and constant vigilance of field parties with respect to both scientific objectives and safety. The grassroots of a positive safety culture are within field parties at the front lines of fieldwork where the risks are most immediate. Forethought and planning, both individual and as a group, can set up field parties for success.

3.2.3.4.1 Medical Health Evaluations and Fitness for Fieldwork

Employers and employees may request and provide medical health evaluations. If there has been a change in health status of an individual field-party member, a medical health assessment may be a regulated prerequisite. Standards are jurisdictional and can require different levels of medical authority. Frequency and content of an assessment can depend on the health status and age of the field-party member, task-hazard analysis, and anticipated risk exposures (e.g., Treasury Board of Canada Secretariat, 2013).

If a field-party member's fitness for fieldwork is in question (e.g., recent injury, illness, and mental or physical change in health status), then it is reasonable to seek a health assessment by a qualified medical professional. Since health status can change suddenly, judicious authority should also be granted to field-party leaders on site to decide whether or not participation in fieldwork should continue. For their own protection and that of their colleagues, field-party members should be obligated to divulge medical conditions, medications, and allergies in advance of fieldwork and take individual responsibility for their personal fitness and health.

3.2.3.4.2 Threats to Health and Safety Within Field Parties

Higher risk is inherent in fieldwork because it represents significant departures from the office, lecture hall, laboratory, and, frequently, regular working hours. Field parties and leaders must recognize the factors that can compromise the health and safety of personnel and use both preventive measures and relief from these potential triggers. They include: fatigue; distraction; pain; cold; heat; thirst; hunger; boredom; loneliness (isolation); discomfort; illness; drugs or alcohol; disorganization; lack of communication; and conflict. Attention to workloads, work pace, schedules, working conditions, stimulating tasks, and compatibility of personnel will help to alleviate many of these challenges, along with regular monitoring of field-party members, verbally or by observation.

3.2.3.4.3 *Balance of Expertise and Compatibility*

Not all field-party members will have the same training, experience, or skill sets. Because fieldwork is an opportunity for instruction, mentoring, demonstration, learning, and practice, a balance of skills must exist between experienced and novice field-party members. Rookie field crew members can learn from resident experts and refine their technical skills. Moreover, novices can be given opportunities to lead field-safety review sessions and make daily safety decisions as practical role-playing experiences in supervised field settings.

The research team will perform better during fieldwork if its scientific goals, methods, equipment, and schedules are compatible. For example, operating high-voltage geophysical instruments on board ship while others in the party are simultaneously handling large volumes of seawater nearby is incompatible from a safety perspective because of the high electrical conductivity of seawater and increased risk of electrocution. Likewise, unreasonable expectations of workloads can lead to greater risk of human error. Well-rested field parties can sustain a reasonable workload and perform better throughout the duration of fieldwork. It is important for the field-party leader and field-party members alike to discuss expectations before and during fieldwork and strive for compatibility.

3.2.3.4.4 *Joint Fieldwork*

Integrated joint fieldwork implies that the safety practices of one organization can affect the health and safety of a partner organization. Legal requirements of joint fieldwork with external organizations or in visiting jurisdictions should be satisfied as a first priority through discussions, joint fieldwork planning, joint safety planning, formal agreements, waivers, visitor-control programs, policies for loaning and sharing of equipment, and host-facility protocols as warranted. The more stringent field-safety requirements of one partner organization should supersede those of its collaborators.

If an organization is hosting a field camp, it must confirm the status of its guests. A field party that includes, for example, students, postdoctoral fellows, volunteers, Emeritus, visiting scientists, or collaborators from industry, governments, or academia, must confirm their equivalent training, qualifications, insurance coverage, responsibilities, and access to shared equipment and supplies. This often takes the form of a written agreement planned in advance with a completed task-hazard analysis, which ideally, has been developed jointly and shared. If an organization is a guest at another field camp hosted by an external partner, a similar arrangement should be in place, along with a task-hazard analysis that is discussed and shared prior to fieldwork.

3.2.3.4.5 *Working Alone*

Policies are usually already in place for the rare occurrence of working alone in the field (e.g., Canada Centre for Occupational Health and Safety, 2006). Working alone is generally avoided as an option for fieldwork because of the inherent increased risk. Working-alone policies typically include detailed itineraries, regular check-in procedures with local contacts, and contingency planning if a check-in is missed after a predetermined period of time. Acceptability of working alone and its conditions must be under mutual agreement between the employer and the employee. Examples of working alone in the field could be downloading data-logging equipment in an urban or suburban setting or asking property owners for permission to gain access at a later date.

3.2.3.5 Communications

Field-safety communications include two types of messages: (1) distress calls for emergency assistance and (2) iterative exchanges of field safety protocols, advice, and feedback. Although the former is urgent during an incident, it requires thorough advance planning. The latter is forward-thinking and essential to an informed safety culture.

3.2.3.5.1 Emergency

As suggested in Section 3.2.3.2.2, field parties must not assume emergency response systems are in place, but proactively identify the relevant system for the field-site location and possible incident types in advance (e.g., poison control, medical evacuation). Local, regional, or centralized jurisdictions and OHS experts may provide assistance, but the emergency response system must be tested or verified in advance of fieldwork to avoid testing it for the first time during a field incident.

Emergency communication devices must be reliable, provide timely service, be easy to use, cover the geographic range of fieldwork, and function under field conditions. Personal locator beacons, with two-way communication and messaging are effective for distress situations. The field party can both put out a distress signal and also receive instructions or information. Other types of communication technologies exist, such as, satellite phones, radio, regular check-ins scheduled in a preplanned itinerary, in addition to other personal communication devices which continue to develop as technology progresses. For example, mapping and monitoring the field route with global positioning systems (GPS) can communicate last known location electronically to emergency responders and facilitate search and rescue efforts. All field-party members should be trained and proficient in the use of emergency communication devices and protocols.

Field parties must also be able to receive pertinent local information. For example, vital field-safety data include: current weather predictions; navigation information; natural disaster warnings; wildfire threats; and alerts to the presence of wildlife through access to news outlets and other resources where available.

3.2.3.5.2 Field Safety

Effective discussion of field-safety issues is a required element of a safety culture which, ultimately, is an informed culture (Reason, 1998). Field-safety issues and requirements can be communicated formally, informally, and at different organizational levels. They must be communicated clearly and uniformly, particularly from management and OHS personnel. Otherwise, poor communication can lead to mixed messages and confusion, particularly when requirements change as new processes or standards are adopted. If the information to be communicated is a change in policy or a requisite, then it must be communicated in a citable format. Regular briefings with service providers (e.g., trainers, equipment suppliers, emergency responders, and OHS personnel) generate feedback as a tool to improve safety performance. When formal opportunities for feedback do not exist, they can be created within an organizational structure that supports a culture of safety.

3.2.3.6 Equipment

Field equipment includes gear for supporting safety, communications, field operations (scientific instruments, camp set up), and transportation. If the equipment is on loan or rented, then either an equipment loan policy or rental agreement would be activated. In addition to keeping field gear in working order and ensuring that it meets safety guidelines, an equipment loan program could also include equipment maintenance, renewal, inventory control, and centralized storage for user groups. Safety items that are used with other equipment can be bundled logically in field kits prepared by equipment providers to ensure compatibility and completeness. For example, a boat that is automatically equipped with a compatible motor (e.g., high transom boat with long shaft motor of suitable horsepower), oars, anchor, rope, life jackets, bailer, and trailer, reduces risk.

Communication and training are closely linked to field equipment. Discussions between field parties can expose new advances in field equipment and instrumentation. Equipment providers can inform field parties of their inventories and suppliers of cutting edge technologies. Similarly, field parties can provide feedback to equipment providers about equipment and service needs and make suggestions to renew or upgrade inventories. OHS personnel can confirm if equipment conforms to relevant safety codes.

In the realm of scientific fieldwork, custom-made or modified instruments are commonly used over many field seasons. If failure of specialized instruments could lead to injury, task-hazard analysis should specify risk mitigation strategies, such as appropriate maintenance and inspection targets, to ensure the equipment is safe to use.

During the approval process of equipment loans, rentals or purchases, proof of training or certification is required. For example, snow machines, ATVs, boats, vehicles, and firearms require licenses. Some preventive training can also be associated with equipment and can make the difference in an emergency situation (e.g., small motor repair). Equipment providers can also take training (e.g., with field parties) for the equipment they stock to become more familiar with it, its use, and client needs. Increasing the equipment suppliers' knowledge of the context of fieldwork, risks encountered by field personnel, and relevant regulatory requirements will improve their services and enhance their ability to become a safety resource to field crews.

Fieldwork creates an occupational work setting that should include portable automated external defibrillators (AEDs) as safety equipment for effective intervention in the chain of survival after sudden cardiac arrest (e.g., Starr, 2002; Mayr et al., 2012). First-aid training typically includes instruction on the use of AEDs, which are becoming available in many workplaces, but not explicitly for fieldwork. Additional examples of field-safety equipment include: first aid kits; survival kits; personal locator beacons; alarm systems; electrified bear fences; firearms; fall protection (e.g., safety harness); respiratory equipment; fire extinguishers; and personal protective equipment (e.g. helmet, hard hat, goggles, safety glasses, gloves, reflective clothing, and steel-toe boots).

3.2.4 LEADERSHIP OF SAFETY CULTURE

Some will automatically defer to the leadership of senior management at the apex of a traditional hierarchy that is strongly linked to organizational culture (Chowdry et al., 2007). Certainly, complacency of senior management who were perhaps not fully apprised of safety risks has been identified in hindsight as a potential causal factor in breaches which led to notorious industrial disasters of the past (e.g., Chernobyl, Union Carbide, Three Mile Island; Cox and Flin, 1998 and references therein; Pidgeon, 1998 and references therein; Reason, 1998). Ideally, organizational leadership must not only commit to making field safety a priority, but also participate rigorously and endorse safety initiatives in an unencumbered climate of continual reflection and improvement to influence employer and employee perceptions and behaviors (e.g., O'Toole, 2002). Cultivating positive feedback loops within an organizational culture requires working frameworks that allow for logical processing of information from relevant sources of expertise, to recommendations, review, and decision-making for improvements to be implemented (Fig. 3.2.4).

A key element of organizational leadership is to engage authoritative expertise with accredited professional OHS designations dedicated to leading field-

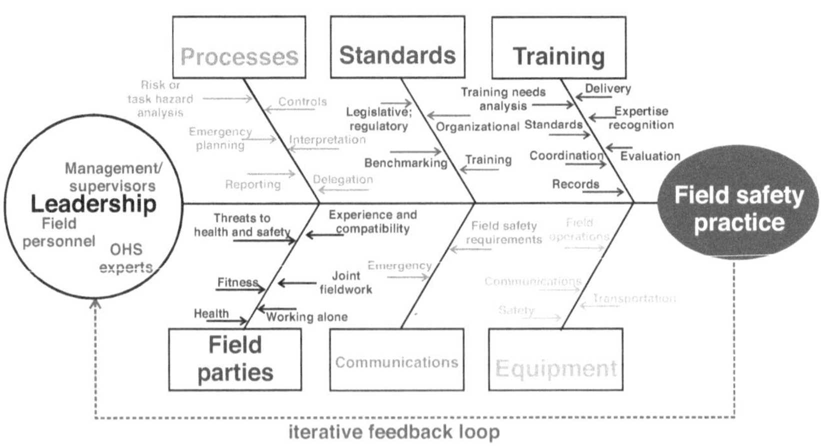

FIGURE 3.2.4 Elements of a dynamic field-safety culture within science organizations

safety programs in collaboration with management and employees (Fig. 3.2.4). Adequate resources must be invested in the key role of relevant OHS expertise to act as a catalyst that drives a health and safety program by establishing, guiding, implementing, and refining all facets of health and safety operations. OHS experts would be able to: (1) provide a resource for compliance with and interpretation of jurisdictional laws, regulations, and directives; (2) consult thoroughly to set organizational minimum standards in support of field safety, (3) provide guidance for, and evaluate task-hazard and training-needs analyses; (4) provide counsel to field personnel and managers as a reliable resource to support preventive behaviors and appropriate responses to field incidents; (5) establish and implement standards and processes in support of field safety; (6) report and assess safety practices and improvements; and (7) engage additional specialized OHS expertise as needed for issues beyond their knowledge. OSH experts must necessarily become familiar with field operations and continue their training to apply OHS principles to their organization's fieldwork. Safety management must include everyone from management personnel to front line fieldworkers with key organizational support necessary to make a positive safety culture thrive.

Managers, supervisors, and OHS experts will need to collaborate with and place value on operational knowledge and experience from field experts to avoid making critical decisions with incomplete information (e.g., Pidgeon, 1998). Although controls are necessary from a regulatory perspective, they do not replace full disclosure of operational needs, critical thinking, and challenging underlying assumptions case by case. An appropriate action can be taken on the basis of a thorough understanding of the context and the personnel. Although the general approach is widely applicable, the plan of action is specific to the context of the fieldwork, task-hazard analysis, and risk management options.

Logically, some leadership falls to regulatory frameworks, organizational governance structures, and controls. None of these should impede the natural thought process of developing task-hazard analyses and proposing risk-mitigation methods for each unique field context. Although controls and reporting enforce proactive responsibility and accountability, they can also be used to assign blame. Blame does not necessarily promote a constructive feedback loop to support a positive culture of safety, particularly if a violation is done out of error rather than premeditation. Since human nature seeks to avoid or deflect blame for fear of reprisal, to impugn indiscriminately or unjustly is to inhibit full investigation of incidents or near misses and will obstruct a safety culture. Similarly, when organizational leadership fails to recognize risk, evades responsibility, or suspends action on known safety deficiencies, latent risk is increased and confidence in the safety culture eroded (e.g., Reason, 1998). Untimely or inadequate action on safeguards damages trust within the organization and discourages personnel to come forward in good faith with constructive recommendations to improve safety.

If minimum standards, controls, and processes are in place, then exceptions should not be accommodated because they will set a precedent that undermines the existing culture of safety and destabilizes the credibility of a just workplace (e.g., Reason, 1998). Although organizations cannot rely on controls to replace critical thinking, controls can serve as a necessary redundancy for employers and employees to confirm minimum compliance with safety criteria when participation in the safety culture is resisted. Inevitably, some forces will seek to breach, disable, or circumvent existing minimum safeguards (Reason, 1998). When safety standards are deliberately violated, rather than by error, consequences are appropriate. Organizational leadership is expected to enforce minimum standards.

Some leadership authority should be decentralized, as suggested by Pidgeon (1998), so that decisions are not made so far up the chain of command, that they are too far removed from the front lines of fieldwork to be either fully informed or timely. Of particular importance is the delegation of authority to field party leaders, who may need to improvise field-safety solutions on site and make rapid decisions without necessarily having the luxury of consulting with a manager or supervisor (e.g., assessing field-party members for fitness to continue fieldwork, hiring local experts or guides for mitigation of unforeseen risk). In fact, Chowdry et al. (2007) have suggested that a model of an organization-wide safety culture may not be the only option; good safety cultures (or subcultures) within individual working groups can be both efficient and effective in identifying and mitigating risk unique to their specific operations.

Leadership in safety practices can come from within field parties. If any member of the field party senses a threat to their personal safety, they have a legal right to refuse work, as long as their withdrawal would not impose imminent risk to their colleagues (e.g., Human Resources and Skills Development Canada, 2009b). The responsibility to declare trepidation with safety planning and risk management lies with everyone in the potential role of whistle-blower with the intention to improve safety practices. Ideally, working out mitigation measures would be done collectively and inclusively within the field team.

An organizational culture must also recognize the need to create time so that more attention is directed to field-safety planning by field parties and their leaders. Others can be delegated to perform supporting field tasks, such as, scheduling training, arranging permits, organizing travel, procuring field equipment, booking accommodations, arranging rentals, and securing on-site support services (e.g., wildlife monitor, camp cook). Increasing pressure for performance, in the form of heavy workloads and tight timelines with few people to drive them, does not usually create a climate where safety planning is a priority (Cox and Flin, 1998 and references therein).

Leadership also means playing the role of devil's advocate in suggesting alternate scenarios and giving voice to concerns which challenge past practices and their underlying assumptions. Past experience can inherently present a bias in decision-making and should be questioned, so that previous acceptance of risk can be re-evaluated for its current pertinence to field safety. Cutting corners in the interest of increased productivity, time saving, or yielding to power relationships can lead to greater latent risk. A decision to compromise one safeguard can lead to another and another and, effectively, escalate the degree of risk regardless of the consequences (e.g., Staw, 1976; Reason, 1998). An illustration is one of the most scrutinized tragedies, the explosion of the NASA space shuttle Challenger in 1986. The rubber O-rings, which were to seal the joints of the solid rocket booster, lost elasticity in the cold weather of launch day, failed, and led to loss of life (Antonsen, 2009 and references therein). NASA and the contractor responsible for the solid rocket booster were both aware of potential problems with the O-rings in previous missions but the missions were allowed to recur and the potential O-ring issue was perceived an acceptable risk. During a teleconference between NASA and the contractor the night before the Challenger launch, the contractor's engineers declared the risk of O-ring failure in cold weather and recommended delay of the launch. NASA officials reacted negatively to the recommendation, given previous acceptability of the risk and the inordinate pressures to proceed with the launch, which included justifying the cost of the space missions. In response, the contractor's senior management withdrew their engineers' recommendations and endorsed the launch. NASA representatives who communicated with the contractor did not inform NASA senior management of the engineers' concerns and the launch proceeded with catastrophic results. Risk escalation was fueled by several factors which suppressed vital information from decision-making long before the launch: funding constraints, scheduling expectations, continued employment, performance pressure, incubation of latent risk (or normalization of deviance), and organizational power dynamics (Pidgeon, 1998 and references therein; Antonsen, 2009 and references therein).

Leadership in safety transcends organizational hierarchy and governance structures. Specialized field-safety expertise resides with field-party members, colleagues, collaborators, and field party leaders, regardless of affiliation. What is essential in organizational leadership is to create an open climate for critical thought, learning, evaluating, resolving conflict, and sharing safety practices, in addition to supporting frameworks with the tenacity to measure safety performance and effect change. Organizational culture can stimulate a culture or subcultures of safety to the point of forming a contagious positive habit which permeates the organization. It is at this level that an organization attains the resilience and reliability to react nimbly to unexpected risks, in addition to those which can be foreseen.

3.2.5 CONCLUSIONS

The current approach to field safety is moving toward a change in attitudes and perceptions from reactive processes and mechanical fulfillment of regulated minimum requirements to a proactive cycle of forethought, problem solving, questioning assumptions of acceptable risk, active monitoring of risk, planning, and evaluation to drive improvement. Organizational leadership can create an open climate with support systems and adequate resources (e.g., standards, processes, OHS expertise) to encourage greater engagement and critical thinking of employers and employees who deliberate practical solutions for field safety. This is a shift from adhering to the letter of the law to recognizing the spirit of the law and taking it further with intention to preserve life, health, property, and the environment. The mere act of cultivating positive cultures of safety and exploring creative solutions will make natural science organizations more resilient to field risks, whether foreseen or unexpected.

ACKNOWLEDGMENTS

The authors wish to thank colleagues at the Geological Survey of Canada (GSC) and within the Field Safety Advisory Group, Earth Sciences Sector at Natural Resources Canada (2010–2014), for innumerable discussions that led to many of the concepts presented. The impetus to compile the cumulative experience of the GSC and policies was part of an effort to establish protocols for field and lab methods under the Geo-Mapping for Energy and Minerals (GEM) Program (2008–2013). This paper benefited from reviews by M. Demuth, R. Boutilier, M.J. Hinton (GSC), and editors of this volume. We thank P. Bobrowsky, A. J-M Pugin, M. Sanborn-Barrie and B. Anderson for providing field photographs. Comments and feedback on the field-safety topics presented in this paper are welcome and can be communicated to the authors. ESS Contribution number 20140208.

REFERENCES

Antonsen, S., 2009. Safety culture and the issue of power. Safety Sci. 47, 183–191.

Canada Centre for Occupational Health and Safety (CCOHS), 2006. Working Alone – General. http://www.ccohs.ca/oshanswers/hsprograms/workingalone.html (accessed 10.09.2014).

Canadian Red Cross Society, 2013. Wilderness and Remote First Aid Trip Plan. http://www.redcross.ca/crc/documents/What-We-Do/First-Aid-and-CPR/Workplace/Online-Tools/Trip-Plan-2013.pdf (accessed 10.09.2014).

Carroll, J.S., 1998. Safety culture as an ongoing process: culture surveys as opportunities for enquiry and change. Work Stress 12 (3), 272–284.

Chowdry, R.M., Fang, D., Mohamed, F., 2007. The nature of safety culture: a survey of the state-of-the-art. Safety Sci. (45), 993–1012.

Cox, S., Flin, R., 1998. Safety culture: philosopher's stone or man of straw? Work Stress 12 (3), 189–201.

Department for Transport, 2012. Carriage of Dangerous Goods: Approved Derogations and Transitional Provisions. Department for Transport, Health and Safety Executive for Northern Ireland, Department of Justice (Northern Ireland). TSO Publishers, United Kingdom pp. 26. http://assets.dft.gov.uk/publications/carriage-of-dangerous-goods-approved-derogations-transitional-provisions/approved-derogations-transitional-provisions.pdf (accessed 10.09.2014).

Earth Sciences Sector (ESS), 2012. Earth Sciences Sector Field Guide. Natural Resources, Canada p. 61.

Earth Sciences Sector (ESS), 2014. Standard Operating Procedures: Firearms Handling, Transportation, and Storage of Unrestricted Firearms. Natural Resources, Canada p. 19.

Edwards, J.R.D., Davey, J., Armstrong, K., 2013. Returning to the roots of culture: a review and re-conceptualisation of safety culture. Safety Sci. 55, 70–80.

Geological Association of Canada, 2011. Field Trips – Policy Statement. https://www.gac.ca/wp/?page_id=2010 (accessed 10.09.2014).

Government of Canada, 2014. Canada Occupational Health and Safety Regulations. SOR/86-304. 246 p. http://laws.justice.gc.ca/eng/regulations/sor-86-304/page-1.html (accessed 10.09.2014).

Health and Safety Executive. 2013. A guide to health and safety regulation in Great Britain. http://www.hse.gov.uk/pubns/hse49.pdf (accessed 10.09.2014).

Human Resources and Skills Development Canada, 2009a. Information on Occupational Health and Safety. Pamphlet 6B. Work Place Health and Safety Committees. p. 4. http://www.labour.gc.ca/eng/health_safety/pubs_hs/committee.shtml (accessed 10.09.2014).

Human Resources and Skills Development Canada, 2009b. Information on Occupational Health and Safety. Pamphlet 4. Right to Refuse Dangerous Work. p. 6. http://www.labour.gc.ca/eng/health_safety/pubs_hs/refuse.shtml (accessed 10.09.2014).

Human Resources and Skills Development Canada, 2010. Information on Occupational Health and Safety. Pamphlet 7. Hazardous Occurrence Investigation Recording and Reporting. p. 6. http://www.labour.gc.ca/eng/health_safety/pubs_hs/hoir.shtml (accessed 10.09.2014).

Mayr, N.P., Mayr, T., Tassani, P., Martin, K., 2012. Use of automated external defibrillators in the occupational setting in Germany – a pilot study. J. Occup. Environ. Med. 54 (7), 789–791.

Ontario Ministry of Labour, 1990. Occupational Health and Safety Act. R.S.O. 1990. http://www.e-laws.gov.on.ca/html/statutes/english/elaws_statutes_90o01_e.htm# (accessed 10.09.2014).

O'Toole, M., 2002. The relationship between employees' perceptions of safety and organizational culture. J. Safety Res. 33, 231–243.

Parliament of Canada, 2000. Second Session, Thirty-sixth Parliament, 48-49 Elizabeth II, 1999-2000, Statutes of Canada, Chapter 20. Canada Labour Code Part II. p. 58. http://www.parl.gc.ca/HousePublications/Publication.aspx?DocId=2330437&Language=e&Mode=1&File=34 (accessed 10.09.2014).

Parliament of Canada, 2003. Second Session, Thirty-seventh Parliament, 51-52 Elizabeth II, 2002-2003, Statutes of Canada, Chapter 21, Bill C-45 of the Criminal Code of Canada. p. 14. http://www.parl.gc.ca/HousePublications/Publication.aspx?Pub=Bill&Doc=C-45&Language=E&Mode=1&Parl=37&Ses=2 (accessed 10.09.2014).

Parliament of the United Kingdom, 1974. Health and safety at work etc. Act 1974. General Public Act, Great Britain. Chapter 37. http://www.legislation.gov.uk/ukpga/1974/37/pdfs/ukpga_19740037_en.pdf (accessed 10.09.2014).

Pidgeon, N., 1998. Safety culture: key theoretical issues. Work Stress 12 (3), 202–216.

Pugin, A.J-M, Pullan, S.E., Hunter, J.A., 2013. Shear-wave high-resolution seismic reflection in Ottawa and Quebec City, Canada. The Leading Edge. Special Section: Urban Geophysics. pp. 935–940.

Reason, J., 1998. Achieving a safe culture: theory and practice. Work Stress 12 (3), 293–306.

Royal Canadian Mounted Police, 2014. Firearm Safety-Training Requirements. p. 2. http://www.rcmp-grc.gc.ca/cfp-pcaf/fs-fd/safety-securite-eng.htm (accessed 10.09.2014).

Starr, L., 2002. Automated External Defibrillation in the Occupational Setting. J. Occup. Environ. Med. 44 (1), 2–7.

Staw, B.M., 1976. Knee-deep in the big muddy: a study of escalating commitment to a chosen course of action. Organ. Behav. Hum. Perf. 16, 27–44.

Treasury Board of Canada Secretariat, 2013. Occupational Health Evaluation Standard. http://www.tbs-sct.gc.ca/pol/doc-eng.aspx?id=12561&section=text (accessed 10.09.2014).

Turner, B.A., Pidgeon, N.F., Blockley, D.I. Toft, B., 1989. Safety culture: its importance in future risk management. Second World Bank Workshop on Safety Control and Risk Management, Karlstad, Sweden. November 1989.

United Nations Economic Commission for Europe, 2013a. Recommendations on the Transport of Dangerous Goods, Model Regulations (Rev.18) http://www.unece.org/trans/danger/publi/unrec/rev18/18files_e.html#c38190 (accessed 10.09.2014).

United Nations Economic Commission for Europe, 2013b. Globally Harmonized System of Classification and Labelling of Chemicals (GHS, Rev.5) http://www.unece.org/trans/danger/publi/ghs/ghs_rev05/05files_e.html (accessed 10.09.2014).

United States Geological Survey (USGS), 2014. Safety and Health for Field Operations Handbook 445-3-H. U.S. Department of the Interior. p. 344. http://www.usgs.gov/usgs-manual/handbook/hb/445-3-h.pdf (accessed 10.09.2014).

United States Department of Transportation, 2014. How to Comply with Federal Hazardous Materials Regulations. Federal Motor Carrier Safety Administration. http://www.fmcsa.dot.gov/regulations/hazardous-materials/how-comply-federal-hazardous-materials-regulations (accessed 10.09.2014).

United States Department of Labor. 1970. Occupational Safety and Health Act of 1970. Title 29. Chapter 12. Public Law 91–596. Occupational Safety and Health Administration https://www.osha.gov/pls/oshaweb/owadisp.show_document?p_table=oshact&p_id=2743 (accessed 10.09.2014).

University of Northern British Columbia (UNBC), 2014. Field Procedures and Safety Manual. UNBC Field Safety Committee, UNBC. p. 83. http://www.unbc.ca/sites/default/files/sections/safety/unbcfieldproceduresandsafetymanual-version1.2march2014.pdf (accessed 10.09.2014).

University of Texas at Austin, Environmental Health and Safety, 2010. Safety Guidelines for Field Researchers. http://www.utexas.edu/safety/ehs/fieldguide/field_guide.pdf (accessed 10.09.2014).

Yukon Geological Survey (YGS), 2012. Health and Safety Program Manual. YGS Occupational Health and Safety Committee, Government of Yukon. 329 p. http://ygsftp.gov.yk.ca/YGSWebsite/YGS_SafetyManual_2012.pdf (accessed 10.09.2014).

Chapter 4

Teaching Geomorphology in the Field

Mary J. Thornbush*[,1], Casey D. Allen**, Faith A. Fitzpatrick[†]
*School of Geography, Earth and Environmental Sciences University of Birmingham, Edgbaston, Birmingham, United Kingdom; **Department of Geography and Environmental Sciences, University of Colorado Denver, Colorado, USA; [†]U.S. Geological Survey, Wisconsin Water Science Center, Middleton, Wisconsin, USA
[1]Corresponding author: E-mail: m.thornbush@bham.ac.uk

The importance of fieldwork's pedagogical potential, especially in such a field-based discipline as geomorphology, should not be overlooked. Although dozens of studies across disciplines outline strategies for success and advantages of learning in field settings, this section focuses specifically on examples from geomorphology, based on research from three distinct angles. First, in Chapter 4.1, "Student Learning Styles," Gomez-Heras offers insight into learning styles that are associated with students' awareness of what it means to conduct fieldwork. He uses VARK (Visual, Aural, Read/Write, Kinesthetic), developed in New Zealand during the 1990s, to distinguish between students' learning styles and applies it to how students engage in fieldwork as a "novelty space." His findings suggest that while most students perceive fieldwork as valuable, their learning styles influence those perceptions. Gaining awareness of student learning styles increases field-based engagement, and incorporating them into pedagogical practices further enhances the already-strong ties students feel they possess while in the field. Fuller and France bring field-based geomorphological pedagogy into the twenty-first century by showing the reader how "Fieldwork Going Digital" (Chapter 4.2) can enhance field-based pedagogy. Utilizing best-practice case studies spanning two continents, they assess not only feasibility but also student engagement in the landscape. To do this, they employ digital video assignments, such as incorporating digital video production into a "Cook's Tour" fieldtrip and evaluating general landscape form and process in relation to management issues using Web 3.0 technology (iPads), to not only gather data but also use real-time social media applications. Their evaluation suggests that field practitioners should include these types of activities and tools in their research, but even more so when it comes to teaching in the field. To round-out this section, Chapter 4.3, "Field-Based Learning in

Developments in Earth Surface Processes, Vol. 18. http://dx.doi.org/10.1016/B978-0-444-63402-3.00006-6

Undergraduate Geomorphology Courses" by Thornbush centers on undergraduate pedagogical issues, specifically drawing on data obtained from three different courses at three different undergraduate levels. Using extensive quantitative and qualitative data, she discusses in depth the challenges undergraduates face when performing fieldwork, including pertinent findings such as younger, non-science students enjoying the experience, even though they struggle with new content. Her findings also suggest that the sooner students engage in fieldwork the better, and gradually adding fieldwork to pedagogy can have strong implications for success in later endeavors such as graduate school. In the end, whether assessing learning styles, integrating new technologies, or evaluating student perceptions of fieldwork across a range of classes and age groups, this section's contribution to geomorphology as a discipline and its ability to enhance field-based pedagogy is invaluable.

Chapter 4.1

Student Learning Styles

Miguel Gomez-Heras*,**,[1], Stephen McCabe**,†

*CEI Campus Moncloa, UPM-UCM, CSIC: ETS Arquitectura, Universidad Politécnica de Madrid and Instituto de Geociencias (CSIC, UCM), Madrid, Spain; **School of Geography, Archaeology and Palaeoecology, Queen's University of Belfast, Northern Ireland, the United Kingdom; †Northern Ireland Environment Link, Belfast, the United Kingdom

[1]Corresponding author: E-mail: miguel.gomez.heras@gmail.com

4.1.1 INTRODUCTION

Krepel and Durrall (1981, p. 7) define a fieldtrip as an activity "in which students go to a place where the material of instruction may be observed and studied directly in their functional setting". This definition applies across all disciplines, but it is perhaps in the case of natural sciences and environmental education where it finds its fullest expression, as the object of study can often be found far from the lecture theater (geographically and sometimes conceptually). Fieldtrips are also relevant in some of the human and social sciences, especially when we want students to gain an understanding of human realities and experiences that are different to those they are normally exposed to. In a discipline such as geography, that combines the study of natural and human environments and their interactions, fieldtrips are, perhaps, the most relevant way to gain a full, and more rounded, understanding of the subject; there are a great number of studies on the utility of fieldwork in geography (e.g., see Kent et al. (1997) for an extensive review on fieldwork teaching in geography). Thus, fieldtrips are commonplace in undergraduate geography courses across the United Kingdom (frequently considered as preparation for final-year dissertations/independent study) and commonly viewed as central elements of the degree (Livingstone et al., 1998).

If designed appropriately, fieldtrips can be the epitome of effective learning because, as Woolnough (1994, p. 24) has stated, the students must play an active role to be able to learn, appreciate, personalize and remember the new information effectively. As such, outdoor education has been shown to increase motivation for learning (Kern and Carpenter, 1984) and to actually improve the performance of students across other curriculum areas (see Lieberman and Hoody, 1998). In the promotion of this important idea, there is an opportunity for the growth of geographical and geomorphological field study through all levels of education. Fieldtrips give students the opportunity of interacting with materials and

Developments in Earth Surface Processes, Vol. 18. http://dx.doi.org/10.1016/B978-0-444-63402-3.00007-8

phenomena that are not available in the classroom; however, with increasingly large groups, there is a danger that fieldtrips may adopt a "show and tell" format (Guertin, 2006) and risk turning into "walking lectures, or even bus tours with commentary, where students become passive" (Habeshaw et al., 1992; p. 117).

Therefore, a great concern is expressed in the literature to design strategies that optimize the learning environment during the fieldtrips and encourage as much as possible the interactivity that should be inherent to the fieldwork experience – the benefit of limited time in the field should be maximized to the student as much as possible. This concern is enhanced given an appreciation of the significant cost of fieldtrips, often involving an extra fee for the students. In this context, there is an onus on the educator or coordinator to make the fieldwork more efficient and effective, as it is regularly "at risk" of disappearing from any teaching agenda due to budgetary restrictions in university departments and the perception that the extra cost for the students can be unnecessarily burdensome.

Orion (1993, 1998) has suggested a model for the learning structure of a fieldtrip module that begins with a "preparatory unit" (taking place in the laboratory/classroom), then goes through the fieldtrip (usually an extended period in the field) to finish with a "summary unit" (essentially a "debrief" again taking place in the classroom). Most of the research on the optimization of the fieldtrip learning environment focuses on the implementation of strategies during the fieldtrip itself. Recently, for example, a number of studies have been presented on the introduction of handheld computers (Guertin, 2006) and portable audio/video systems (Elkins and Elkins, 2006) to increase the interactivity, improve the student-led learning and reduce the periods where there is no involvement in learning activities during the fieldtrips. However, the current authors suggest that more emphasis be placed on what happens before students venture beyond the laboratory/classroom – the study on the impact of student learning during the "preparatory unit" and its effect on the subsequent learning during the fieldtrip itself is often neglected in the educational research of fieldwork.

Students need to be thoroughly prepared for the fieldtrip to maximize the benefits of the potentially unique learning environment – it is very difficult for them to be placed into a new learning environment without first going through a process of "equipping". A relationship occurs between student capability to learn in the fieldtrip and the novelty of the field environment (Falk et al., 1978). The main aim of the preparatory unit is therefore to "reduce the effect of the "novelty space" of the outdoor event" (Orion, 1993; p. 163). Orion and Hofstein (1994) identified three factors in this "novelty space": the "geographical novelty" (i.e., the place is new) and the "cognitive novelty" (i.e., the information and concepts are new), which relate to the "declarative knowledge" (Biggs, 2002, p. 3), and also the "psychological novelty" (i.e., students may see fieldtrips not as a learning activity and experience, but, e.g., as an opportunity for socializing).

A key point is that the influence of the different preferred learning styles of students on how they overcome these different kinds of novelty spaces related to fieldwork, and thus how they are able to learn effectively during the fieldtrip itself, are aspects that have not been explored in the educational literature in relation to fieldwork.

There is wide recognition of the impact of students' learning styles in education in general (e.g., Entwistle, 1981; Entwistle, 1991; Honey and Mumford, 1992; Fleming, 1995; Marriott and Marriott, 2003; Goldfinch and Hughes, 2007) and how tailoring teaching to fit the needs of students with different preferred learning styles is usually associated with improved learning and performance (Fleming, 1995).

Several models through which the preferred learning styles in adults can be rationalized exist. Some of them focus on the preference for various stages of the learning cycle (e.g., Honey and Mumford, 1992), whereas others focus on the preferred mode of information presentation (e.g., Fleming, 1995). In this chapter, the latter approach was selected to assess the students' preferred learning styles. Fleming (1995) identified four types of learning styles: Visual (V), Aural (A), Read/write (R) and Kinaesthetic (K). Preference will depend on how students receive information through graphs, charts, pictures (Visual), speech (Aural), text (Read/write) or through multi-sensorial experiences (Kinaesthetic). Fleming (1995) developed a questionnaire at Lincoln University, New Zealand, which, although it has some recognized limitations (Fleming and Baume, 2006), is still a good way to gauge preferred learning styles in the context of education "to help to overcome the predisposition of many educators to treat all the students in a similar way" (Fleming, 1995).

In light of these considerations, the aim of this study is to explore the impact that preferred learning styles have on students' perceptions of their performance during fieldwork and on overcoming the issues of novelty space described above (with the aim of optimizing their performance). In the same way that a "traditional lecture" may favor only certain types of learners, fieldwork may resonate with particular individuals. Therefore is important to take into account the impact of learning style in relation to fieldwork both at the preparatory stage and during the fieldtrip itself so that all learners and preferences are catered for.

To pursue this aim, the perceptions of a group of second year geography students taking up the module "Field and Research Techniques in Geographical Practice" at the School of Geography, Archaeology and Palaeoecology (Queen's University Belfast) were surveyed. This module comprised a "preparatory unit" with 20 x 1h lectures and one short field research project (that took place locally), one week of fieldwork in Mallorca and a "summary unit" consisting of group preparation of two written reports on different aspects of the physical and human geography of Mallorca.

4.1.2 METHODOLOGY

For the purposes of this study, a questionnaire was circulated to students attending a follow-up session after the fieldtrip to Mallorca (see Appendix 4.1.1). Students filled in the questionnaire voluntarily after receiving an explanation of the purpose of the questionnaire, and were informed of the possibility that the results were going to be published, complying with the Revised Code of Ethics of the British Educational Research Association (BERA, 2004). The questionnaire began with some introductory questions (gender, preference for physical or human geography and specific group in which they undertook fieldwork) to test the representativeness of the results. Following this section, the questionnaire was divided into two parts. First, a section aiming to find out the preferred learning styles of the students with a version of the VARK questionnaire of learning styles (Fleming, 1995; outlined above), and second, a series of questions about the students' perceptions of different aspects of fieldwork. This series of questions was divided into four different blocks. The first one related to whether the students felt prepared or not to carry out the fieldwork, i.e., how they perceived the extent to which they had overcome the "geographical novelty" and "cognitive novelty" during the preparatory unit. The second one focused on how valuable they found fieldwork as a learning experience in the overall context of their degree pathway. The third one related to "psychological novelty" and how the students perceived social interaction (with their peers and others) to have had an impact on their performance during the fieldtrip. The fourth and final block of questions dealt with how students felt about some specific aspects related to learning during the fieldwork, such as feedback and assessment.

The first section of the questionnaire consisted of 10 multi-option questions in which, according to the original design of Fleming (1995), the students could choose none, one or several options for each question. This part of the questionnaire was shortened from the original VARK-model study, to prevent students' survey fatigue (in the context of the second section of the questionnaire also being undertaken during the same session). This second part consisted of 39 questions: 35 affirmations/statements to which the students had to show their degree of agreement with a scale from 1 (strongly disagree) to 5 (strongly agree), and 4 questions related to the students' social behavior. Questions were expressed both in positive and negative ways to check the consistency of replies.

For each of the questionnaires, the preferred learning style for each student was calculated as the style that scored the highest in the VARK test. When a student showed either bimodal or multimodal learning styles (i.e., two or three styles obtained the same score) it was counted as belonging to all the relevant groups. The total percentage of respondents who selected each one of the different degrees of agreement to the proposed affirmation (or the average value when the question required a numerical answer) was calculated for each one of the questions. The responses were then split into groups depending on the preferred learning styles presented by the respondents and, for each one of the learning styles, the percentage was calculated of each reply over the total body of students showing that preferred learning style.

4.1.3 DISCUSSION OF RESULTS

The results (see Appendix 4.1.2) present the total percentage of students showing each one of the different learning styles (V, A, R, and K), as well as the percentage of students that indicated unimodal, bimodal and multimodal learning styles. The total percentage, Tot (%), of respondents for each one of the degrees of agreement to the affirmations can be seen alongside the percentage of students within a preferred learning style group (V, A, R, and K) and their degree of agreement to the various affirmations. In the appropriate cases, the results are also summarized in brackets in the form: (showing some degree of disagreement/undecided/showing some degree of agreement).

There were 60 respondents to the questionnaire, which accounted for 55% of the students enrolled on the module. The sample can be considered representative; the gender distribution was representative of the total of the class, as well as the representation of three different groups on the fieldtrip for logistical purposes (37%, 32% and 43%). Overall, Aural (A) was the preferred learning style for the highest percentage of students (48.3%), followed by Kinaesthetic (K, 35.0%), Visual (V, 28.3%) and Read/write (R, 25.0%). Preferred learning style seemed to have an influence on the degree of interest toward physical or human geography, with the A-learners showing an interest well above the average on human geography (66% of A-learners showed some degree of preference toward human geography in comparison to a 58% of the whole group of students showing some degree of preference toward human geography), whereas R-learners showed the highest rate of interest toward physical geography (60% of R-learners showed some degree of preference toward physical geography in comparison to 35% for the whole group of students showing some degree of preference toward physical geography). Seemingly no significant relationship exists between preferred learning style and gender of the respondents; however the percentage of females (76%) among the V-learners was significantly above the total average (65%).

4.1.3.1 Student perceptions of "novelty spaces"

In relation to the perception on how the "geographical" and "cognitive" novelty was overcome during the preparatory unit, the R-learners consistently perceived themselves to have the highest level of preparation before the fieldtrip. In all cases, they showed above-average degrees of agreement to the questions related to the adequacy and perception of the learning material presented, and an above-average disagreement to the affirmations that suggested the introduction of new material or activities to the preparatory unit of the module. K-learners would be placed on the other end of the spectrum, as they tended to perceive the preparatory unit as unhelpful, and an above-average percentage of K-learners perceived the need for further preparatory material to be delivered. V- and A-learners demonstrated a more mixed response, with V-learners showing the highest percentage of appreciation of the preparatory lectures (71% in comparison to a 65% for the whole group of students) and the elective project (59% in comparison to a 51% for the whole group of students) as a tool that made the main fieldwork easier for them. A-learners showed the highest percentage of perception of learning from the written information received during the course (48%), with similar values to the R-learners (47%). However, when asked about their familiarity with carrying out a field research project, the response of A-learners was very varied, with the highest percentage in comparison to the average corresponding to a perception of unfamiliarity. A-learners also demonstrated the highest percentages expressing that they would have liked to have more demonstrations (58% in comparison to 50% for the whole group of students) and longer local fieldtrips as a preparation for the main fieldwork in Mallorca (65% in comparison to a 52% for the whole group of students).

In regard to how valuable students found fieldwork as a learning experience, the majority of the students in all groups showed a very high acceptance of fieldwork as a learning experience (affirming the overall value of the approach), and specifically of fieldwork overseas. K- and V-learners were the ones who showed the highest percentages of appreciation of the value of fieldwork as a resource for geographical education, and also the ones who showed the highest level of acceptance to having the fieldwork take place in an international destination (i.e., Mallorca). On the other hand, R-learners were the group with the lowest levels of appreciation of the value of fieldwork (27% of R-learners showed disagreement or were undecided when given the statement "I think fieldwork is essential for geographical research", in comparison to only 8% of the whole class).

In terms of the students' perception of the social interaction and how it affected their performance during fieldwork, no major differences occur between the different groups of learners, although R-learners showed a slightly "lower-profile" tendency in relation to "night outings" and "rest time" (i.e., the average number of outings for R-learners was slightly lower and rest time slightly earlier). Having said that, there were important differences in the students' perceptions of how these social activities influenced their learning experience, with R-learners perceiving more intensely that the night outings and lack of rest had a potentially negative influence on their ability to learn from the fieldwork. A-learners demonstrated the highest degree of appreciation of night outings and the lowest degree of perception that these had a negative impact on their performance during fieldwork. In general, although few students showed agreement to the affirmation "I think night outings have negatively influenced in my learning" (V: 12%, A: 0%, R: 7%, K: 14%), in all cases a high percentage thought they suffered from sleep deprivation over the course of the Mallorca fieldtrip (V: 48%, A: 44%, R: 60%, K: 52%). 56% of the whole body of respondents agreed that night outings negatively influenced the ability of other members of the group to learn, and so it is probably appropriate to deduce that, although they do not recognize it when asked directly, the students were in general negatively influenced to some extent by night outings. In any case, the students showed a high degree of agreement for the need to take more disciplinary actions for those who did not take the fieldwork seriously (83% of the whole group).

The last group of questions showed very different responses for each of the learner groups in relation to different aspects of the learning experience during the fieldtrip. V-learners showed the highest degree of appreciation for the poster presentations that students carried out in some evening sessions during the fieldwork and thought that they should form part of the assessment (65% in comparison to a 46% for the whole group of students). A-learners showed a higher degree of disagreement with the level of feedback received (21% in comparison to a 15% for the whole group of students). R-learners thought that their learning experience was less effective than in the lecture theatre (20% in comparison to 11% for the whole group of students), and the majority of the K-learners struggled to compile a meaningful field notebook (52% in comparison to a 34% for the whole group of students).

4.1.4 EDUCATIONAL IMPLICATIONS

These results show that the majority of students perceive fieldwork as a very positively learning experience and support previous research on this matter using different evaluation instrument (e.g., Orion et al., 1997). They support the view that outdoor environmental study is an extremely valuable component of geographical and geomorphological education, providing students with a different (and hopefully innovative) learning interface that ultimately aims to produce a more rounded understanding of the subject matter in students. In addition, the present results highlight how different students' preferred learning styles (Visual, Aural, Read/write, Kinaesthetic) condition different perceptions of fieldwork as a learning experience, and how they perceive their degree of preparation to carry it out.

With regard to overcoming "novelty spaces" (new places, new information and concepts, the perception of fieldwork as more of a social rather than learning experience), a relationship exists between the different groups of learners and their appreciation of preparation for fieldwork in the laboratory/classroom. In the cases of "geographical and cognitive novelty", although the majority of students (62%) agreed that the level of information given before the fieldtrip was adequate, only 40% thought of themselves as being familiar with how to carry out a fieldwork project. R-learners were the group that perceived themselves as best prepared to carry out the fieldwork. This is not surprising as the majority of the preparatory unit is based on a "traditional lecture" format. However, R-learners were the ones least represented on the module with just 25% of the students preferring this learning style. There is a paradox here; the group of learners who felt more prepared is also the one which showed lower values of appreciation of fieldwork as a learning experience – but this is not unexpected given that this group prefers to learn from books. This underlines that the degree of perceived preparation for the fieldwork, that is, the perception of having overcome the novelty space, does not necessarily relate directly to the learning experience during the fieldwork.

In the same way that "traditional lectures" are a better "fit" for R- and A-learners, fieldwork is an activity that tends to suit V- and K- learning styles. However, as Fleming and Baume (2006) recognize, VARK styles are not a measure of personality *per se*, and preferred learning styles do not necessarily correspond to personal strengths. Therefore, these aprioristic relations have to be interpreted carefully. One example of this it is shown in the preferences cited by the sample of students toward physical and human geography. Although one prior assumption may be that physical geography would suit multi-sensorial K-learners with a preference for hands-on approaches to learning, and human geography be more aligned with R-learners, the results in fact show the opposite trend, with the majority of R-learners preferring physical geography and the majority of K-learners inclining toward human geography. Regardless, the results from this simple study suggest that fieldwork deserves a considered approach – it does not naturally resonate with all students. Preparation for fieldwork could be improved significantly by introducing or increasing the presence of modifications and innovations in teaching and assessment methods that cater for all learning styles.

In relation to the "psychological novelty", the majority of students perceived the social experience of fieldwork in a positive light, but only 38% placed more emphasis on social aspects than educational value. The results show that the students were in general negatively influenced to some extent in terms of learning experience by night time outings. Therefore, this is an issue that needs to be addressed to improve the fieldwork learning experience. As Orion et al. (1997) have said, "the preparation should employ concrete activities to reduce the effect of the "novelty space"", in this case explaining clearly beforehand the "ground rules" and impact that social activities can have to their performance in the field and enforcing disciplinary actions, which is a point that the majority of students were in favor of.

4.1.5 CONCLUSIONS AND RECOMMENDATIONS

As a way of summary, the following conclusions could be drawn regarding the most effective teaching strategies in relation to the different learning styles and how these could improve the geomorphological fieldwork experience for the students:

- V-learners very strongly perceived presentations as a learning tool. They showed the highest levels of positive perception of the "elective projects" during the preparatory unit as a learning tool (these projects have to be presented in class and are assessed on the basis of the presentation) and they also showed the highest levels of appreciation of the poster presentations that the groups made in the evening sessions during the fieldtrip to Mallorca. Therefore, teaching activities that involve students presenting results will be suitable for V-learners and will enhance their learning experience.

- The Aural learning style was the most common among the students of the module studied. A-learners showed a preference for having the concepts and methods used in the fieldtrip explained either on-site or beforehand in a very tangible and concrete way. They would have liked to have more demonstrations of methods and more local fieldtrips before the fieldtrip to Mallorca, as well as a higher level of feedback during the fieldtrip. Teaching strategies involving a lecturer talking them through the methods before and during the activities, or the use of portable audio/video systems such as the ones suggested by Elkins and Elkins (2006), would be useful for A-learners to improve their learning experience.
- R-learners tend to feel more comfortable in the classroom, and correspondingly feel more unsure of learning in a fieldwork context. The preparatory unit is crucial for R-learners, as they are the group who gained most from it. In the same way, R-learners are the ones who learned more effectively from writing reports to be submitted after fieldwork. However, this means that they need to be monitored, perhaps given special attention, during the fieldwork itself to guarantee that they do not miss the richness of the interactive learning experience. This can be done by highlighting the "written" aspects of the fieldtrip, for example, the translation of sensorial observation into meaningful notes in their field notebooks – and by showing how those notes can form the basis of any subsequent report.
- K-learners tend to be a more natural "fit" for fieldwork, feeling more at ease with the overall experience. K-learners are keen to interact with the environment and with local culture. However, this is the group for which the preparatory and summary unit presented more difficulties. K-learners' perception of the preparatory unit as unhelpful was higher than average and they were unlikely to perceive report writing as helpful for their learning. Most of their learning took take place in the field. However, over 50% of K-learners struggled to produce a meaningful field notebook. As most of the assessment of the module is based in both the reports and the field notebook, this means that is especially important for these students to receive specific support and prompt feedback in relation to the record of their sensorial learning in written format.

Those of us who are familiar with fieldwork may take the "overcoming of novelty space" for granted. This study demonstrates that such spaces do exist, and that beneath the attractive idea of using fieldwork as an educational tool/learning experience for students (which undoubtedly has value), there are subtleties that require consideration. Whereas educators may be tempted to treat all students in the same way, to fully maximize the benefit of fieldwork as a learning experience for the student, it is valuable to consider the various ways in which different preferred learning styles in students may be engaged and catered for. Then the potential of fieldwork and engagement with the outdoor environmental can be realized more fully – both for the benefit of the discipline as a whole and the positive development of the students with whom we interact.

ACKNOWLEDGEMENTS

We wish to thank Dr. Fleming for granting permission to use the VARK questionnaire. Thanks are also due to Dr. Roe for her support in carrying out this research and the coordinators of the PGCHET – Queen's University Belfast at the time (L. Carey & J. Milliken) for their inspiration. Above all, thanks go to Prof. Bernie Smith – a man who was passionate about seeing geomorphology in the field rather than just the armchair, and who was always an encouragement, supporting our teaching careers.

REFERENCES

BERA, 2004. Revised Code of Ethics. http://www.bera.ac.uk/files/guidelines/ethica1.pdf (accessed 14.09.2014)

Biggs, J., 2002. Aligning the curriculum to promote good learning. Constructive Alignment in Action: Imaginative Curriculum Symposium. LTSN Generic centre. http://www.palatine.ac.uk/files/1023.pdf (accessed 14.09.2014).

Elkins, J.T., Elkins, N.M.L., 2006. Improving student learning during travel time on fieldtrips using an innovative, portable audio/video system. J. Geosci. Educ. 54 (2), 147–152.

Entwistle, N., 1981. Styles of Learning and Teaching. John Wiley & Sons, New York.

Entwistle, N., 1991. 'How students learn and why they fail'. In: Radford, J. (Ed.), Talent, Teaching and Achievement. Jessica Kingsley, London, pp. 77–96.

Falk, J.H., Martin, W.W., Balling, J.D., 1978. The novel fieldtrip phenomenon: adjustment to novel setting interferes with task learning. J. Res. Sci. Teach. 75, 127–134.

Fleming, N.D., 1995. I'm different; not dumb. Modes of presentation (VARK) in the tertiary classroom. In: Zelmer, A. (Ed.), Research and Development in Higher Education, Proceedings of the 1995 Annual Conference in the Higher Education and Research Development of Australasia (HERDSA), HERDSA, vol. 18, pp. 308–313.

Fleming, N.D., Baume, D., 2006. Learning styles again: VARKing up the right tree! Educational Developments. SEDA Ltd 7, 4–7.

Goldfinch, J., Hughes, M., 2007. Skills, learning and success of first-year undergraduates. Active Learning in High. Education 8 (3), 259–273.

Guertin, L.A., 2006. Integrating handheld technology with field investigations in introductory-level geoscience courses. J. Geosci. Educ. 54 (2), 143–146.

Habeshaw, S., Gibbs, G., Habeshaw, T., 1992. 53 Problems with Large Classes: Making the Best of a Bad Job. BPCC Wheatons, Exeter.

Honey, P., Mumford, A., 1992. Manual of Learning Styles. Honey Publications, Maidenhead.

Kent, M., Gilberts, D.D., Hunt, C.O., 1997. Fieldwork in Geography Teaching: a critical review of the literature and approaches. J. Geogr. Higher Educ. 21 (3), 313–332.

Kern, E.L., Carpenter, J.R., 1984. Enhancement of student values, interests and attitudes in earth science through a field-oriented approach. J. Geol. Educ. 32, 299–305.

Krepel, W.J., Durrall, C.R., 1981. Fieldtrip: A guide for planning and conducting educational experience. NSTA, Washington, DC.

Lieberman, G., Hoody, L., 1998. Closing the Achievement Gap: Using the Environment as an Integrated Context for Learning. State Education and Environment Roundtable, California.

Livingstone, I., Mathhews, H., Castley, A., 1998. Fieldwork and Dissertations in Geography. GDN – Cheltenham & Gloucester College of Higher Education, Cheltenham.

Marriott, N., Marriott, P., 2003. Student learning style preferences and undergraduate academic performance at two UK universities. Int. J. Manage. Educ. 3 (1), 4–13.

Orion, N., 1993. A practical model for the development and implementation of fieldtrips as an integral part of the science curriculum. School Sci. Math. 93, 325–331.
Orion, N., 1998. Development of a high-school geology course based on fieldtrips. J. Geol. Educ. 37, 13–17.
Orion, N., Hofstein, A., 1994. Factors that influence learning during a scientific fieldtrip in a natural environment. J. Res. Sci. Teach. 31, 1097–1119.
Orion, N., Hofstein, A., Tamir, P., Giddings, J., 1997. Development and validation of an instrument for assessing the learning environment of outdoor science activities. Sci. Educ. 81, 161–171.
Woolnough, B., 1994. Effective Science Teaching. Open University Press, Milton Keynes.

APPENDIX 4.1.1 (QUESTIONNAIRE)S

This questionnaire aims to examine how successful you perceive has been your learning experience in the Mallorca fieldtrip and how it correlates with your preferred learning strategies. This questionnaire is anonymous and its results will be part of a research within the frame of the PGCHET course and might eventually form part of a paper in an education journal. *The questionnaire takes about 10-15 min to be completed.* If you have any query in how the data will be handled, please email to *Dr. Miguel Gomez-Heras*:

Some general information on you and the group you attended (circle as appropriate):

Week 1	Week 2
Alcudia	Palmanova
Male	Female

What aspect of geography do you feel more interested in?:

Physical (strongly)	Physical (weakly)	Do not have clear preference	Human (weakly)	Human (strongly)

Had you been in Mallorca before?

YES	NO

The following set of questions aims to evaluate what kind of learner you are. Choose the answer which best explains your preference and circle the letter next to it. Please circle more than one if a single answer does not match your perception. Leave blank any question that does not apply. (Adapted from the VARK questionnaire © Copyright Version 7.0 (2006) held by Neil D. Fleming and Charles C. Bonwell)

You are helping someone who wants to go to your town centre or railway station. You would:

a. go with her;
b. tell her the directions;

c. write down the directions;
d. draw, or give her a map.

You are planning a holiday for a group. You want some feedback from them about the plan. You would:

a. describe some of the highlights;
b. use a map or website to show them the places;
c. give them a copy of the printed itinerary;
d. phone, text or email them.

You are going to cook something as a special treat for your family. You would:

a. cook something you know without the need for instructions;
b. ask friends for suggestions;
c. look through the cookbook for ideas from the pictures;
d. use a cookbook where you know there is a good recipe.

You are about to purchase a digital camera or mobile phone. Other than price, what would most influence your decision?

a. Trying or testing it.
b. Reading the details about its features.
c. It is a modern design and looks good.
d. The salesperson telling me about its features.

Remember a time when you learned how to do something new. Try to avoid choosing a physical skill, e.g., riding a bike. You learned best by:

a. watching a demonstration;
b. listening to somebody explaining it and asking questions;
c. diagrams and charts – visual clues;
d. written instructions – e.g., a manual or textbook.

You have a problem with your knee. You would prefer that the doctor:

a. gave you a web address or something to read about it;
b. used a plastic model of a knee to show what was wrong;
c. described what was wrong;
d. showed you a diagram of what was wrong.

I like websites that have:

a. things I can click on, shift or try;
b. interesting design and visual features;
c. interesting written descriptions, lists and explanations;
d. audio channels where I can hear music or interviews.

Do you prefer a lecturer who uses:

a. demonstrations, models or practical sessions;
b. question and answer, talk, group discussions or guest speakers;

c. handouts, books or readings;
d. diagrams, charts or graphs.

You have finished a test and would like some feedback. You would like to have feedback:

a. using examples from what you have done;
b. using a written description of your results;
c. from somebody who talks it through with you;
d. using graphs showing what you have achieved.

You are going to choose food at a restaurant or café. You would:

a. choose something that you have had there before;
b. listen to the waiter or ask friends to recommend choices;
c. choose from the descriptions in the menu;
d. look at what others are eating or look at pictures of each dish.

The following set of questions aims to know what are your perceptions on how well prepared you were to carry out the fieldtrip. Please circle one of the numbers to express the degree of agreement you have on the following statements, knowing that:

1 = strongly disagree 2 = disagree 3 = undecided 4 = agree 5 = strongly agree

I would have liked to have more talks on the "mallorquin" culture and lifestyle beforehand.

1 2 3 4 5

I learned a lot from the written information and handouts I received before the fieldtrip.

1 2 3 4 5

I was familiar with how to carry out a field research project when I arrived to Mallorca.

1 2 3 4 5

Fieldwork is more difficult if you do not attend to all the preparatory lectures.

1 2 3 4 5

The elective project was useful to gain understanding on how to carry out a field research project.

1 2 3 4 5

I would have liked more demonstrations of methods and concepts before going on the fieldtrip.

1 2 3 4 5

I would have liked more written information before the fieldtrip.

1 2 3 4 5

I think it would be useful to have more long local fieldtrips to be fully prepared for continental fieldwork.

1 2 3 4 5

I think the level of information given before the fieldtrip was adequate to the work I had to carry out in the field

1 2 3 4 5

*The following set of questions aims to examine how valuable you find fieldwork for your "geographical" learning experience***:** *Please circle one of the numbers to express the degree of agreement you have on the following statements, knowing that:*

1 = strongly disagree 2 = disagree 3 = undecided 4 = agree 5 = strongly agree

I think fieldwork is essential for geographical research.

1 2 3 4 5

Fieldwork gave me a unique perspective of the physical environment.

1 2 3 4 5

Fieldwork gave me a unique perspective of human geographies.

1 2 3 4 5

I feel I could now carry out a field research project on my own.

1 2 3 4 5

I could have learned the same things in a fieldtrip in the UK/Ireland.

1 2 3 4 5

Travelling abroad has made me more appreciative of different geographies and landscapes.

1 2 3 4 5

I would like to take a similar continental fieldtrip again at level 3.

1 2 3 4 5

I am now better aware of the economic cost and human resources necessary to carry out fieldwork.

1 2 3 4 5

I would prefer to go to Mallorca (student cost £300) rather than a destination in the UK/Ireland, staying in a youth hostel or field station, at a slightly lower cost (£220).

1 2 3 4 5

The following set of questions aims to understand your social experience during the fieldtrip: *Please, answer the questions with an approximate figure.*

How many nights did you go out for dinner?

______

How many nights did you go out to pubs/discos?

______

What was the average time you went to bed?

______

How many times did you try local food during the week?

______

What was approximately your expenditure during the field trip?

______

Please circle one of the numbers to express the degree of agreement you have on the following statements, knowing that:

1 = strongly disagree 2 = disagree 3 = undecided 4 = agree 5 = strongly agree

This experience has made me feel more appreciative of the local culture.

1 2 3 4 5

I would have liked to have more interaction with local people.

1 2 3 4 5

I felt challenged by the cultural difference.

1 2 3 4 5

I think night outings have negatively influenced in my learning.

1 2 3 4 5

I felt very tired during the day because of lack of sleep.

1 2 3 4 5

I believe night outings negatively influenced the learning of some members of the group.

1 2 3 4 5

The social aspect of the fieldtrip was more valuable for me than the academic work I carried out.

1 2 3 4 5

I think more disciplinary actions should be taken for those not taking the fieldwork seriously.

1 2 3 4 5

Going out at night time has improved my work relation to my class mates.

1 2 3 4 5

The resort made me feel safe as I felt it was like home.

1 2 3 4 5

The following set of questions aims to better understand aspects on your learning experience during the fieldtrip. Please circle one of the numbers to express the degree of agreement you have on the following statements, knowing that:

1 = strongly disagree 2 = disagree 3 = undecided 4 = agree 5 = strongly agree

The level of feedback in the field was adequate.

1 2 3 4 5

I felt supported in my work by the group.

1 2 3 4 5

I would prefer to have all the assessment in the field.

1 2 3 4 5

I would have liked to have the presentations made in the fieldwork assessed.

1 2 3 4 5

I found myself lost and challenged trying to make a meaningful field notebook.

1 2 3 4 5

I often felt disoriented and not knowing what to do.

1 2 3 4 5

I think I learn more in lectures theatre than in the field.

1 2 3 4 5

As a part of this research, I would like to carry out some group focus research interviews. If you would like to participate in one of these, please let me know after finishing the test.

Thank you for your time!

APPENDIX 4.1.2 (RESULTS FROM THE QUESTIONNAIRE)

Preferred learning styles:

V:	28.3%	A:	48.3%	R:	25.0%	K:	35.0%

Unimodal:	68.3%	Bimodal:	26.7%	Multimodal: 5%

What aspect of geography do you feel more interested in?:

	Physical (strongly)	Physical (weakly)	Do not have clear preference	Human (weakly)	Human (strongly)
V(%)	24	24	6	6	35
A(%)	24	3	7	21	45
R(%)	33	27	0	7	33
K(%)	14	24	10	10	43
Tot(%)	18	17	8	15	42

	Male	Female
V (%)	24	76
A (%)	34	66
R (%)	33	60
K (%)	33	67
Tot (%)	33	65

The following set of questions aims to know what are your perceptions on how well prepared you were to carry out the fieldtrip.

1 = strongly disagree 2 = disagree 3 = undecided 4 = agree 5 = strongly agree

1. I would have liked to have more talks on the "mallorquin" culture and life-style beforehand.

	1	2	3	4	5	(disagree:47/undecided:24/agree:30)
V (%)	6	41	24	24	6	
A (%)	10	21	31	31	7	(31/31/38)
R (%)	13	27	20	40	0	(40/20/40)
K (%)	5	19	19	43	14	(24/19/57)
Tot (%)	7	25	27	33	8	(32/27/41)

2. I learned a lot from the written information and handouts I received before the fieldtrip.

	1	2	3	4	5	
V (%)	12	29	12	47	0	(41/12/47)
A (%)	7	21	24	45	3	(28/24/48)
R (%)	13	20	20	47	0	(33/20/47)
K (%)	5	24	33	33	5	(29/33/38)
Tot (%)	8	22	25	24	3	(30/25/45)

3. I was familiar with how to carry out a field research project when I arrived to Mallorca.

	1	2	3	4	5	
V (%)	12	24	24	35	6	(36/24/41)
A (%)	7	31	31	28	3	(38/31/31)
R (%)	13	0	40	40	7	(13/40/47)
K (%)	10	19	24	48	0	(29/24/48)
Tot (%)	7	23	30	38	2	(30/30/40)

4. Fieldwork is more difficult if you do not attend to all the preparatory lectures.

	1	2	3	4	5	
V (%)	12	0	18	47	24	(12/18/71)
A (%)	3	17	17	41	21	(20/17/62)
R (%)	7	13	13	40	27	(20/13/67)
K (%)	5	19	10	48	19	(24/10/67)
Tot (%)	3	15	17	45	20	(18/17/65)

5. The elective project was useful to gain understanding on how to carry out a field research project.

	1	2	3	4	5	
V (%)	6	18	18	53	6	(24/18/59)
A (%)	3	21	28	28	21	(24/28/49)
R (%)	0	27	20	27	27	(27/20/54)
K (%)	5	24	24	43	5	(29/24/48)
Tot (%)	3	22	23	38	13	(25/23/51)

6. I would have liked more demonstrations of methods and concepts before going on the fieldtrip.

	1	2	3	4	5	
V (%)	12	29	24	29	6	(41/24/35)
A (%)	3	10	28	41	17	(13/28/58)
R (%)	7	27	13	53	0	(34/13/53)
K (%)	0	33	14	38	14	(33/14/52)
Tot (%)	3	25	22	37	13	(28/22/50)

7. I would have liked more written information before the fieldtrip.

	1	2	3	4	5	
V (%)	12	35	6	35	12	(47/6/47)
A (%)	3	31	28	28	0	(34/28/28)
R (%)	7	47	13	33	0	(54/13/33)
K (%)	0	29	24	33	14	(29/24/57)
Tot (%)	3	35	23	32	7	(38/23/39)

8. I think it would be useful to have more long local fieldtrips to be fully prepared for continental fieldwork.

	1	2	3	4	5	
V (%)	18	18	6	35	24	(36/6/59)
A (%)	3	10	21	55	10	(13/21/65)
R (%)	20	27	13	27	13	(47/13/40)
K (%)	14	29	14	29	14	(43/14/43)
Tot (%)	12	20	17	40	12	(32/17/52)

9. I think the level of information given before the fieldtrip was adequate to the work I had to carry out in the field.

	1	2	3	4	5	
V (%)	6	12	24	53	6	(18/24/59)
A (%)	7	7	31	48	7	(14/31/55)
R (%)	0	13	13	67	7	(13/13/74)
K (%)	10	10	14	57	10	(20/14/67)
Tot (%)	7	7	25	55	7	(14/25/62)

The following set of questions aims to examine how valuable you find fieldwork for your "geographical" learning experience:

1 = strongly disagree   2 = disagree   3 = undecided   4 = agree   5 = strongly agree

10. I think fieldwork is essential for geographical research.

	1	2	3	4	5	
V (%)	6	0	0	24	71	(6/0/95)
A (%)	3	3	0	41	52	(6/0/93)
R (%)	7	7	13	33	40	(14/13/73)
K (%)	0	5	5	14	76	(5/5/90)
Tot (%)	2	3	3	32	60	(5/3/92)

11. Fieldwork gave me a unique perspective of the physical environment.

	1	2	3	4	5	
V (%)	6	0	0	29	65	(6/0/94)
A (%)	3	3	7	34	52	(6/7/86)
R (%)	7	0	7	47	40	(7/7/87)
K (%)	0	5	5	24	67	(5/5/91)
Tot (%)	2	2	7	35	55	(4/7/90)

12. Fieldwork gave me a unique perspective of human geographies.

	1	2	3	4	5	
V (%)	0	0	0	35	65	(0/0/100)
A (%)	3	3	10	41	41	(6/10/82)
R (%)	0	13	7	40	40	(13/7/80)
K (%)	5	5	0	24	67	(10/0/91)
Tot (%)	2	5	5	38	50	(5/5/88)

13. I feel I could now carry out a field research project on my own.

	1	2	3	4	5	
V (%)	0	0	6	65	29	(0/6/94)
A (%)	3	0	10	59	28	(3/10/87)
R (%)	0	0	13	40	47	(0/13/87)
K (%)	10	5	14	38	33	(15/14/72)
Tot (%)	3	2	10	57	28	(5/10/85)

14. I could have learned the same things in a fieldtrip in the UK/Ireland.

	1	2	3	4	5	
V (%)	18	47	18	18	0	(65/18/18)
A (%)	10	38	38	7	7	(48/38/14)
R (%)	7	40	33	13	7	(47/33/20)
K (%)	14	52	14	14	5	(66/14/19)
Tot (%)	12	42	28	13	5	(54/28/18)

15. Travelling abroad has made me more appreciative of different geographies and landscapes.

	1	2	3	4	5	
V (%)	0	0	0	41	59	(0/0/100)
A (%)	0	3	10	48	38	(3/10/86)
R (%)	0	7	0	53	40	(7/0/93)
K (%)	0	0	5	38	57	(0/5/95)
Tot (%)	0	2	5	47	47	(2/5/94)

16. I would like to take a similar continental fieldtrip again at level 3.

	1	2	3	4	5	
V (%)	6	12	6	18	59	(18/6/77)
A (%)	3	7	10	17	62	(10/10/79)
R (%)	13	7	20	27	33	(20/20/60)
K (%)	5	0	14	5	76	(5/14/81)
Tot (%)	3	5	12	18	62	(8/12/80)

17. I am now better aware of the economic cost and human resources necessary to carry out fieldwork.

	1	2	3	4	5	
V (%)	0	6	6	71	18	(6/6/89)
A (%)	7	0	3	62	28	(7/3/90)
R (%)	0	13	7	53	27	(13/7/80)
K (%)	5	5	10	52	29	(10/10/81)
Tot (%)	3	3	5	65	23	(6/5/88)

18. I would prefer to go to Mallorca (student cost £300) rather than a destination in the UK/Ireland, staying in a youth hostel or field station, at a slightly lower cost (£220).

	1	2	3	4	5	
V (%)	6	0	0	12	82	(6/0/94)
A (%)	7	0	7	17	69	(7/7/86)
R (%)	7	0	13	13	67	(7/13/80)
K (%)	0	0	5	19	76	(0/5/95)
Tot (%)	3	0	7	17	73	(3/7/90)

The following set of questions aims to understand your social experience during the fieldtrip:

19. How many nights did you go out for dinner?

V. 4.6 **A. 5.0** **R. 4.2** **K. 4.2**

20. How many nights did you go out to pubs/discos?

V. 3.6	**A. 3.6**	**R. 3.3**	**K. 3.6**

21. What was the average time you went to bed?

V. 1:40	**A. 1:45**	**R. 1:10**	**K. 1:55**

22. How many times did you try local food during the week?

V. 1.8	**A. 0.7**	**R. 1.3**	**K. 1.6**

1 = strongly disagree   2 = disagree   3 = undecided   4 = agree   5 = strongly agree

23. This experience has made me feel more appreciative of the local culture.

	1	2	3	4	5	
V (%)	6	0	0	65	29	(6/0/94)
A (%)	3	0	10	52	31	(3/10/83)
R (%)	7	7	13	53	20	(14/13/73)
K (%)	0	0	10	48	43	(0/10/89)
Tot (%)	2	2	8	55	32	(4/8/87)

24. I would have liked to have more interaction with local people.

	1	2	3	4	5	
V (%)	6	12	47	24	12	(18/47/36)
A (%)	3	10	28	52	3	(13/28/55)
R (%)	7	20	20	47	7	(27/20/54)
K (%)	0	5	19	52	24	(5/19/76)
Tot (%)	2	10	28	48	10	(12/28/58)

25. I felt challenged by the cultural difference.

	1	2	3	4	5	
V (%)	24	12	29	29	6	(36/29/35)
A (%)	10	17	34	21	14	(27/34/35)
R (%)	27	20	40	13	0	(47/40/13)
K (%)	10	24	38	10	19	(34/38/29)
Tot (%)	12	18	38	18	12	(30/38/30)

26. I think night outings have negatively influenced in my learning.

	1	2	3	4	5	
V (%)	29	53	6	12	0	(82/6/12)
A (%)	31	34	31	0	0	(65/31/0)
R (%)	7	47	40	7	0	(54/40/7)
K (%)	24	43	19	14	0	(67/19/14)
Tot (%)	27	42	23	7	0	(69/23/7)

27. I felt very tired during the day because of lack of sleep.

	1	2	3	4	5	
V (%)	18	29	6	24	24	(47/6/48)
A (%)	7	28	17	34	10	(35/17/44)
R (%)	7	20	13	27	33	(27/13/60)
K (%)	10	24	14	14	38	(34/14/52)
Tot (%)	12	28	13	25	20	(40/13/45)

28. I believe night outings negatively influenced the learning of some members of the group.

	1	2	3	4	5	
V (%)	12	18	12	29	29	(30/12/58)
A (%)	14	14	21	28	21	(28/21/49)
R (%)	0	27	0	27	47	(27/0/74)
K (%)	14	14	5	38	29	(28/5/67)
Tot (%)	12	17	13	33	23	(29/13/56)

29. The social aspect of the fieldtrip was more valuable for me than the academic work I carried out.

	1	2	3	4	5	
V (%)	6	18	35	24	18	(26/35/42)
A (%)	3	14	34	38	7	(17/34/45)
R (%)	7	13	47	33	0	(20/47/33)
K (%)	0	10	62	14	14	(10/62/28)
Tot (%)	2	12	47	28	10	(14/47/38)

30. I think more disciplinary actions should be taken for those not taking the fieldwork seriously.

	1	2	3	4	5	
V (%)	0	24	18	29	29	(24/18/58)
A (%)	0	14	10	34	38	(14/10/72)
R (%)	0	20	7	40	33	(20/7/73)
K (%)	10	19	10	38	24	(29/10/62)
Tot (%)	3	18	15	33	28	(21/15/61)

31. Going out at night time has improved my work relation to my class mates.

	1	2	3	4	5	
V (%)	0	6	12	24	59	(6/12/83)
A (%)	0	7	3	28	59	(7/3/87)
R (%)	0	7	20	20	53	(7/20/73)
K (%)	0	5	10	29	57	(5/10/86)
Tot (%)	0	5	10	25	58	(5/10/83)

32. The resort made me feel safe as I felt it was like home.

	1	2	3	4	5	
V (%)	6	12	24	47	12	(18/24/59)
A (%)	3	17	31	28	17	(20/31/45)
R (%)	0	13	27	53	7	(13/27/60)
K (%)	0	5	14	62	14	(5/14/76)
Tot (%)	3	10	28	42	13	(13/28/55)

The following set of questions aims to better understand aspects on your learning experience during the fieldtrip.

1 = strongly disagree 2 = disagree 3 = undecided 4 = agree 5 = strongly agree

33. The level of feedback in the field was adequate.

	1	2	3	4	5	
V (%)	6	12	6	65	12	(18/6/77)
A (%)	14	7	31	38	10	(21/31/48)
R (%)	7	0	20	67	7	(7/20/74)
K (%)	5	10	29	52	5	(15/29/57)
Tot (%)	7	8	25	50	10	(15/25/60)

34. I felt supported in my work by the group.

	1	2	3	4	5	
V (%)	6	0	0	41	53	(6/0/94)
A (%)	3	3	10	41	41	(6/10/82)
R (%)	7	0	20	67	7	(7/20/74)
K (%)	5	10	29	52	5	(15/29/57)
Tot (%)	5	3	8	45	38	(8/8/83)

35. I would prefer to have all the assessment in the field.

	1	2	3	4	5	
V (%)	6	24	6	29	35	(30/6/64)
A (%)	3	17	21	28	31	(20/21/59)
R (%)	7	33	20	33	7	(40/20/40)
K (%)	5	24	5	33	33	(29/5/66)
Tot (%)	3	20	17	32	28	(23/17/60)

36. I would have liked to have the presentations made in the fieldwork assessed.

	1	2	3	4	5	
V (%)	0	24	12	41	24	(24/12/65)
A (%)	7	28	10	17	38	(36/10/55)
R (%)	13	60	13	7	7	(73/13/14)
K (%)	10	24	14	29	24	(34/14/53)
Tot (%)	8	30	15	23	23	(38/15/46)

37. I found myself lost and challenged trying to make a meaningful field notebook.

	1	2	3	4	5	
V (%)	12	29	35	18	0	(41/35/18)
A (%)	0	34	31	24	7	(34/31/31)
R (%)	7	40	13	27	7	(47/13/34)
K (%)	10	19	19	38	14	(29/19/52)
Tot (%)	5	28	32	27	7	(33/32/34)

38. I often felt disoriented and not knowing what to do.

	1	2	3	4	5	
V (%)	18	59	6	18	0	(77/6/18)
A (%)	17	52	14	17	0	(69/14/17)
R (%)	13	47	13	27	0	(60/13/27)
K (%)	19	33	24	19	5	(52/24/24)
Tot (%)	18	47	15	18	2	(65/15/20)

39. I think I learn more in lectures theatre than in the field.

	1	2	3	4	5	
V (%)	59	35	0	0	6	(94/0/6)
A (%)	31	41	10	10	7	(72/10/17)
R (%)	20	20	40	13	7	(40/40/20)
K (%)	48	19	29	5	0	(67/29/5)
Tot (%)	37	32	20	8	3	(69/20/11)

Chapter 4.2

Fieldwork Going Digital

Ian Fuller*[,1], Derek France**
*Physical Geography Group, Institute of Agriculture and Environment, Massey University, Palmerston North, New Zealand; **Department of Geography and Development Studies, University of Chester, Chester, United Kingdom
[1]Corresponding author: E-mail: I.C.Fuller@massey.ac.nz

4.2.1 INTRODUCTION

The field is, "any arena or zone within a subject where, outside the constraints of the four walls classroom setting, supervised learning can take place via first-hand experience (Lonergan and Andreson, 1988, p. 64)." Fieldwork has long been perceived to lie at the heart of geography and its companion/subdiscipline of geomorphology (Sauer, 1956; Gold et al., 1991; Haigh and Gold, 1993; Kent et al., 1997; Pawson and Teather, 2002; Bracken and Mawdsley, 2004; Stoddart and Adams, 2004; Dummer et al., 2008; Herrick, 2010; Fuller, 2012). Fieldwork has been identified as geography's "signature pedagogy" (Hovorka and Wolf, 2009, p. 99) and every geomorphologist surely learns their trade "through the soles of their feet" (McEwen, 1996, p. 379). Fieldwork provides demonstrably valuable learning experiences (Kern and Carpenter, 1984; Kern and Carpenter, 1986; Fuller et al., 2003, 2006, 2010; Scott et al., 2006), opportunities to acquire "hands-on" subject-knowledge in the "real world," and the glue that bonds together a student cohort (Fuller, 2006). Incorporation of technology *per se* into field teaching has been perceived to enhance fieldwork learning for a range of applications, as reviewed by Welsh et al. (2013). However, Welsh et al. (2013) also recognize that technology needs to be used effectively if learning is to be enhanced. Surprisingly, Welsh et al. (2013) suggest that technology is in fact used least when students are actually out in the field, being more likely to be utilized during the stages of data analysis and write-up. In this chapter we demonstrate the application and incorporation of digital technology as a means of enhancing field teaching of physical geography and geomorphology at the tertiary level in a range of diverse geomorphic settings.

4.2.2 A COOK'S TOUR OF GEOMORPHOLOGY: NORTH ISLAND, NEW ZEALAND

Underpinning Sauer's often quoted presidential address to the American Association of Geographers in 1956: "the principal training of geographers should

Developments in Earth Surface Processes, Vol. 18. http://dx.doi.org/10.1016/B978-0-444-63402-3.00008-X

come, wherever possible, by doing fieldwork" (Sauer, 1956, p. 296) was a slow excursion, punctuated by leisurely halts on vantage points across landscapes (Fuller, 2012). This was the Cook's Tour' fieldtrip, so named after the nineteenth century travel agent, Thomas Cook, who pioneered tours of extended regions. Kent et al. (1997) identified this as the simplest and most traditional form of observational fieldwork and suggested that it was the predominant mode of field teaching in the 1950s and 1960s, when Sauer was speaking. It was also quickly recognized as the most boring mode of fieldwork because students become disengaged (Brown, 1969). However, a place remained for such observational fieldtrips (or at least as a component part of a fieldtrip) and in the mid-1990s Kent et al. (1997) described a variety of approaches that foster engagement of students on such visits, including critical assessments, map production, worksheets, and self-paced guides, essentially meshing together "look-see" and experiential fieldwork, as implemented by several practitioners (Keene, 1982; Higgitt, 1996a; Fuller et al., 2000). These revisions to the straightforward "Cook's Tour" fieldtrip have fostered student engagement (Fuller et al., 2000), recognizing the importance of the pedagogically established notion that students learn best "by doing" (Race, 1993) in an active learning setting (Higgitt, 1996b), adopting experiential learning approaches (Kolb, 1984; McEwen, 1996). Geography fieldwork has developed in the decades since Sauer's address. Kent et al. (1997) plotted the changing approach to fieldwork from 1956 to 1997. The nature of fieldwork in the modern undergraduate degree continues to change and embrace the pace of modern life, with each hour accounted for fieldtrips crammed with a plethora of learning activity and action to satisfy the consumer-driven, value-for-money demands of today's fee-paying students (Fuller, 2012). This development is essential if fieldwork is to remain a central, fundamental component of modern undergraduate degrees as so many recognize. In this context, can there still remain any place for the Cook's Tour, observation-based fieldtrip, particularly when fieldwork is identified as, "a key sphere of learning in which subject-specific and generic skills are developed" (Wall and Speake, 2012, p. 433)? We evaluate the incorporation of digital video assignments into a primarily observation-based, multicenter fieldtrip run in New Zealand's North Island to address this question.

4.2.2.1 The Fieldtrip

The field course takes the form of a 7-day road trip, which encompassed a range of environments, topics, and issues as part of a third year undergraduate paper in applied field geomorphology running around the highly erodible terrains of the North Island's East Coast and through its volcanic Central Plateau regions. The intention and genre of the trip are observational, rather than hands-on measurement of process, with an emphasis on the relationship between evaluation of process and critical assessment of landscape management approaches within a broad spectrum of geomorphic topics (e.g., landslides, flooding, coastal erosion, volcanic hazards). Hands-on technical skills are taught elsewhere in the curriculum in this degree (see field experiments below), thus an observational approach is entirely appropriate to add diversity and breadth to geomorphology teaching and fieldwork. However, there is a danger that students undertaking such a field course become passive learners (Kent et al., 1997). To help foster engagement, students are required to complete field notebook entries at each site visited; it forms a compulsory form of assessment that is assessed on a daily basis. Students also take ownership of the content of the course by becoming the local expert at key sites: each student prepares a talk accompanied by illustrative material that is presented at the relevant (their chosen) location in the field course. Their material is informed by recently published material (journal articles and "gray" literature, that is, council reports) that they read and digest prior to the fieldtrip. Handouts they prepare are distributed to the group as a whole for reference in the field. In effect, the student becomes the "knowledgeable expert" *sensu* Coe and Smyth (2010) and their learning more active and student centered. In addition, students prepare for the fieldtrip by writing an essay discussing the larger-scale (tectonic and climatic) variables conditioning the geomorphology of the region.

However, there remains a need to foster engagement further and at a deeper level because note taking and field sketching, and listening to others' presentations can still prove somewhat passive. Furthermore, over the course of a 7-day road trip, the novelty of visiting a site, discussing processes and management, and questioning peers can wear off, as became evident in the first iteration of the field course in 2009. To address this issue, a digital video component was introduced. Jarvis and Dickie (2010) and Kemp et al. (2012) reviewed a range of uses for podcasts in student learning, among which is the suggestion that they provoke reflective thought (Fisher and Baird, 2006), which in turn fosters deeper levels of learning and engagement (Jarvis and Dickie, 2010; Kemp et al., 2012). At each site, following discussion led by the appropriate "knowledgeable expert," students worked in groups between four to six to produce a piece to camera which summarized key processes and issues at the site. Accordingly, this approach generated a series of knowledgeable experts at each site, feeding off the primary presentation and discussion. Known colloquially as the "video diary," this was modeled on the "digital story" of France and Wakefield (2011). Underpinning the educational value of this approach is that to communicate clearly and effectively a succinct piece to camera requires that the students have recognized and understood the processes and issues at each site. For assessment, each group selected 10 video pieces, with at least one from each day, covering the breadth of course content. These were edited together into a coherent whole of 10-min duration and presented to the class on the final evening of the fieldtrip. Marks were awarded on the basis of clarity of video and audio, as well as content (grasp of subject matter) and overall quality (e.g., use of captions). Inevitably the variability of New Zealand's weather in the last week of August (late Winter/early Spring) renders live video impossible at times, but video diaries could comprise a sequence of stills and/or pan shots with voice-overs.

In addition, each group member was to appear in front of camera as the knowledgeable expert at least once in the final assessed sequence. Therefore, the final production required students to discuss and negotiate what to include and what to omit, requiring a review of content and assessment of quality. To assist with production and editing, students received briefings from staff, as well as instruction using France and Wakefield (2011). Students were able to refer to their production, as well as the complete set of peer-prepared material in order to inform the final piece of assessment for the course, which comprised a report critically appraising the management at a site of their choice visited during the fieldtrip.

4.2.2.2 Learning Evaluation and Discussion

The extent to which the introduction of digital video assessment improved understanding and engagement during this field course was assessed using a range of approaches (Tables 4.2.1 and 4.2.2), which are elaborated by Fuller and France (2015).

The introduction of a digital video assessment component to a primarily observational "Cook's Tour" field course proved to be effective in fostering engagement and encouraging participation in this fieldwork genre. A traditionally dull, stale form of practicing fieldwork in geomorphology has been transformed

TABLE 4.2.1 Methods of Evaluation

	Cook's tour	Field Experiments		
Evaluation method	Response rate 2011 (%)	Response rate 2010 (%)	Response rate 2011 (%)	Response rate 2012 (%)
Pre-fieldwork information sheet	97	–	–	–
Pre-fieldwork "video bus"	–	62	–	–
On-fieldwork "video bus"	60		–	–
Post-fieldwork "video bus"	–	100	–	–
Post-fieldwork questionnaire	80	92	75	83
Focus group discussion	67	77	–	–
Post-fieldwork follow-up sheet	93	–	–	–
Marks analysis 2009–2012	N/A	–	–	–

TABLE 4.2.2 Marks Analysis: Cook's Tour

		Mean mark (grade)		
	n	Essay	Report	Difference
2009	21	69 (B)	72 (B+)	3
2010	16	65 (B)	72 (B+)	6
2011	33	62 (B−)	70 (B+)	8
2012	25	69 (B)	72 (B+)	3
2013	21	62 (B−)	72 (B+)	10

NB: The same markers were used to assess the work in each year over the course of this study.

into an enjoyable, engaging, fun means of learning in the approach adopted here (Fuller and France, 2015). Kemp et al. (2012) suggest that a similar approach incorporating podcasting into fieldwork was also enjoyed by students and motivated their study. The spin-off to such enjoyable learning is that digital video assessment has improved levels of understanding of geomorphology. Students learn best when "doing": production of video diaries engaged students in active learning in an otherwise passive, knowledge-transfer situation. When such active learning is enjoyed and becomes fun, its effectiveness is improved. This is supported by Oblinger and Oblinger (2005) and Prensky (2009), who argued that students prefer technology-led learning.

Marks analysis (Table 4.2.2) suggests that there was a tangible effect in student performance among some cohorts. In 2009, before the introduction of video, a slight increase occurred in marks between the prefieldcourse essay and the post-trip report. Such an increase was reasonable, since following time spent in the field, students' understanding of processes and the environment ought to be enhanced. The improvement appeared to be more marked in both 2010 and 2011, although not statistically significant when applying a *t*-test (*p*-value 0.078, thus no significant difference occurred between the marks increase between essay and report after video component was introduced), there was nevertheless a 6–8 point improvement in mean marks – more than twice the improvement. However, this improvement was not repeated in 2012. The 2012 cohort appeared to be stronger than the two preceding years, remarkably similar to 2009 in fact. The greatest improvement was seen most recently in 2013 and the difference between the marks for the essay and the report was statistically significant (*p*-value 0.006), thus H1 was upheld, the marks for the post fieldtrip report in 2013 were significantly higher than the prefieldtrip essay. The 2013 cohort had the lowest pretrip essay score (61.9%). This suggests that engaging students using digital video had the greatest impact among weaker students (i.e., those whose grades in the initial assessment were below average), with less impact among stronger students. This is intuitive, since stronger students tend to be more engaged with the learning process (which is why they are stronger), whereas weaker students are in need of greater help and encouragement. As such, the approach adopted here could be argued as a powerful learning tool within geomorphology, facilitating engagement and learning of less able students in particular. As such, digital video strategically incorporated into geomorphology fieldwork ought to enrich the learning space and enhance the student-learning experience. Stokes et al. (2011) have recently affirmed findings that show active participation in fieldwork promoted retention and recall of subject-specific information, which thereby, and perhaps subtly, has improved certain cohort performances in the final assessment in this course.

4.2.3 FIELD EXPERIMENTS IN PROCESS GEOMORPHOLOGY

Since 2010, digital video technology has also been incorporated into two final year undergraduate papers in process geomorphology that have a strong emphasis on field-based experimentation. The experiments in these papers introduce students to the use of both sophisticated technology and standard field sampling procedures to acquire data in order to test a range of hypotheses. Traditionally, students find some of these approaches dull and repetitive and/or perceive them to be complex and involved. The intention of incorporating digital video into these field experiments was twofold: (1) to increase the level of engagement and enjoyment involved in tedious data collection; and (2) to improve the level of understanding of the method employed, particularly where new and unfamiliar equipment was being introduced and deployed.

4.2.3.1 Approach to Teaching and Assessment

Field experimentation deployed in this study was concerned with the investigation of fluvial processes and process geomorphology in an alpine setting. The former were studied by students in a series of short, 2–3 h long field experiments using a stream local to the university campus in Palmerston North, New Zealand. These experiments run in the normal course of the teaching day and deployed equipment including electromagnetic flow meters, a total station, cone penetrometer, and standard grain size sampling tools (sieves and pebble template). Experiments were designed to measure stream flow and resistance, bed sediment characteristics, pool-riffle morphology, and channel morphodynamics. Process geomorphology was studied as part of an 8-day residential fieldtrip at Fox Glacier, South Island New Zealand, a site remote from the university and in a highly contrasting, alpine environment. In addition to equipment described above, field experimentation used a high-precision Global Positioning System using Real Time Kinematic survey (RTK-dGPS) to measure glacier velocity using surveyed ablation stakes. Cross-profiles of the Fox Glacier sandur and its sediment assemblages were also measured, as were flow properties in the proglacial stream. In both courses, students work in self-selected groups of c.4 and had the responsibility of setting up and running the equipment to derive data that they processed and interpreted in their own time or during the evenings of the residential fieldtrip. The intention of the fieldwork in these courses was to provide detailed, hands-on technical experience measuring key parameters in rivers and alpine landscapes, providing an introductory research experience (see Fuller et al., 2014). Thus, students become familiar with technology used in these contexts, as well as having the opportunity to learn "by doing," which according to Race (1993) is when students learn best, particularly in such an active learning setting (Higgitt, 1996b), adopting these experiential learning approaches (Kolb, 1984; McEwen, 1996).

These theatres of learning thus deploy best pedagogic practice (Fuller, 2012), but there is a need to keep pace with opportunities to develop and improve practice. Accordingly, from 2010, digital technology was introduced in the form of video cameras (digital cameras with a video function and flip videos) to report on methods deployed and environments studied by each student group in these courses. These "digital stories" (*sensu* France and Wakefield, 2011) were intended to increase levels of engagement by the group as a whole, because each student had to appear and/or narrate a part of each digital story produced. This approach also removed the need to write tediously about the nuances of equipment setup or at descriptive length about the nature of the environment. The video sequence became a means of assessing the students' understanding of the methods deployed as well as their competency in execution. To be able to successfully narrate a piece to camera required familiarity with the equipment and methodological approach that cannot be easily glossed over. Furthermore, incorporation of digital video can add some interest to tedious operations such as measuring 100 stones to characterize the grain size of riverbeds and bars. Jarvis and Dickie (2010) and Kemp et al. (2012) reviewed a range of uses for similar podcasts in student learning, among which is the suggestion that they provoke reflective thought (Fisher and Baird, 2006), which in turn fosters deeper levels of learning and engagement (Jarvis and Dickie, 2010; Kemp et al., 2012).

Video productions were awarded marks on the basis of clarity of video and audio, as well as content (grasp of methodology and approach) and overall quality (e.g., use of captions). Inevitably as with the Cook's Tour described above, the variability of New Zealand's weather may render live video impossible at times, but video sequences could comprise a sequence of stills and/or pan shots with voice-overs. To assist with production and editing, students received briefings from staff, as well as (more latterly) instruction using France and Wakefield (2011).

4.2.3.2 Learning Evaluation and Discussion

The extent to which the introduction of digital video assessment improved understanding and engagement with the methods taught during these courses was assessed using a combination of approaches (Table 4.2.1), elaborated by Fuller and France (in review). They found that incorporating digital video into field-based experiments fostered engagement and interest among student groups, whether they were on campus or farther afield at Fox Glacier. In both settings, the approach incorporating digital video as part of field experiments in process geomorphology proved enjoyable and improved understanding of methods in particular, as well as processes being studied. In addition, as with use of video diaries in the North Island tour, this deployment of digital technology has helped students work together in a group, having fostered a greater sense of group identity and cooperation. Potential pitfalls, however, are to be avoided. Notably adequate time should be allowed for students to play and experiment with the equipment when presented with "challenging new experiences" (student feedback recorded by Fuller and France, in review). Nevertheless, the overwhelming perception of students' experience of digital technology was positive in a social, developmental, and learning perspective. This was in accord with findings of the use of similar digital technology (podcasts) by, for example, Kemp et al. (2012). Mavroudi and Jöns (2011, p.18) also recognized similar benefits of incorporating video documentaries into their assessment of fieldwork and concluded that "video documentary works extremely well to stimulate students' interest and critical thinking." Overall, the learning opportunities offered by video and the additional skill-set development can be best summarized by Dando and Chadwick (2014, p. 6) who noted that "Students gain enhanced geographic knowledge as they apply concepts they have learned and practice communicating them to others.... Both geographic education and media literacy lead to enhanced critical thinking skills. Even if the videos are not perfect, learning happens."

4.2.4 WEB 3.0 TECHNOLOGY

Web 3.0 is sometimes called the Semantic Web (Ohler, 2008) and the emergence of Web 3.0 technology is enabling users to share, collaborate, and work more effectively through enhanced Web-based platforms. These websites utilize outputs created on "Web 2.0" devices such as iPads (Cook and Kelly, 2013). Therefore, students outside the traditional lecture-based setting can, with the aid of a digital device, integrate a range of resources, whether multimedia-, data-, or text-based to potentially produce a new product. This gives students the opportunity to develop their own Personal Learning Environments, which Educause (2009) described "as the tools, communities, and services that constitute the individual educational platforms learners use to direct their own learning and pursue educational goals." The complexity of information transfer around such "learning spaces" within an Educational Space is further elaborated in France et al. (2013).

4.2.4.1 Mobile Technologies and Fieldwork

The arrival of the iPad in 2010 augmented engagement and access with the digital world, with the use of terms like "apps" and "the cloud" becoming more ever-present. Academics are now beginning to consider how mobile devices can be used to enhance their students' field-based investigations. Smartphones are becoming more affordable and ubiquitous (Melhuish and Falloon, 2010) and can be used effectively in the field (Welsh et al., 2013), whether internet connected or not. Williams et al. (2011) noted that earlier iPad trials with paramedic students were not as successful, and intermittent internet connectivity posed real access issues. Smartphones can still be better utilized than laptops when conducting research in the field because of their portability (Welsh and France, 2012). Smartphones and iPads can be used to record data on spreadsheets or with specialized applications or with "add on" hardware (e.g., microscope, thermometer, pH meter, and/or still or video cameras). Data can be collected in digital notebooks, digitally processed in the field, shared, and backed up through distributed computer (on "the cloud"). From simple search tools such as *Google Scholar*, the bibliographic data obtained can be used in open applications like *Mendeley* and *PaperShip*. However, a recent study of undergraduate students (Woodcock et al., 2012) found that many students who own smartphones were "largely unaware of their potential to support learning" (p.1). More importantly, however, students were found to be "interested in and open to the potential as they become familiar with the possibilities" (p.1).

4.2.4.2 Case Study: Application of Mobile Devices to Residential Fieldwork

The Geography and Development Studies Department at the University of Chester has incorporated the use of digital technologies into a range of fieldwork opportunities for a number of years. Whether a first year residential field course to Devon (the United Kingdom) or final year field course to the Bay of Naples (Italy), both provided the basis for geography students to enhance their own research skills through the use of mobile devices.

Encouraging students to use their own mobile devices and/or departmental iPads in conjunction with a wide array of inexpensive mobile apps (see France et al., 2013) has fostered student engagement within the small-group research project and field environments. Social media enabled students to share day-to-day fieldwork reflections through 140-character Tweets on *Twitter*. The use of *Twitter* has led to discussions in the evenings fuelled by the day,s Tweets. At the end of the residential fieldwork Tweets were collated and shared through *Storify* (see the Bay of Naples example http://bit.ly/1lbg1zp). One student has suggested those discussions might not have been as productive had they confined their thoughts to traditional hardcopy notebooks.

Students recorded primary data through spreadsheet apps such as *Numbers*, and displayed the analysis directly in the field through *Keynote*. Geo-referenced sampling positions can be recorded using *Geospike*, and later exported and mapped on software such as *Google Earth*. Research data and photographs were shared through a number of cloud-based data-storage facilities, such as *Dropbox*. A key benefit for the students was the ease with which geo-tagged photographs could be captured on their own mobile device, and through applications such as *Flickr*, and synchronized automatically with their cloud-based account and shared with a wider research group. The *Fotobabble* app enabled students to take a photograph and attach their audio comments to the picture for further reference and research on return to university. Students were able to record and edit short videos of their research projects or reflections on their fieldwork experiences using *Splice*, or *iMovie*.

Some students were concerned that the iPads were cumbersome, difficult to share at times, and may even break (which did not happen). Overall, however the students' feedback was very positive, for example,"[the iPad] brings together several useful applications in one place, GPS, photo/video/internet, so we do not need 3–4 pieces of equipment" and"[the iPad] was exceptionally useful for the fieldwork; instant note-taking; recording and manipulating data instantly." (student feedback recorded in Whalley et al. (2014, p.108)).

4.2.5 RECOMMENDATIONS

Digital technology has considerable potential to enhance field teaching and learning in geomorphology. This section outlines some key recommendations for going digital on fieldwork.

4.2.5.1 Transferable Skills Development for Geomorphologists

Digital video and deployment of Web 3.0 Technology actively engages students in a range of fieldwork environments and contexts, contributing to development of a skill set in geomorphology, such as enhanced appreciation of processes and methodologies. Furthermore, digital technology as reported here also develops new communication and presentation skills in their own right. These are key transferable skills, which students recognize as having lasting value (Fuller and France, in review). In addition, a core value of using digital video, whether on a tour of North Island geomorphology, or detailed measurement of process, was perceived by the group as a whole to be in fostering group work (Fuller and France 2015; in review). The exercises required group negotiation, discussion, compromise, and to be effective, team working was required, which is a valuable transferable skill to learn and develop. This agrees with the assertion of Fisher and Baird (2006) that collaborative learning is fostered using podcasting. Accordingly, the approach adopted here ensures an updated version of a Cook's Tour, as well as field experiments and deployment of Web 3.0 Technology, provide a key sphere of learning by developing both subject specific (learning about the landscape) and generic skills (Wall and Speake, 2012). We suggest this is a powerful incentive for adoption of digital technology in the field teaching of future geomorphologists.

4.2.5.2 Limitations and Cautions

Inevitably, not all students appreciated the opportunity for participation and learning facilitated by the introduction of digital video into these field courses.

A very small minority expressed this in terms of hating using the video camera and perceived it as a waste of time (cf. Fuller and France, 2015). To be fair, this might be true of the best students, who probably benefit least from its use, because they are already engaged and stimulated to learn. Nevertheless, the overwhelming majority believed that producing a digital video was a valuable and helpful exercise (Fuller and France, 2015; in review). Interestingly, students did not necessarily appear to recognize the links between their video production and other written assessments (report). The deeper holistic learning connections between the use of video and later assignments that build on knowledge generated were not always perceived by students (Fuller and France, 2015). However, this may also reflect the need for clearer instruction and communication of the linkages between assessments by staff. Links between various components in a course cannot be assumed and need to be clear. Stokes et al. (2011, p.138) suggest that, "to be an effective learning environment, the purpose of a fieldwork event must be made explicit to students." On one hand, the relationship between digital video, Web 3.0 components, and final assessment may not be recognized by the student as being tangible. On the other hand, given the enhancement to the learning experiences reported by Fuller and France (2015; in review), this work contends that contrary to student perceptions, preparation for subsequent assessments were helped by the level of engagement in the field fostered by video production. This is in agreement with the suggestion of Stokes et al. (2011) that active participation promotes information retention and recall.

Mature, or distance/extramural students also ought not to be neglected in their need for being brought up to speed with technology that may be new to them. Indeed, as a whole, adequate time ought to be allowed for students to experiment with the equipment when presented with "challenging new experiences" (Fuller and France, 2015). This would ensure learning is not compromised by any unfamiliarity with technology. Accordingly, some staff technical ability is essential to troubleshoot technical difficulties that inevitably arise, and no student should be disadvantaged, whatever their background prior to geomorphology fieldwork.

4.2.6 CONCLUSIONS

Digital video production on a Cook's Tour style of fieldtrip enhanced the learning experience for students on a multiday, multisite geomorphology field course. Engagement and enjoyment fostered deeper levels of learning that improved student performances in assessment for weaker students in particular. Video production fostered understanding of the environment, group cooperation and made the touring fieldtrip enjoyable. Videos used in this Cook's Tour provided a means to both actively engage students in a range of fieldwork environments and develop new and transferable communication and presentation skills. In field experiments, incorporating digital video fostered subject and methodological understanding, and group coordination and cooperation in a range of field experiments in two courses over 3 years. These examples demonstrated that digital video can be strategically incorporated into a variety of geomorphology fieldwork (home or away) to enrich the learning space and enhance the student learning experience, and develop new communication and presentation skills. Mobile technologies can provide an efficient and accurate way to capture, store, analyze, and share data during fieldwork. These technologies can help build cohesive research teams and provide a means to establish good working relationships between staff and students. Tablets and smartphones can provide a personal learning environment for any undergraduate research project, whether for recording data, storing and manipulating field notes (written, audio, and images), or "traditional" use of the web and cloud-based applications. Therefore, the authors wholeheartedly recommend the adoption and use of a range of digital technology to enhance field teaching in geomorphology and physical geography.

ACKNOWLEDGEMENTS

Thanks to the classes of 2010, 2011, 2012, and 2013 for their willingness to contribute their time and opinions. Thanks also to Drs Martin Brook, Alastair Clement, Rob Dykes, Kat Holt, Sam McColl, and Jon Procter for their assistance in teaching the New Zealand residential fieldcourses at one time or another. The authors would also like to acknowledge the contribution of the project team members involved in the HEFCE-funded UK National Teaching Fellowship Project, Enhancing Fieldwork Learning, including Drs Alice Mauchline, Victoria Powell, and Katharine Welsh, and Professors Julian Park and Brian Whalley.

REFERENCES

Bracken, L.J., Mawdsley, E., 2004. "Muddy glee": rounding out the picture of women and physical geography fieldwork. Area 36, 280–286.

Brown, E.H., 1969. The Teaching of Fieldwork and the Integration of Physical Geography. In: Cooke, R.U., Johnson, J.H. (Eds.), Trends in Geography: An Introductory Survey. Heinemann, London, pp. 70–78.

Coe, N.M., Smyth, F.M., 2010. Students as Tour Guides: Innovation in fieldwork assessment. J. Geogr. Higher Educ. 34, 125–139.

Cook, M., Kelly, S.Q., 2013. Connecting the dots: Web 3.0 and Interdisciplinary Freshman Research. CUR Quarterly 34, 8–16.

Dando, C.E., Chadwick, J.J., 2014. Enhancing geographic learning and literacy through filmmaking. J. Geogr. 113, 78–84.

Dummer, T.J.B., Cook, I.G., Parker, S.L., Barrett, G.A., Hull, A., 2008. Promoting and assessing "deep learning" in geography fieldwork: an evaluation of reflective field diaries. J. Geogr. Higher Educ. 32, 459–479.

Educause, 2009. 7 things you should know about Personal Learning Environments. Available at: http://net.educause.edu/ir/library/pdf/ELI7049.pdf. (accessed 23.02. 2014).

Fisher, M., Baird, D.E., 2006. Making mLearning work: Gen Y, learning and mobile technologies. J. Educ. Technol. Systems 35, 3–30.

France, D., Whalley, W.B., Mauchline, A., 2013. Using mobile devices to enhance undergraduate field research. CUR Quarterly 34, 38–42.

France, D., Wakefield, K., 2011. How to produce a digital story. J. Geogr. Higher Educ. 35, 617–623.

Fuller, I.C., 2006. What is the value of fieldwork? Answers from New Zealand using two contrasting undergraduate physical geography field trips. NZG 62, 215–220.

Fuller, I.C., 2012. Taking students outdoors to learn in high places. Area 44, 7–13.

Fuller, I.C., France, D. 2015. Securing field learning using a 21st century Cook's Tour, J. Geogr. Higher Educ., (in press).

Fuller, I.C., France, D. Does digital technology enhance student learning in field-based experiments and develop graduate attributes beyond the classroom? J. Geogr. Higher Educ. (in review).

Fuller, I.C., Rawlinson, S.L., Bevan, J.R., 2000. Evaluation of student learning experiences in Physical Geography fieldwork: paddling or pedagogy? J. Geogr. Higher Educ. 24, 199–215.

Fuller, I.C., Gaskin, S., Scott, I., 2003. Student perceptions of Geography and Environmental Science fieldwork in the light of restricted access to the field, caused by Foot and Mouth Disease in the UK in 2001. J. Geogr. Higher Educ. 27, 79–102.

Fuller, I.C, Edmondson, S., France, D., Higgitt, D., Ratinen, I., 2006. International perspectives on the effectiveness of Geography fieldwork for learning. J. Geogr. Higher Educ. 30, 89–101.

Fuller, I.C., Brook, M.S., Holt, K.A., 2010. Linking teaching and research in undergraduate Physical Geography papers: the role of fieldwork. NZG 66, 196–202.

Fuller, I.C., Mellor, A., Entwistle, J., 2014. Combining research-based student fieldwork with staff research to reinforce teaching and learning. J. Geogr. Higher Educ. 38, 383–400.

Gold, J.R., Jenkins, A., Lee, R., Monk, J., Riley, J., Shepherd, I.D.H., Unwin, D.J., 1991. Teaching Geography in Higher Education. Blackwell, Oxford.

Haigh, M., Gold, J.R., 1993. The problems with fieldwork: a group-based approach towards integrating fieldwork into the undergraduate geography curriculum. J. Geogr. Higher Educ. 17, 21–32.

Herrick, C., 2010. Lost in the field: ensuring student learning in the "threatened" geography fieldtrip. Area 42, 108–116.

Higgitt, D.L., 1996a. The effectiveness, of student-authored field trails as a means of enhancing geomorphological interpretation. J. Geogr. Higher Educ. 20, 35–44.

Higgitt, M., 1996b. Addressing the new agenda for fieldwork in higher education. J. Geogr. Higher Educ. 20, 391–398.

Hovorka, A.J., Wolf, P.A., 2009. Activating the classroom: Geographical fieldwork as pedagogical practice. J. Geogr. Higher Educ. 33, 89–102.

Jarvis, C., Dickie, J., 2010. Podcasts in support of experiential field learning. J. Geogr. Higher Educ. 34, 173–186.

Keene, P., 1982. The examination of Pleistocene sediments in the field: a self-paced exercise. J. Geogr. Higher Educ. 6, 109–121.

Kemp, J., Mellor, A., Kotter, R., Oosthoek, J., 2012. Student-produced podcasts as an assessment tool: an example from geomorphology. J. Geogr. Higher Educ. 36, 117–130.

Kent, M., Gilbertson, D.D., Hunt, C.O., 1997. Fieldwork in geography teaching: a critical review of the literature and approaches. J. Geogr. Higher Educ. 21, 313–332.

Kern, E., Carpenter, J., 1984. Enhancement of student values, interests and attitudes in Earth Science through a field-orientated approach. J. Geol. Educ. 32, 299–305.

Kern, E, Carpenter, J., 1986. Effect of field activities on student learning. J. Geol. Educ. 34, 180–183.

Kolb, D., 1984. Experiential Learning: Experience as the Source of Learning and Development. Prentice Hall, London.

Lonergan, N., Andreson, L.W., 1988. Field-based education: some theoretical considerations. Higher Educ. Res. Dev. 7, 63–77.

Mavroudi, E., Jöns, H., 2011. Video Documentaries in the Assessment of Human Geography Field Courses. J. Geogr. Higher Educ. 35, 579–598.

McEwen, L., 1996. Fieldwork in the undergraduate geography programme: challenges and changes. J. Geogr. Higher Educ. 20, 379–384.

Melhuish, K., Falloon, G., 2010. Looking to the future: M-learning with the iPad, Computers in New Zealand Schools. Learning, Leading. Technology 22, 1–16.

Oblinger, D.G., Oblinger, J.L. 2005. Educating the Net Generation. Educause. Available online at http://www.educause.edu/research-and-publications/books/educating-net-generation. (accessed 23.02.2014).

Ohler, J. 2008. The Semantic Web in Education. Educause Quarterly, 4, 7–9. Available online at https://net.educause.edu/ir/library/pdf/EQM0840.pdf. (accessed 2.02.2014).

Pawson, E., Teather, E., 2002. "Geographical Expeditions": assessing the benefits of a student-driven fieldwork method. J. Geogr. Higher Educ. 26, 275–289.

Prensky, M., 2009. H. Sapiens digital: from digital immigrants and digital natives to digital wisdom. Innovate 5, 1–9.

Race, P., 1993. Never Mind the Teaching Feel the Learning. SEDA Paper 80.

Sauer, C.O., 1956. The education of a geographer. Ann. Assoc. Am. Geogr. 46, 287–99.

Scott, I., Fuller, I.C., Gaskin, S., 2006. Life without fieldwork: some lecturers' perceptions of geography and environmental science fieldwork. J. Geogr. Higher Educ. 30, 161–171.

Stoddart, D.R., Adams, W.M., 2004. Fieldwork and Unity in Geography. In: Matthews, J.A., Herbert, D.T. (Eds.), Unifying Geography: Common Heritage, Shared Future. Routledge, London, pp. 46–61.

Stokes, A., Magnier, K., Weaver, R., 2011. What is the use of fieldwork? Conceptions of students and staff in Geography and Geology. J. Geogr. Higher Educ. 35, 121–141.

Wall, G.P., Speake, J., 2012. European Geography Higher Education Fieldwork and the Skills Agenda. J. Geogr. Higher Educ. 36, 421–435.

Welsh, K., France, D., 2012. Spot light on…Smartphones and Fieldwork. Geography 97, 47–51.

Welsh, K., Mauchline, A., Park, J.R., Whalley, W.B., France, D., 2013. Enhancing Fieldwork Learning with technology: practitioner's perspectives. J. Geogr. Higher Educ. 37, 399–415.

Whalley, W.B., France, D., Park, J.R., Mauchline, A.L., Walsh, K.E., Powell, V., 2014. In: Souleles, N., Pillar, C. (Eds.), Ipad Use in Fieldwork: Formal and Informal Use to Enhance Pedagogic Practice in a Bring Your Own Technology World. Proceedings of the first International Conference on iPads in Higher Education, pp. 100–112.

Woodcock, B., Middleton, A., Nortcliffe, A., 2012. Considering the Smartphone Learner: an investigation into student interest in the use of personal technology to enhance their learning. SEEJ 1, 1–15.

Williams, P., Wong, W., Webb, H., Borbasi, S. 2011. Mobile Technologies in the Field: iPads – Rescuer or Rescuee? Proceedings ASCILITE, Hobart. Available at http://www.ascilite.org.au/conferences/hobart11/downloads/papers/Williams-concise.pdf. (accessed 23.02.2014).

Chapter 4.3

Field-Based Learning in Undergraduate Geomorphology Courses

Mary J. Thornbush[1]
School of Geography, Earth and Environmental Sciences University of Birmingham, Edgbaston, Birmingham, United Kingdom
[1]*Corresponding author: E-mail: m.thornbush@bham.ac.uk*

4.3.1 INTRODUCTION

Most of all we must bear in mind that the laboratory for the study of field sciences is the 'field'; anywhere else the experience becomes second-hand, out of context and relatively meaningless. (Fisher, 2001, p. 94)

The purpose of this chapter is to examine student reactions to the deployment of field-based learning throughout the undergraduate curriculum. To this end, an introductory geomorphology course along with an upper-year course in environmental geomorphology and a field school are considered in order to examine the response of students to the inclusion of a varied level of field exposure in undergraduate courses. Students were tested based on field study from weekly assignments, to a major topical research project, to major field-based data collection based at a variety of landscapes. The level of exposure increased along with the proportion of corresponding assessments allotted to fieldwork (from 15% to 65% of the final course grade). Based on an assessment of the results, recommendations are set forth in this chapter for employing field-based learning at various levels in the undergraduate level. This chapter begins with a literature review of field-based learning, particularly as it has been executed at the undergraduate level (in addition to what is already presented in the Introduction of this volume); it then considers case studies from different levels of the undergraduate experience, and discusses the pedagogical implications, specifically gauging the perception of fieldwork by (undergraduate) learners.

Geography is a subject area where visuals are a crucial aid for student learning. Some have argued that learning through the physical senses can reinforce conventional learning. For instance, Auer (2008) attested to the importance of sight, taste, touch, and such external senses for learning in environmental studies.

Developments in Earth Surface Processes, Vol. 18. http://dx.doi.org/10.1016/B978-0-444-63402-3.00009-1

He posited that sensory perception triggers primitive learning pathways that become apparent in post-secondary outdoor education through coursework as part of experiential learning (learning by doing), which involves deep learning. In his sensory-intensive approach to learning, the five senses aid pedagogy by exciting the curiosity (sparking interest) in ways not apparent with abstract learning. Audio-visual resources (AVRs) were examined by McKendrick and Bowden (1999), who discovered that television programs and slides were almost universally deployed, with some integration of videos, for learning about field locations and techniques, and some minor use of videos for learning laboratory techniques. AVRs were found to be more predictably used by academics in their third to fifth year of teaching, when they broaden their resource base in learning. A majority of students (80%) were satisfied with the way in which AVRs were used, especially in order to prepare them for fieldwork techniques. AVRs were seen favorably for supporting student-centered, experiential, and active learning in geography, with the potential for enhancing student motivation (McKendrick and Bowden, 2000). It has also been found that video podcasts used in support of experiential learning appealed to students who were visual learners (Jarvis and Dickie, 2010). Field students have been known to establish better site visualization by building a geographic context (Stumpf et al., 2008). Photographs have been employed in geographical fieldwork for contemporary human geography, where cheap disposable cameras were deployed during a 5-day fieldtrip to Barcelona, Spain (Sidaway, 2002). Moran and Round (2010) used visual sources to overcome the linguistic limitations of participants (non-Russian speakers) in a fieldwork module in Moscow.

Whereas the virtual fieldtrip experience can be a cost-effective alternative to the fieldtrip in order to teach simple terms and concepts in introductory courses, fieldwork is required for learning difficult concepts that require an association with place. As demonstrated for the Dark Peak website, even though virtual fieldwork compliments real fieldwork, it cannot replace it (McMorrow, 2005). Similar findings were reported by Scott et al. (2006), who interviewed nine lecturers at five institutions and discovered that they perceived fieldwork as providing an essential engagement with the external or "real" world. A strong emotional connection develops between field scientists and fieldwork over computer-based "field" exercises. They found that: "Lecturers use fieldwork as a means to help students understand theory, to inject reality into their teaching and to teach subject-specific skills" (p. 169). Drennan and Evans (2011) provided physical models for three-dimensional visualization in order to assist student learning and comprehension in the course and field. Some consider the use of videos, photographs (slides), and lectures to be part of a passive learning approach that does not promote a deeper learning based on "experience" (e.g., Jones, 2000). Direct experience through fieldwork can challenge students in ways beyond the class-based mode of learning, as in the challenging of preconceptions (Hope, 2009). This can be achieved through an emotional connection and deeper understanding of "the other" (in human geography). It is also noteworthy that different people will respond differently to the same fieldtrip, so that fieldwork may not be an equally appropriate mode of learning. However, it is possible to deepen learning through direct experiences in the field and the potential for vivid encounters eliciting emotional responses (an affective response). According to Hope (2009, p. 180):

Fieldwork is important because at its heart lies a direct, active encounter with 'the other'– others who call us to attention with a jolt; who challenge us to rethink our preconceptions; who draw a sense of fellowship from us – and a new attentiveness that aids a deeper understanding. It is this direct encounter that makes fieldwork challenging and at times difficult and it is these features that, in my view, continue to make it a valuable mode of learning for human geography.

Perhaps one of the reasons for a preference for virtual learning, in addition to disability, distance, and the inconveniences and dangers associated with executing fieldwork, is the expense. Long-haul travel is particularly costly, and one of the reasons why local fieldtrips may be becoming increasingly popular. Examples of local geographical studies exist, such as on green school grounds in southern Ontario, Canada (Dyment, 2005), although for the elementary through to high-school level. At the post-secondary level, Hudak (2003) developed three field campus exercises for the study of basic Earth processes relating to rock weathering, retaining wall stability, and stream flow. His students found the outdoor work to be more realistic, interesting, and interactive than indoor exercises. This indicates that conducting fieldwork may be worth the associated costs, effort, and time. He advocated that keeping students in the outdoor campus environment could be used to facilitate teaching fieldwork in large introductory courses. However, this also poses a problem of large classroom size (for instance in North America and Ireland) for first-year excursions in terms of logistics, cost, and overall feasibility.

Nevertheless, regardless of the cost, research shows that fieldwork continues to offer opportunities for learning that are unique (and worth the effort) in comparison to classroom teaching. Salvage et al. (2004), for instance, used watershed testing of hydrology, geochemistry, and geophysics as a teaching tool in environmental geology. Environmental hydrology was a component of this curriculum that was inquiry-based and alternated between field-based "laboratory" classes. Fieldwork was executed in teams, but individual final reports were submitted for grading. A research question was set in order to focus watershed analysis. An inquiry-based approach to students meant doing more fieldwork. Their students reported that they learned more in the field than they would in a lecture-based classroom. It was hoped that by implementing the watershed focus and field exercises at lower undergraduate levels that this will foster the development of needed research skills required for independent research at the upper-year level. Jennings and Huber (2003, p. 190) performed campus-based teaching of undergraduate geography through a hands-on participant approach and discovered that students appreciated practical (concrete) work in a real-world context. They found that these courses are especially appropriate for upper-level courses and can bridge the way to a job or graduate school because these students are able to understand and address real-world problems. To this end, the authors propose that institutions "should work within the constraints of their surroundings and develop some portion of their curriculum as field oriented classes."

Others, such as Hirsch and Lloyd (2005), found that a field-based (hands-on) perspective to learning (acquired through field schools in Vietnam, Laos, and Thailand) can often stimulate self-awareness through the adoption of preparation and debriefing, presentation, and projects, which allowed students to prepare and reflect (enhancing learning), as well as relate their reading along with course-based theory. Bednarz et al. (2008) have examined case studies from Australia, New Zealand, the UK, and the US, finding that community engagement could serve to apply course content and provide students with future research opportunities. They called for a response of geographers in other parts of the world to disclose their practice. Exotic locations have been found to be effective for recruiting students, while at the same time providing them with meaningful experiences (with deep learning) in an active (real world) geography (Robson, 2002).

Studies have shown that students prefer to have conceptual learning paired with practical application. For example, Carlson (2007) used a questionnaire to gauge student satisfaction with a field-based GIS inquiry approach. Students commented on the "reality-based" part of exercises as bridging class and reality, preferring practical application when paired with theoretical concepts. Questionnaires in this research have commonly taken the form of (generally anonymous) feedback about part or entire coursework. Birdsall (2003) received feedback from 85% of students about an assignment used to test for flexible thinking involving reassessing landscape. Experiential (or constructivist) learning can enhance critical learning in ways that challenge preconceptions and provide opportunity for new experiences through reconstruction (Elwood, 2004).

Several studies have shown the lack of a systematic assessment (or empirical evaluation) of the effectiveness of fieldwork as a learning strategy (Kent et al., 1997). Indeed, Fuller et al. (2003) explicated that since geographers (and Earth and environmental scientists) value fieldwork, they commonly do not contribute to objective experimentation to ascertain its impact on the student-learning experience. Their research indicated that fieldwork withdrawal (due to the foot and mouth disease in 2001) was perceived by students to be detrimental to student understanding of subject matter (in geography and environmental science) and to their experience of university modules. It was ascertained that fieldwork motivated students and had other positive effects on their learning and performance. More specifically, students perceived fieldwork as mostly positive, affecting their experience of reality and subject comprehension, learning of methods and techniques (technical learning), developing transferable (and holistic) skills and interpersonal skills with peers and lecturers. Overall, the withdrawal of fieldwork was perceived as negative by final-year students from the five institutions that they examined in the UK. This work, however, comprised of fieldwork across two disciplines, specifically geography and environmental science.

This issue of interdisciplinarity is a critical consideration, since field schools can draw from a greater student pool than just geographers. For example, undergraduates and seventh graders involved in concept maps in the area of weathering science could draw from non-expert knowledge of form-process by employing knowledge of biology, chemistry, geology, archaeology, as well as physical geography to address the cultural heritage resource of rock art (Allen and Lukinbeal, 2010). A course involving archaeogeophysical survey and mapping was offered as an interdisciplinary course to undergraduates and graduates, with several departments cooperating in its execution, including programs in geography, geology, and archaeology (Barrett et al., 2004). This approach made it difficult to find a suitable textbook, and an eclectic set of course readings was used instead. Students lacking familiarity with the research project utilized in this study were dissatisfied with its "disorganization" in the context of a field survey. The course was also constrained by the operation of equipment in certain weather conditions, such as with frozen ground (because the probes need to be put into the ground) and in cold weather (since sensors become unaligned and require constant rebalancing). Nevertheless, similar research in (residential) field courses across the disciplines of geography, Earth, and environmental sciences by Boyle et al. (2007) conveyed that affective responses can trigger higher motivation and deeper learning (as outlined in their Table 1, p. 301). According to Elton (2001, p. 53), pedagogic research serves various purposes, including informing teaching practice; improves this teaching practice by providing evaluation; as action research that academics could use in their own teaching; and for teachers (non-academics) who also engage in pedagogic research.

Students are known to respond positively to fieldwork as a means to "hands-on" learning, including the acquisition of experience with instruments, research design, and data analysis (Fuller et al., 2006). This positive perception of fieldwork (for developing social skills, such as teamwork) is particularly notable for Caucasian-dominated institutions, and more work is needed to test whether it is culturally-constrained, as for example at Asian universities. More work could be performed in gauging student perception; for instance, of the relevance of research to learning (Healey, 2005). This could be in an inquiry-based learning approach (IBL) or in its more specialized form of problem-based learning (PBL).

According to Pawson et al. (2006), the former approach (IBL), such as problem-solving, requires knowledge acquisition in-advance of the problem-solving process, whereas the latter (PBL) a problem (generally set by the instructor) defines learning. Where curriculum content takes this latter approach it is considered to be pure PBL, but where only a course or assignment is involved then it appears in its hybrid form organized around problem scenarios (rather than subjects or topics). PBL characteristically requires that disciplines collaborate, because problems may not respect traditional disciplinary boundaries. What is important in this approach is the ability to learn how to learn (metacognition) rather than the material (subject matter, topics, etc.). Students may react to the messiness of real-world PBL scenarios. Nevertheless, the PBL approach is suited to geography because of its inherent interdisciplinary. "It usually takes place in group environments where the focus is on attempting to resolve problems, or to work through scenarios, with the aim of developing lifelong learning skills that are transferable to career situations (Pawson et al., 2006, p. 114).

Inquiry-based learning (IBL) involves an active approach to learning; is question-driven (or research-focused); an inductive approach; student-centered (with the instructor acting as facilitator); is facilitated or scaffolded learning; and constructivist (Spronken-Smith et al., 2008). It could also involve group and individual learning; be university-focused; involve the community; comprise a field-based activity; be resource-based learning; and have a multidisciplinary or interdisciplinary focus (see their Table 1, p. 72). In their model, these authors conceived of problem-based learning as a subset of inquiry-based learning, which is in turn a part of active learning (see their Fig. 2, p. 74). Inquiry-based learning can constitute entire courses at more advanced levels, acting in a progression towards preparation for dissertations or theses (at the graduate level). They encountered problems in upper-year courses based on inquiry-based learning that stemmed from the time commitment and group work involved. They also encountered problems with workload issues in inquiry-based learning, which provoked anxiety in students who are fully engaged in these courses (at the expense of other courses). These authors found inquiry-based learning courses to be less common in geography, and stated that these courses have the potential to engage students in authentic problems that may be relevant to the workplace and could stimulate and motivate students, at the same time more fully developing their research skills. They (Spronken-Smith et al., 2008, p. 83) argued that: "[Teachers] using IBL can benefit from a strengthening of teaching-research links and take pride in the clear gains in student engagement and learning. IBL teaching is often perceived to be fun and enjoyable."

A problem-based learning (PBL) approach was taken by Dadd (2009), who incorporated real workplace scenarios in the teaching of a third-year course in environmental geology at Macquarie University in Australia. Choosing a class size of 20-25 students divided into five groups (which could be facilitated by a single faculty member), forming "company groups" with four-five members each. Students were able to form their own groups; however, there was still some dissatisfaction expressed pertaining to group composition when the unit was evaluated – even though students thought that group work had aided their learning. Students also disliked the workload associated with PBL, although this is unavoidable in this approach to learning, which requires student involvement and self-sufficiency (disallowing learners to be passive). It was also possible that final-year students are concerned about weaker students not contributing, or providing lower quality work that could bring down their course grade. More specific group problems concerned contribution, attendance, and personality conflicts among group members. Although most studies report difficulties with group work, one study by Mooney (2006) relayed a positive experience with group work, involving groups of five students, which included gaining a deeper level of understanding and learning. Students identified improved team skills in terms of leadership as well as compromise. Only 38% responded negatively to group work, reporting problems with work allocations, disagreements, and difficulties reaching compromises. Most (over 80%) responded that they would prefer to work with friends, and less than 10% indicated that they preferred working alone. However, the random allocation of groups could improve the skills representation within groups. In addition, a group report has the benefit of cementing groups post-fieldwork (although non-active members could lead to problems with group reports). Students commonly report that they need more guidance in this PBL approach, particularly at the outset of the project; they require a clear definition of expectations (expected outcomes). Even though "hollow" assignments can trigger discomfort, they can also stimulate critical thinking and creativity (Lev, 2004). Problem-based learning poses various concerns for instructors, according to Hung et al. (2003), as they elicit depth (versus breadth of curriculum); higher-order thinking (versus factual knowledge acquisition); long-term effects (versus immediate learning outcomes); roles as tutors (versus traditional professors); and the initial discomfort of students (versus positive attitudes). An integral part of PBL is for students to solve realistic problems (that they may meet in their professional careers) by using a teaching method that deploys suitable problems for them to solve and they are assessed on their solution (Biggs, 1999, p. 71). This is based on the presumption that: "To [practice] as a particular professional requires solving problems that belong to that profession." This allows for a focus on what students should be doing (rather than on the responsibilities of teachers and administrators).

Project-based learning is a type of PBL approach (and research-based learning approach), which was deployed at the second-year level in an introductory field-based course (Eppes, 2009). Students were evaluated based on their participation (50%) and field notebooks and worksheets (50%). This approach (in a soil-based project requiring sedimentological, geomorphological, hydrological, and ecological observations beyond the coverage of soil taxonomy typically included in introductory geology textbooks) received positive feedback from students with regards to a semester-long project. Some students even indicated that this experience allowed them to become employed in geotechnical and environmental consulting firms. A dendrochronology exercise, delivered at upper-level to undergraduates who participated in a field geography course, was designed to provide concrete experience that complemented lectures and reading assignments (more abstract forms) and found to be an effective teaching tool (Lafon, 2005). It was discovered that students developed a deeper understanding than through conventional (lecture-based) learning. However, students may not always possess the necessary background (prerequisites) required to perform well in project-based courses, such as an adequate statistical background (Resler and Kolivras, 2009). This meant that visual inspection of plots and conveying patterns (trends) in data substituted statistical analysis. Debriefing was executed through a group poster presentation that included both quantitative and qualitative results. Activity reports were also submitted to track individual contributions to group work. Nevertheless, students were excited to work together in a problem-solving learning experience that provided opportunities for independent thought and self-direction.

4.3.2 METHODS

4.3.2.1 Introductory Geomorphology

Three sets of student satisfaction surveys were considered in the assessment of a second-year introductory geomorphology course in geography and environmental studies. The course was held in the winter term (January to April) of 2008 and held subsequently in two fall terms (September to December) in 2008 and 2009. The outline of the course included the following handbook description: "An examination of terrestrial and marine geomorphological processes, and the effects of structure, time, and climatic change on the development of landforms. Laboratory work will include the analysis of landforms from maps and aerial photographs." The prerequisite comprised the first-year course offering in an introduction to geographical issues or the permission of the instructor. The required textbook was *Geomorphology: A Canadian Perspective* by Trenhaile (2007, 2010). In addition, some recommended books included *Global Geomorphology* by Summerfield (1991) as well as *The Human Impact on the Natural Environment* by Goudie (2005).

The course content did not vary substantially during this time, comprising lectures, seminars, laboratory sessions, and fieldtrips (Table 4.3.1). Some minor changes in the grades scheme were made in 2009, with the removal of a laboratory session and an increase in the worth of the remaining (two) laboratory sessions (from 5% to 10% each) and an increase in the worth of field exercises (also from 5% to 10% each). In addition, the mid-term test was removed in the final term (in 2009). Seminars offered an opportunity to discuss with students some relevant (recent) articles published in geomorphology and were assessed through a written response. The laboratory sessions essentially covered the basics of introductory geomorphology, including remotely-sensed data; air photo analysis; and geomorphological mapping. The former was removed in the last year. Finally, fieldwork was assessed through a completed exercise administered in the local area in an urban setting. This consisted of an excursion to a local churchyard for weathering studies, a local beach for coastal geomorphology, and a nearby port/marina in order to examine local environmental geomorphological issues, as pertaining to land-use. One of the factors that caused some changes in the course from the first session in the winter of 2008 to the two subsequent fall sessions is that they occurred in different seasons, which affected the quality of fieldtrips.

Lectures covered all major parts of geomorphology. This included address of the following topics: introduction to geomorphology and its approaches;

TABLE 4.3.1 Grade Allocations for all Three Courses

Course (year)	Assessment	Weightings	Field-Based
Introductory geomorphology (2008)	Seminars	25% (5 @ 5% each)	Fieldtrips = 15%
	Labs	15% (3 @ 5% each)	
	Fieldtrips	15% (3 @ 5% each)	
	Mid-term test	20%	
	Final exam	25%	
Introductory geomorphology (2009)	Seminars	25% (5 @5% each)	Fieldwork = 30%
	Labs	20% (2 @ 10% each)	
	Fieldwork	30% (3 @ 10% each)	
	Final exam	25%	
Environmental geomorphology (2010)	Seminars	30% (6 @ 5% each)	Field project = 30%
	Field project	30%	
	Presentation	10%	
	Test	30%	
Field school (2009)	Field assignments	30% (3@ 10% each)	Field assignments + Research project = 65%
	Equipment demonstrations	20% (2 @ 10% each)	
	Research proposal	15%	
	Research project	35%	

historical development of geomorphology as a discipline; the Earth's surface; global tectonics; planetary geomorphology; weathering; karst; lakes; oceanic coasts; fluvial processes; fluvial landforms; glacial sediments and landforms; periglaciation; soils and sediments; deserts; natural hazards; mass-wasting; human impacts on landscapes; climate and environmental change; and new developments for the future.

Students enrolled in the course in all three terms were asked to complete a student satisfaction survey. The questions in this survey varied somewhat over the course of the three terms (compare Tables 4.3.2a and 4.3.2b). It originally (in winter and fall terms of 2008) was comprised of 11 questions in Part A to be assessed on a six-point scale; however, this changed to 23 questions in the final year (in the fall term of 2009). The same (three) open-ended questions were asked on the back of the sheet. It was originally divided into three parts (Parts A–C), with a total of 17 questions, but in the final version, only two parts (Parts A and B) were used, with a total of 26 questions. Students were asked for details of the course they were assessing and evaluated it anonymously in class, but in the absence of the instructor, at the end of the course (specifically, in April and November 2008 and November 2009). For the protection of these pupils and the university the

TABLE 4.3.2a Questions for Parts A–C in the Original Survey in 2008 for Introductory Geomorphology

Part	Question number	Question
A	1	The course outline and other descriptive information (e.g., goals, reading list, topics covered, assignments, exams, due dates, grade weightings, etc.) provided enough detail about the course.
	2	The course as delivered followed the course outline and other descriptive information.
	3	The course material was presented in a well-organized manner.
	4	Student questions and comments were responded to in a knowledgeable and professional manner.
	5	The course content was communicated effectively.
	6	Opportunities for course assistance were available.
	7	The evaluation methods used for determining the course grade were fair.
	8	The instructor made the course interesting.
	9	The support materials used in this course helped me learn.
	10	The instructor demonstrated knowledge of the subject.
	11	The overall quality of instruction was satisfactory.
B	12	For you, this course/course component is:
	13	The percentage of class sessions you attended in this course/course component is:
	14	Compared to other courses of this type and at this level, the work load for the course/course component being surveyed is:
C	15	What did you like about this course?
	16	What suggestions do you have for improving this course?
	17	Additional comments

institution where this course was offered and evaluated will not be identified. Readers should note, however, that it is a second-year undergraduate course in geography and environmental studies offered at a Canadian university.

4.3.2.2 Environmental Geomorphology

This upper-year (third-year) undergraduate course in geography and environmental studies was designed to include: "An examination of applied geomorphological studies in selected fields and their use in determining the physical

TABLE 4.3.2b Questions for Part A in the Changed Survey Administered in 2009 for Introductory Geomorphology

Part	Question number	Question
A	1	A comprehensive course outline was provided (for example, the outline included: office hours, course topics, evaluation methods, etc.).
	2	Learning objectives were clearly presented.
	3	The course content matched what was presented on the course outline.
	4	The course materials were useful in promoting learning of the course concepts.
	5	The instructor encouraged student participation.
	6	The instructor responded to all questions in a respectful manner.
	7	The instructor respected diversity among the students (e.g., differences in: age, gender, race, ethnic background, sexual orientation, etc.).
	8	The instructor clearly explained the course concepts.
	9	The instructional method(s) enhanced student learning.
	10	The instructor monitored student learning in order to make any necessary adjustments to the pace of delivery.
	11	The instructor appeared enthusiastic about teaching the course.
	12	The instructor was accessible to students outside of the class.
	13	The instructor appeared well prepared for class.
	14	I found the course appropriately challenging.
	15	I found the course stimulating.
	16	I have acquired new knowledge as a result of taking this course.
	17	The knowledge from this class can be applied to a variety of situations.
	18	The instructor provided opportunities to further develop my analytical skills.
	19	Graded materials were aligned with the learning objectives for the course.
	20	Feedback on graded materials was timely.
	21	Feedback on graded materials was valuable.
	22	Overall, I would recommend this instructor.
	23	Overall, I would recommend this course.
B	24	What did you like about this course?
	25	What suggestions do you have for improving this course?
	26	Additional comments:

impact of human activities and assessing environmental quality." This was the first time that this course was delivered by the instructor at this university, where students were enrolled in a teacher-training program and were not geography majors. The prerequisite consisted of an introductory geomorphology course that had comprised material about glacial sediments and landforms. Lectures were delivered 2 h per week, with an additional 2 h laboratory session.

Geomorphology and Global Environmental Change (Slaymaker et al., 2009) was the recommended, optional textbook. Text-based topics were covered during one lecture each week and included environmental geomorphology (an introduction); land use and land cover change; land use change; water control; coasts; risk-based prediction and adaptation; human impacts; climate change; aeolian systems; mass wasting; permafrost regions; and new geomorphology agendas. The second weekly lecture focused on case studies for supplementary learning. The lecture-based sessions were assessed as outlined in Table 4.3.1.

In order to guide students through the research laboratories, regular meetings were held to address: the field study topic, location, design, equipment, and set-up; field monitoring; equipment retrieval; and presentations. Field monitoring occurred during an extended period spanning 6 weeks. Students were assigned permanent groups before field monitoring began. They were to work in these five groups (of three or four people each) to collect data for their field study, write up their report, and deliver the presentation (see Table 4.3.1). The report was to include field logs (worth 20% of the research laboratory grade).

The approach to research in the laboratory sessions was student led, with the instructor acting as research director and facilitator. Background was given to students pertaining to the selected research topic, location, study design, equipment, and set-up. Specifically, the chosen topic for this course, because it was administered in a winter term was the appearance of icicles locally (in an urban setting) and their potential as a hazard in the content of climate change (with the onset of greater temperature fluctuations and potentially warmer winters). Students collectively specified research questions to investigate during their term research project. Help was provided with the selection of equipment used to monitor outdoor temperature and relative humidity. Students were also asked to photo document during monitoring, and to record observations of local outdoor conditions (e.g., wind, cloud cover, etc.) at the field site in order to substantiate their weather information.

Advice was provided by the instructor on stylizing the final report that included a project title page, overview, introduction, methods and instruments, results, discussion, conclusion, references, and appendices. References were collected during regular laboratory sessions, because they included a review of relevant literature provided by the instructor. Students were assessed on the basis of their inclusion of papers from laboratory sessions, but more particularly, for their own extended search for literature.

At the end of the course, standard, anonymous, student-evaluation forms were administered (in the absence of the instructor) in order to evaluate the

TABLE 4.3.2c Questions on the Six-Point Scale in the Student Satisfaction Survey for Environmental Geomorphology

Question number	Question
1	A comprehensive course outline was provided (for example, the outline included: office hours, course topics, evaluation methods, etc.).
2	Learning objectives were clearly presented.
3	The course content matched what was presented on the course outline.
4	The course materials were useful in promoting learning of the course concepts.
5	The instructor encouraged student participation.
6	The instructor responded to all questions in a respectful manner.
7	The instructor respected diversity among the students (for example, differences in: age, gender, race, ethnic background, sexual orientation, etc.).
8	The instructor clearly explained the course concepts.
9	The instructional method(s) enhanced student learning.
10	The instructor monitored student learning in order to make any necessary adjustments to the pace of delivery.
11	The instructor appeared enthusiastic about teaching the course.
12	The instructor was accessible to students outside of the class.
13	The instructor appeared well prepared for class.
14	I found the course appropriately challenging.
15	I found the course stimulating.
16	I have acquired new knowledge as a result of taking this course.
17	The knowledge from this class can be applied to a variety of situations.
18	The instructor provided opportunities to further develop my analytical skills.
19	Graded materials were aligned with the learning objectives for the course.
20	Feedback on graded materials was timely.
21	Feedback on graded materials was valuable.
22	Overall, I would recommend this instructor.
23	Overall, I would recommend this course.

instructor and course at large. The front half of the questionnaire comprised questions to be answered on a six-point scale: (1) "Strongly Agree;" (2) "Agree;" (3) "Neither Agree Nor Disagree;" (4) "Disagree;" (5) "Strong Agree;" and (6) "Not Applicable," and totaled 23 questions (Table 4.3.2c). On the back of the page, an additional three open-ended questions were included, such as: (1) What did you like about this course?; (2) What suggestions do you have for improving this course?; and (3) Additional comments. In addition to this evaluation, which all students had to complete, the team leaders of the five groups in the laboratory sessions (selected by the instructor) were asked to complete a separate questionnaire about individual members of their group for their level of engagement. Space was also provided in their evaluation forms for outlining any problems with individual participation and contribution.

4.3.2.3 Field School

Upper-level (also at the third-year) undergraduates enrolled in this course, some without any previous background study of physical landscapes, in order to obtain course credit in an applied (outdoor) setting in a geographic area that was local-regional and in proximity to their university. The course was equivalent to a half-year special topic in geography entitled Geography Field School in Simcoe County, where the course was mostly based. Nine students enrolled for this summer course. Lectures, providing students with the necessary (lecture-based) background, were held every weekday for 2 h, with Fridays being allotted to field-based studies. For the latter, the course instructor arranged all access requirements and transportation was organized through car-pooling to the various field sites. The only prerequisite was the introductory geography course or the permission of the department, so that some students had not even taken an introduction to the field. The required textbook for the course was *Field Techniques in Glaciology and Glacial Geomorphology* by Hubbard and Glasser (2005), because it provided students with a sufficient introduction to field techniques, some (but not all) of which would be relevant to glaciology and glacial geomorphology. The recommended textbook was *The Physiography of Southern Ontario* (third edition) by Chapman and Putnam (1984).

The field-based assessment of this course was based on the completion of three field assignments and a research project (see Table 4.3.1), with up to 65% of the course content being field-based. The course ran from the end of July to the middle of August in the summer term. The course spanned these 3 weeks, with introductory (background) material being provided in Week 1, including an introduction to field-based research and the landscapes of southern Ontario, along with provision of county-specific literature pertaining to rivers. The instructor also provided demonstrations of field equipment used to study rivers. At the end of this first week, a field study was executed at the Nottawasaga River. The topics for Week 2 were coasts and karst environments. Lectures included a consideration of the possibilities for field-based research in Simcoe County in order to assist students with the selection of research topics for their research projects. Furthermore, the landscapes of Simcoe County were introduced alongside relevant published research in Simcoe County, plus an instructor demonstration of field equipment for coasts and karst environments. On the Friday, the class headed to Wasaga Beach and Blue Mountain for field study on these respective topics. Finally, in Week 3, quantitative field-based research was discussed in terms of a suitable protocol for studies and assistance was provided in generating a proposal to conduct field study. This time, the published research in Simcoe County and the instructor demonstration of field equipment regarded glaciology. The final field research study session was held at Awenda Provincial Park, where studies conducted their own research studies under the guidance of the instructor.

The purpose of the final fieldtrip was to expose students to overnight camping and include some coverage of camping skills, which should be a part of informal geographical skills. However, some students in this course were mature students who were not able to participate in this extended trip. Other students had work commitments and could also not remain the entire time. For this reason, the overnight part of the course was canceled for this class and students completed their projects during the day of the field excursion to Awenda Provincial Park, but were not asked to remain overnight. Moreover, not all students could make it the day of this fieldtrip and were provided with a non-field project based at the National Archives (examining aerial photographs), and the instructor assisted setting this up for a couple of students. However, this replacement assignment for the research project in these two cases is not considered here, because this chapter focuses on field-based learning (rather than desk-based study).

Standard student-evaluation forms were completed in the final week of classes (in mid-August), on an anonymous basis and with the instructor absent. More specifically, these student satisfaction surveys for courses comprised only a single page (front and back), where students were required to complete their department/school offering the course; the faculty; instructor; subject; date; course number; and section. Students were informed that the form would be given to the instructor after the submission of the final grades. They were asked for the type of course: lecture/seminar/teleconference; lab/practicum/studio; or other (please specify). There were three parts (Parts A, B, and C) to be completed, including 11 six-point scale responses, where students was asked to tick the most appropriate response to a specific questions as: (1) "Strongly Agree;" (2) "Somewhat Agree;" (3) "Neutral;" (4) "Somewhat Disagree;" (5) "Strongly Disagree;" and (6) "Not Applicable." The next section, Part B, continued with the above style of questioning, but on an altered scale that was question specific; three questions were in this part. On the back of each survey was Part C, which comprised three open-ended questions, for a total number of 17 questions in this survey (see Table 4.3.2d).

4.3.3 RESULTS

The overall findings convey that students generally agreed with positive statements made about the courses in the evaluation forms. This feedback relays that undergraduates either "Strongly Agree" or "Somewhat Agree"/"Agree" with

TABLE 4.3.2d List of Questions in the Undergraduate Satisfaction Survey for the Field School

Part	Question number	Question
A	1	The course outline and other descriptive information (e.g., goals, reading list, topics covered, assignments, exams, due dates, grade weights, etc.) provided enough detail about the course.
	2	The course as delivered followed the course outline and other descriptive information.
	3	The course material was presented in a well-organized manner.
	4	Student questions and comments were responded to in a knowledgeable and professional manner.
	5	The course content was communicated effectively.
	6	Opportunities for course assistance were available.
	7	The evaluation methods used for determining the course grade were fair.
	8	The instructor made the course interesting.
	9	The support materials used in this course helped me learn.
	10	The instructor demonstrated knowledge of the subject.
	11	The overall quality of instruction was satisfactory.
B	12	For you, this course/course component is: A program requirement; or An option or elective
	13	The percentage of class sessions you attended in this course/ course component is: 0–49%; 50–74%; 75–100%; or Not applicable.
	14	Compared to other courses of this type and at this level, the workload for this course/course component being surveyed is: lower; about the same; or higher.
C	15	What did you like about this course?
	16	What suggestions do you have for improving this course?
	17	Additional comments:

such positive statements. As the level of field exposure increases, as is evident in the introductory geomorphology course, responses still remained generally positive, but to a lesser extent. This is likely linked to an applied pedagogy, which students may have trouble grasping at this (introductory) level, and the increased workload associated with active learning. The specific quantitative (scale-based) and qualitative results follow for each course.

4.3.3.1 Introductory Geomorphology

As conveyed in Table 4.3.3a–Table 4.3.3c, students mostly viewed the course positively. Their modal responses indicate that they generally agreed with positive statements made in Part A on the six-point scale. There were some split modes; however, overall responses were most frequently positive. The first survey conveys, for instance, some multimodal frequencies (for the most popular responses), particularly for Questions 5 (effectiveness of course content communication), 9 (support materials), and 11 (overall quality of instruction), where students were divided in the most popular opinion. This dispersion in the scores could be potentially linked with student aptitude and performance, but this is notspecifically considered here.

The breadth of response (variation) in the six-point scale varied. Students never chose "not applicable" and the most certain narrow responses (most often popular) were Questions 10 (in the 2008 survey) and 1 (in the 2009 survey). The first concerned the knowledge of the instructor (which was scored high) and the provision of a detailed course outline (which was also scored high). Moreover, for the fall session in 2008, there was a high level of agreement in answering Questions 4 (responses to student questions) and 5 (communication of the course content); these responses were scored high on the scale. The questions that had a low level of agreement in the classes included Questions 7 (fairness of grading methods), 9 (support materials), and 11 (overall quality of instruction) in the first survey. None of the questions in the second survey were spread

TABLE 4.3.3a Most Frequent (Modal) Responses for Part A in April 2008 for Introductory Geomorphology

Question number	Strongly agree	Somewhat agree	Neutral	Somewhat disagree	Strongly disagree	Not applicable
1	2	4	–	2	–	–
2	3	4	1	–	–	–
3	3	3	1	2	–	–
4	4	1	1	2	–	–
5	2	1	–	2	2	–
6	4	3	1	–	–	–
7	1	3	1	1	2	–
8	2	3	2	1	–	–
9	2	2	1	2	1	–
10	7	–	1	–	–	–
11	2	2	1	2	1	–
Total	32	26	10	14	6	0

TABLE 4.3.3b Most Frequent (Modal) Responses for Part A in November 2008 for Introductory Geomorphology

Question number	Strongly agree	Somewhat agree	Neutral	Somewhat disagree	Strongly disagree	Not applicable
1	4	4	–	–	–	–
2	5	3	–	–	–	–
3	4	3	1	–	–	–
4	6	1	1	–	–	–
5	2	6	–	–	–	–
6	5	2	1	–	–	–
7	4	2	2	–	–	–
8	3	4	1	–	–	–
9	2	3	2	1	–	–
10	7	1	–	–	–	–
11	5	2	1	–	–	–
Total	47	31	9	1	0	0

out across five scales the way these were. However, the last survey showed a similar spread of responses, including for Questions 3 (matching of course content with course outline), 4 (course materials), 5 (encouragement of student participation), 6 (responding to questions in a respectful manner), 8 (clear explanation), 10 (monitoring of student learning), 17 (application of knowledge), 18 (further opportunities for analysis), 20 (timely feedback on graded materials), 21 (valuable feedback), 22 (recommendation of instructor), and 23 (recommendation of course).

These responses reflect students who were mainly taking this as a program requirement. This included 63% of respondents in the winter of 2008, 50% in the fall of 2008, and 68% in the fall of 2009. All students in the first couple of surveys had attended most sessions (75–100%). Most of the respondents (63%) in the first survey found the workload to be higher than their other courses at the same level and 50% found it so in the second survey. The overall response rate for these surveys was good, although not perfect, including some 89% (small classes <10 students) for the first couple of surveys and 71% (larger class of 35 students) for the final survey.

The written comments in the open-ended questions (summarized in Appendix 4.3.1 at the end of this chapter) reflect student diversity. Some students indicated that they enjoyed the subject matter, especially the laboratory sessions. Even though they generally enjoyed the course content and the outdoor experience, they needed guidance at this (introductory/second-year)

TABLE 4.3.3c Most Frequent (Modal) Responses for Part A in November 2009 for Introductory Geomorphology

Question number	Strongly agree	Agree	Neither agree nor disagree	Disagree	Strongly disagree	Not applicable
1	13	12	–	–	–	–
2	7	12	5	1	–	–
3	12	9	2	1	1	–
4	8	11	1	3	2	–
5	5	11	7	1	1	–
6	8	8	3	3	2	–
7	14	7	4	–	–	–
8	3	12	4	4	1	–
9	3	13	7	–	2	–
10	4	12	6	2	1	–
11	11	9	5	–	–	–
12	8	11	4	2	–	–
13	10	14	1	–	–	–
14	9	12	4	–	–	–
15	5	14	4	2	–	–
16	11	12	1	–	1	–
17	6	9	8	1	1	–
18	3	13	6	1	2	–
19	5	14	4	–	2	–
20	8	12	2	2	1	–
21	5	13	2	3	2	–
22	5	15	1	2	2	–
23	6	13	2	2	2	–
Total	169	268	83	30	23	0

level. However, they appreciated that real-world materials (from the author's own research and fieldwork) were examined. One student suggested offering the course in another term, which did occur in subsequent years. Often, students felt rushed and that the pace of delivery was too fast. Nevertheless, some students also felt that they learned a considerable amount. Most of them lacked a solid background in the physical sciences and this may have hindered their performance.

Students repeatedly stated that they enjoyed the subject matter presented in the course, and also referred to gained hands-on experience in different course components (and not just during fieldwork). They also felt, however, that the required textbook was inadequate. This is one of the reasons why developing textbooks for field-based learning is critical at this time. Some students clearly struggled with the content. However, such students (who struggled) still enjoyed the material covered and did not blame the instructor for their difficulties. Nevertheless, some of those students who struggled with the workload (but did not blame the instructor), lashed out at the university instead.

In the fall survey for 2008, some students had already heard about the subject matter and had some indication of what to expect. They still, however, struggled with the heavy workload; although, this was not the case for all students. Some students preferred more casual approaches to learning than the traditional lecture hall. One respondent shared that they appreciated the breadth of coverage, which others would envision as too much information. This student also made specific comments about course improvement, which the instructor took seriously. They clearly enjoyed the course. It was interesting how memorable the field experiences were to students; they preferred a varied delivery. Some adjusted well to any novelty in the course. This student had a positive experience that they wished to share. Another student conveyed their appreciation of a varied style of delivery. There were additional negative comments regarding the selected textbook. Again, stressing the need for a fieldwork textbook for undergraduates that comprised of a variety of landscapes and methods utilized for learning geomorphology.

Not all students in the final survey completed the open-ended questions at the back of the survey. Yet, there was still sufficient feedback due to the larger size of this class. Again, students enjoyed applied learning. Yet, students felt that they lacked enough background for the course. Continued reference was made to their enjoyment of the diversity of course delivery. Some concern was expressed among this group, however, about no mid-term test. Some indication was made here that these students resisted reading articles. Several concerns also were expressed over group work. Nevertheless, students felt that fieldwork offered them opportunities to improve their performance. Another student felt that the textbook did not help. Students preferred direct outdoor learning. Some felt the instructor to be a hard marker. Some of these students expressed interest in learning about their local environment. Nevertheless, they felt that some exposure outside of this area would be welcome. More comments were made about the breadth of information. However, they expressed problems with group work. Some positive comments about the instructor were gleaned. Nevertheless, another student (who communicated that they learned a lot) felt that the instructor was unapproachable. Several students struggled with the load of material on the final examination due to a lack of a mid-term test. Other students' positive comments about the course and fieldwork and actually requested more fieldwork. They commended the instructor for having done a great job. One student even responded positively to group work, although problems were identified, as with equally sharing the workload.

4.3.3.2 Environmental Geomorphology

A total of 17 questionnaires were completed by students in this portion of the assessment (see Table 4.3.2c). Counts of ticks appear in Table 4.3.3d.

TABLE 4.3.3d Frequency Counts of Responses on the Six-Point Scale for Environmental Geomorphology

Question	Strongly agree	Agree	Neither agree nor disagree	Disagree	Strongly disagree	Not applicable
1	4	10	2	1	–	–
2	2	7	6	2	–	–
3	2	11	3	1	–	–
4	–	7	4	5	1	–
5	–	10	5	1	1	–
6	4	7	4	2	–	–
7	7	6	2	–	–	2
8	1	3	7	5	1	–
9	–	7	7	2	1	–
10	1	1	9	3	3	–
11	–	13	2	2	–	–
12	5	9	2	1	–	–
13	2	9	5	1	1	–
14	1	8	5	2	–	–
15	1	3	8	5	–	–
16	2	10	2	2	–	–
17	3	5	6	2	1	–
18	3	6	6	2	–	–
19	1	2	6	4	4	–
20	–	1	6	2	4	4
21	–	1	5	–	3	8
22	2	9	3	3	–	–
23	–	6	4	6	1	–
Total	41	151	109	54	21	14

The most frequently checked responses on the six-point scale are generally in agreement with the statements, except for a very strong agreement with Question 7 as well as some questions for which students have no comment (such as Questions 8–10, 15, 17, 19–20, with a split for Question 18), where not applicable has been used most frequently (Question 21) and where there is an equal split between "Agree" and "Disagree" (namely, Question 23). The most agreed upon was Question 11, where 13 students agreed that the instructor was enthusiastic to teach the course. High overall agreement pertaining to Questions 1, 3, 5, and 16 also occurred. All students provided some level of written feedback, but most (59%) did not have any additional comments to make.

Some negative modal responses were made to the questions in the first part of the questionnaire. Specifically, modes tended to be more negative for seven questions or 30% of questions (namely, Questions 8, 10, 15, 19–21, and 23). These questions relate to explication of the content, the pace of delivery, stimulation, grading and feedback, and recommendation of the course. The remainder of questions (70%) was positive.

The second portion of the survey was useful to gain specific commentary on course improvement and to recognize student complaints (Appendix 4.3.2). One student, for instance, expressed a concern about grades going into their professional year (in response to the second open-ended question. This student did, however, appreciate the laboratory sessions, saying that it enabled them to apply lecture-based content. Indeed, students commonly referred to the field portion of the course, mostly enjoying the experience.

Comments about enjoying the field experience are also expressed by other students. A student, for instance, mentioned how they appreciated that a continuation was made from the introductory geomorphology (lower-year, second-level) course. Another student also expressed this preference for a continuation of material from the introductory course. Other students struggled with the topic and preferred a greater distribution of the course grade. Some students realized the gain of transferable skills, such as report-writing skills.

Students were able to compare this learning approach to a traditional (lecture-based) course. In addition to the traditional lecture-based approach, students also appreciated the field learning experience. Conversely, others enjoyed that there was a major assignment. Some students suggested that the course be delivered at another time of the year. This follows from previous statements expressing concern over the selected topic. On the other hand, an instructor cannot please everyone in the course. So, while some students struggled with the topic and subject matter of the major project, some actually enjoyed it. This latter statement is also revealing about how students prefer receiving instruction from an experienced fieldworker and researcher. They did, however, have a problem with group work and explained why (see comments included in Appendix 4.3.2).

4.3.3.3 Field School

A total of five surveys were collected, processed, and eventually submitted to the course instructor, making the response rate 56%. This small sample size did not allow for any statistical analysis, but provides sufficient feedback for a qualitative consideration of the (first) impression to fieldwork of new students taking a physical landscapes field school. The most frequently used (modal) responses were considered for questions in Parts A and B (Table 4.3.3e), rather than average responses used for course evaluation statistics. Modal responses to questions in Part A were mostly in the "Somewhat Agree" option, since "Agree" was not on the scale and students did not "Strongly Agree" and were not "Neutral." Exceptions to this general trend occurred in Questions 1 (where most answered "Somewhat Disagree"); 2 (split between "Neutral" and "Not Applicable"); 7 ("Neutral"); 8 ("Strongly Agree"); and 9 ("Somewhat Disagree").

Part B conveys that most students took this summer field school as an option or elective, with one taking it as a program requirement. Students attended a majority of class sessions in the course (75–100%). Most of these students (80%) found the workload of the course to be higher than other courses of this type at a similar level. Only one student considered this course to be at about the same level as other courses in terms of its workload, but none thought it to be lower.

TABLE 4.3.3e Counts of Tick Marks in Responses on a Six-Point Scale for the Field School.

Question	Strongly agree	Somewhat agree	Neutral	Somewhat disagree	Strongly disagree	Not applicable
1	1	1	–	3	–	–
2	–		2	1	–	2
3	–	3	1	1	–	–
4	1	3	–	1	–	–
5	–	3	2	–	–	–
6	1	3	1	–	–	–
7	1	–	3	–	–	1
8	3	1	1	–	–	–
9	–	1	–	4	–	–
10	2	3	–	–	–	–
11	2	3	–	–	–	–
Total	11	21	10	10	0	3

The written comments in Section C contained positive remarks about the field experience (Appendix 4.3.3). Students described the field experience as fun. They appreciated the hands-on (active) learning experience, along with its applied (realistic) nature. One respondent commented on practical problems in the field, such as having to wait to use field equipment. Another student remarked on how they considered this a better way to learn.

Suggestions for improvement were also provided. One student suggested a lower-level introductory course addressing the use of equipment. Students felt they needed more preparation before going into the field. Considerable dissatisfaction occurred with the required textbook selected for this course, and a couple of students suggested that the instructor compile a course-specific manual to replace the textbook. The latter point expands on an earlier frustration expressed by a student with regards to having to wait for equipment in the field. One student suggested that: "Instead of many [fieldtrips] with many studies, it should be one large project, linking equipment use and research." This suggestion was taken seriously by the author for subsequent course design. However, it should be noted that the course aimed to cover a breadth of landscapes, and so an accordingly diversity of field sites (and, thus, techniques and equipment) had to be selected. There were three further remarks, including one that referred to the high workload of this course. Another respondent reinforced their opinion of the chosen course textbook, which was poorly received.

4.3.4 DISCUSSION

4.3.4.1 Introductory Geomorphology

Even though geomorphology is known to be a field-based discipline, the amount of time spent on fieldwork can perhaps influence trends in long-term learning. This course comprised of short field sessions (normally an hour each) integrated into the course outline. Others have employed fewer fieldtrips, but for longer durations at a time. For instance, Ellis and Rindfleisch (2006) required that students be present at a field location (a beach) for approximately 4 h. This would differ for research-based courses, where fieldwork takes dominance in course time. Active learning can be promoted through research-led (inquiry-based) learning. As, for instance, when a specialized topic is provided to students that requires a research-intensive approach to learning (Harris and Tweed, 2010). These researchers ran a field course (field-based workshop) over the course of 4 days at Glen Etive in the Scottish Highlands, where students applied the knowledge that they gained in the course. Using a standard Staffordshire University end-of-module feedback form with potential for comments as well as a second questionnaire, it was possible to gauge that the opinion of nine students, who suggested more lectures particularly with face-to-face teaching. Nevertheless, it was found that research-led learning can trigger deeper learning in students and cause them to take control of their learning experience. Even in short field sessions, students who are new to geomorphology (and the course content) require guidance in performing fieldwork, especially in the early sessions.

Field-based learning is active and gives students opportunities for self-discovery. Arreola (2001, 485), for instance, considered discovery to be "a pivotal quality of a successful field outing" where the field can be used as a teaching laboratory. In the field, poor class-based students can improve their performance and, conversely, those students who perform well in a classroom setting can demonstrate less proficient field skills. According to these authors (Hefferan et al., 2002, 189): "In many ways, the [fieldtrip] experience is a microcosm of life in the workplace, requiring preparation, effort and logistical consideration, data collection, peer collaboration, multidisciplinary interaction, improvisation and more effort." It can help undergraduates' transition to graduate studies or the workplace, as they gain confidence in their ability to continue applied research. They also gained valuable transferable skills, which could be employed in the workplace. A specifically PBL approach was appreciated by students who worked in groups on (authentic) problems relevant to the workplace, because it developed their teamwork skills; however, concerns were expressed about group size, especially when more than seven to eight students were in a group (Spronken-Smith, 2005). In order to be successful, PBL needs to be preceded by fieldwork training that includes, for example, learning research methodology in advance (Panelli and Welch, 2005). The field experience has also been known to strengthen relationships (the professional bond) between professors and their students in a comfortable (casual) setting, where they can ask about graduate school and careers (Paradis and Dexter, 2007). At the introductory level, however, students appreciate a hands-on approach to learning so that they can better visualize course content and actively engage with it. This provides young scholars with an appreciation for the course as well perhaps as developing situational interest that can develop into individual interest and lead them to pursue further study in the area, for instance in upper-year courses, graduate school, or the workplace.

4.3.4.2 Environmental Geomorphology

It is not surprising that several students overwhelmingly conveyed their enjoyment of field-based study. Students are known to respond positively to fieldwork as a means to "hands-on" learning, including the acquisition of experience with instruments, research design, and data analysis (Fuller et al., 2006). This positive perception of fieldwork (for developing social skills, such as teamwork) is particularly notable for Western-dominated institutions and more work is needed to test whether it is culturally constrained. More work could be performed in gauging student perception; for instance, of the relevance of research to learning (Healey, 2005), as in the current study. This could be in an IBL approach or in its more specialized form of PBL. The project-based approach taken as part of in this study is a type of PBL approach.

Some students did recognize that they were gaining useful skills through applied learning. One student, for instance, mentioned report-writing skills. However, they failed to comprehend the greater impact of this approach in a student-centered, learning approach executed during laboratory sessions. The close supervision provided by the instructor in this course provided students with an understanding of actual field-based (modern) instruments, such as iButtons. So, the students gained technical skills (whether they were aware of it or not). They also developed their ability to visualize data through graphs and statistical analysis, conveyed to the instructor during presentations and in submitted research reports. Even though students struggled with group work, this exposure will help them to improve their ability to work as members of a team (e.g., in the workforce).

The problem with assessments that are based on student course evaluations is that they are often affected by student perception and are often highly subjective. Some would argue that student evaluations of instructors, for instance, are "...primarily popularity contests and not measures of learning" (Wieman, 2007, p. 15). Moreover, studies that engage students in research-led learning should not be exploitative (e.g., as noted by Healey, 2005), and student involvement should at very least be acknowledged (Kent et al., 1997). The students in these courses were not geography majors and lacked any formal research training (e.g., at the graduate level). For this reason in particular, student findings are not conveyed in this paper. What is important here, rather, is the project and (the research-led, problem-solving) approach taken to evoke active learning in these upper-year undergraduates. It is interesting to attain student evaluations of this project- and field-based active approach to learning, which does not require any real training in the way that scientific research demands.

It is noteworthy that active learning does not always need to be conducted in the field and has been executed, for example, as part of a map-based exercise in the classroom (Drennan and Evans, 2011). Video podcasts (vodcasts) have also been employed to support work beyond the classroom towards the acquisition of geographical methods and techniques in physical geography and GIScience (Jarvis and Dickie, 2010). Nevertheless, the importance of fieldwork has been defended, for instance by Herrick (2010), who saw its merit in experiential and active learning that stimulates deep learning. Malam and Grundy-Warr (2011) similarly argued that a problem-solving approach could stimulate applied research skills, including critical engagement and independent research skills among others. Moran and Round (2010) identified the use of "on-street" teaching alongside visual sources in order to help overcome linguistic problems associated with doing fieldwork in a non-English-speaking country, such as Moscow, Russia. However, students need not travel too far even for effective results, as for instance revealed by Rathburn and Weinberg (2011), who had students complete campus-based (hands-on) field-based exercises related to surface and groundwater quantity and quality. Others, for instance Winder and Lewis (2010), have approached regional geography in a field course. Research suggests that learning through the affective domain may be a valuable pedagogical approach (van der Hoeven Kraft et al., 2011), and so meaningful local environments may enhance (field-based) learning.

The feedback from students in this study conveyed that not all students found the course to be stimulating, as most respondents agreed with Question 15. This is also evidenced by the bimodal distribution of response frequency for Question 23. Several reasons occur for this; for instance, students did not appreciate the selected topic; they were unaccustomed to a project-based active approach; they did not work well in their groups (e.g., if only a few people did most of the work); etc. This was most likely attributable to a lack of comfort with the approach (and possibly content in geomorphology, because these were not geography majors) and possibly geographical techniques. Most of these students were training to become elementary school teachers; they struggled with science and mathematics and would also have experienced difficulty with a technical (quantitative monitoring) approach. Follow-up interviews would have benefited the research in understanding the reason for some of these (negative) student reactions. The set questions in the questionnaire (which was a standardized form and not designed specifically for the study) did not serve to target questions relevant for this study. However, students were able to add further comments in the open-ended questions. They may, however, have held back from fear of the instructor recognizing their handwriting. Nevertheless, it is extremely difficult to avoid such bias (subjectivity) when gauging personal perception (reactions).

Additionally, the author has some final reservations about standard assessments as the end-of-term evaluation process. Often, students express hostility and nonconstructive remarks, which can be aggressive and ineffective for further development of the course. Students who lack a proper background, in particular, lash out at professors, and future studies should investigate whether or not a direct correlation occurs with student performance and student satisfaction assessments, which is beyond the scope of this study. Moreover, course evaluation summaries tend to use averages, which can hide trends like those evident in this study using modal (most frequent or popular) responses. Courses, such as more applied upper-year courses, can stimulate negative evaluations for instructors that are supporting learning and attempting to give students a genuine, deep learning experience. Field-based studies are not easy, especially where class size exceeds eight to 10 pupils, as in this study. Also, much of the geography-specific content of this course, which geography specialists would appreciate, was lost in this teacher-oriented group.

4.3.4.3 Field School

The qualitative open-ended questions were more informative then the more quantitative responses on the six-point scale. The main reason for this was that "Agree" and "Disagree" were not clearly indicated as choices on this standard scale, forcing students to choose between "Strongly Agree" or "Somewhat Agree" and "Somewhat Disagree" or "Strongly Disagree." Their comments on the written responses provided a better overall understanding of their attitude towards the course and their suggestions for future improvements. These comments indicate that students did enjoy fieldwork, but found it challenging due to their lack of a background, because some students were admitted into the course without any previous field exposure to the study of landscapes (e.g., an introduction to physical geography or geomorphology or field course).

Other authors have also uncovered that undergraduates enjoy field-based studies. For instance, first-year undergraduates in the United Kingdom demonstrate an overall more positive perception of the field excursion experience than a negative attitude; however, more had an average attitude to fieldwork (see Table 3 in Dalton, 2001, p. 388). Assessments of fieldwork in physical geography convey that it clearly an enjoyable experience for undergraduates and a valuable learning experience compared to class- and desk-based learning (Fuller, 2006). The author concluded that field-based activities enhance student learning by experiencing the real world, as a hands-on application, that can help students gain transferable skills (as through the use of equipment and techniques) and experience analyzing data. Field courses certainly fulfill the last of three student interests, including the self-interest to perform well on examinations; their possible interest in the subject; and in developing their abilities (Elton, 2001). Harackiewicz et al. (2008) addressed interest, including individual and situational, that can develop over the course of an introductory course. The former (individual interest) is evident for students that have some familiarity with the course and its content, with the potential for a deepening interest through the establishment of value or personal relevance or meaning in the latter (situational interest), so that the course holds and develops their interest. Experiences throughout a course may deepen individual interest if situational interest is held, as with mastery goals.

A mixture of individual and group work assignments were given in this course so that students did not find this problematic. It seems that working in groups helped to support individuals with little previous experience with field-based learning. Afterall, fieldwork provides opportunities for group-based learning and for the development of social interaction and the potential to enhance understanding and deeper learning. Positive learning experiences can be gleaned by undergraduates engaging in conservation research, for instance, which can lead them to gain confidence at the same time as they develop specialized expertise and their naturalistic intelligence (Hayes, 2009). The author also stressed the importance of mentorship for undergraduates through fieldwork, as this provides students with a glimpse of their mentor's professional life and is a type of social education.

The small size of the group enabled students to openly ask questions and request help. This helped to establish positive interactions between the instructor and students. This was particularly evident in the final field assignment at the Awenda Provincial Park, when student groups were closely supervised by the instructor and received much direction in the completion of their fieldwork. This may have not been the case with a larger class size. The improvement of teacher-student rapport does not always progress beyond a fieldtrip, especially in large classes, as conveyed by (Lai, 1999) for a free-learning environment (partaken of by secondary students in Hong Kong). Students required much direction and supervision in this course because they were learning new techniques, as well as working in outdoor environments that they had not formerly approached from an educational (research-oriented) perspective. They found the textbook to be unhelpful, because of its focus on glacial environments, and could not rely on it for assistance. For this reason, they approached the instructor more frequently with questions and concerns about the material, including assignments. The small-group dynamics also provided opportunities for interactions with the instructor as well as within and between groups. Students still expressed frustrations with having to wait for equipment because of the expense of purchasing field-portable instruments.

However, if not the shortage of equipment and unsuitability of the textbook, students could have experienced other concerns, as with overnight fieldtrips. In New Zealand, for instance, undergraduates participating in overnight fieldtrips experienced sex-, gender-, and cultured-related conflicts with sleeping arrangements, meaning that the experience is not always a positive one when these factors are taken into account (Nairn, 2003). However, this did not affect this field school due to the cancellation of the camping portion of the course. A mentoring approach in higher education, however, encouraged undergraduates to develop their ability to prepare and plan for field research, as well as engage in collaborative learning in addition to enabling them to integrate classroom material into their fieldwork (Hefferan et al., 2002). Arreola (2001) found that informal mentoring by graduate students helped undergraduates partaking of a 3-day (2-night) field experience offered through Arizona State University of the Mexican-American borderland. The direction provided by the instructor in this course was beneficial for student learning and enhanced the student experience of the course. Students became active participants in their learning experience and were, overall, keenly involved in learning the techniques for geographical fieldwork in different landscapes. Applied exercises to encourage deep education through field-based learning can take as much as just a 1-day exercise, and can be a problem-based task that requires application of in-class learning, such as a consultant exercise

(Thomas, 2000). However, time in the field is not the only variable that is important for effective student learning. For instance, a study examining the traditional description-explanatory mode of learning against the analytical-prediction approach found that the former (traditional approach) was more effective for first-year undergraduates (Fuller et al., 2000). The reason for this is direction, which is required for inexperienced students new to a discipline. The latter PBL approach has been found to be challenging by students, although it promotes deep learning, which should benefit further study, including dissertations and problem solving, and promotes self-confidence (Simm and David, 2002). Evaluations for an inquiry-based undergraduate geography course running five years (between 2000 and 2005) ranged from very good to excellent, with a marking scheme comprising two assignments (on hazards and pollution) each worth 30% and an exam worth 40% (Spronken-Smith and Kingham, 2009). The authors found that by engaging in research projects, students were able to experience actual research and become better acquainted with staff's research areas. This engagement in (environmental geography) research enables students to develop deep learning of the subject matter. These authors encouraged such a research-based approach to deep learning because it allows students to acquire experience with (applied) statistics, which becomes part of their transferable skills set.

4.3.5 CONCLUSIONS

Even though students seemed to struggle with the new content, especially as non-science majors in a concurrent education program, they appreciated the learning experience and the fieldwork component in particular. At the introductory (second-year) level, undergraduates relied heavily on a textbook for the course in order to direct their learning. They were mainly dependent on the instructor (and this textbook) to learn new material. At this level, students required much traditional lecture-based instruction to guide their learning. A slow to medium pace of delivery is preferred and lectures should not be rushed as to not overwhelm these novices. Engaging in fieldwork, even at the introductory level, however, does seem appropriate because it exposes students to opportunities for deeper learning than can occur in the lecture hall. Students appreciate seeing concepts and other information learned in class in the wider world, especially in their local environment, which they encounter everyday and enjoy understanding and seeing through the lens of the scientist and geomorphologist. Fieldwork also breaks the monotony of teaching introductory courses in geomorphology, which are laden with terminology. For example, the churchyard study addressing headstone weathering was a popular excursion that students seemed to appreciate, even though it was a walking distance from the classroom and in their everyday setting. Finally, listening to student feedback provided by end-of-term evaluations is useful in order to improve the future enjoyment of the learning experience.

Even though field-based projects are difficult to execute in large groups, such as in an upper-year undergraduate science course, some noted benefits occur with such an approach. Some students enjoyed learning in the outdoors, as in the local area. They may have been adverse to winter topics (including icicles as a subject matter), but this did not apply to all students. Despite close supervision of field-based learning, even upper-year undergraduates struggled with some aspect of the course, be it in the application of lecture-derived knowledge or learning techniques in the field. Indeed, some students can resist an experiential approach, even at the third-year level. However, research (including the results of this study), as addressed in the literature review of this study, suggests that a research-led, problem-oriented approach works at the upper undergraduate level, as for the acquisition of transferable skills, and this chapter postulates this even when learners are non-geographers. Field-based learning was demonstrative and provided a contextualized framework for learning.

Finally, this study also portrays a positive student opinion of an upper-year undergraduate field school based in southern Ontario, Canada. Undergraduates understood the value of outdoor learning for a hands-on understanding of physical environments in Simcoe County. Even though the workload was high due to the level of this (third-year) course, students attended classes regularly and participated in field-based assignments. It was not possible to have an overnight (camping) component in this course because students had other obligations, such as work and family (in the case of mature students, for instance, those who were married and/or had children). However, students were able to apply class-based material, including the handling of equipment, in field assignments, as well as in the final research project. Fieldwork manuals should be tailored to meet student needs in such a field school, rather than relying on broadly spanning textbooks.

The overall impression at the end of this assessment is that field-based undergraduate learning should begin early in order to get students acquainted with the demands of field-based research and learning. In this way, students are able to acquire necessary skills to better engage in graduate study, which characteristically has a greater research-based component. By incrementally increasing the dosage of field exposure, it is possible to prepare students for graduate work, as well as slowly developing in students an appreciation for applied (experiential) field-based learning that eventually blooms into a research-led approach to learning that will accompany students the rest of their academic careers.

ACKNOWLEDGMENTS

The author is grateful to those students who completed the course student satisfaction surveys. She also wishes to note that this study was not executed at her current institution.

APPENDIX 4.3.1 APRIL 2008

Question 15	Question 16	Question 17
Subject matter covered; labs allowed for hands-on application of material covered in lecture.	A lab environment that would allow for better work environment; field report – a few more details on expectations related to length and formatting.	Very fair with evaluations – worked with students to ensure we could do as well as possible in the course. Instructor was enthusiastic in delivering lectures – very knowledgeable on the material.
Nice that labs had some samples available; ability to read others research gave a different perspective (sometimes language used was confusing, though).	Offer it in Fall term, so that students can actually see examples instead of pictures; more fieldtrips (esp. if in Fall); more lab materials available; sometimes material was presented too quickly.	–
I found it interesting and I did learn a lot of new information.	More detailed descriptions for assignments and lab component. I found the report difficult as we never done anything like it before and felt there could have been more detailed expectations.	–
I really enjoyed the subject matter and the "hands on" during labs.	A different textbook [with] a glossary would be helpful.	More variety of types of questions on exams.
The content was interesting.	Don't expect students to know as much as you when lecturing/giving labs. Provide descriptions/assistance during lab sessions. Don't do outdoor experiments of soil/water when there is 3 ft. of snow. Describe definitions and not just read them, i.e., give examples. [That's] what text is for.	Potential to be a really good course.

(Continued)

Question 15	Question 16	Question 17
I liked that the prof was very knowledgeable about the subject.	Make labs different (not the same one over and over again). Teach at an introductory level – the prof assumed us to know a lot more coming into this course than we did and did not change her teaching style to [accommodate] this. More Canadian content.	[The author] knows her material, but she needed to alter the way she taught it, as we are students with no experience in [geomorphology]. She needed to teach us how to do the labs, not just present us with them and expect us to know what to do – more feedback would have helped. Needed to alter her expectations of us and be more flexible.
I thought that most of the material covered was interesting.	I felt that at times this was not taught as an introduction course making some of the concept hard to get. I also felt that the work load was way higher than my other courses in geography. I also believe that the expectations for students was high [considering] we had never been introduced to most of these concepts before.	One thing that is clear is that [the author] is an expert in her field. She gets excited about the material and that energy transfers to her students.
	This course really needs an introduction course even though it is already one. The jargon used in this course made it impossible to [grasp] topics. The workload was way [too] high for a student with four other classes to manage. The choice of topic for assignments was not suitable with the time of year it was being taught in.	Although [the author] was very nice and approachable she expected us to already have a background in the field which we didn't and thus much of the course material I found very hard to understand it. Also this school needs a physical geography lab so that the Prof. can actually use hand-on aids to demonstrate theories better.

APPENDIX 4.3.2 NOVEMBER 2008

Overall, have you been satisfied with the course, why/why not?	Do you have any suggestions for improvement, either of teaching style or material covered in the course?
Melissa B.: I was satisfied with the Geomorphology course. I enjoyed the hands on [fieldtrips], seminars, and [in-class] lectures. I enjoyed learning about different types [of] landscapes and processes [within] the earth.	Geomorphology was a good choice of course offered. I would recommend any [upper-year] course offered, either 2nd or 3rd year Geomorphology course. I would suggest more [fieldtrips] and less labs for more hands on experience.
This was a good course. I really enjoyed the [fieldtrip] components. Thank you for providing a more scientific approach to geography here at [the University].	During the [fieldtrip] to the cemetery it would have been nice for everyone to be shown how to do the measurements then for the students to be released in different groups to take the measurements on their own. I realize that we would have needed more equipment to do this but each group could possibly be responsible for a specific type of measurement or observations.
If you are going to put pictures for identification on the final exam, please make sure that you either provide the pictures, links to the pictures, or a time to review the pictures in class (or your office) prior to the exam. I am a student who does not learn visually...it actually distracts m to look at the slides on the screen and try to process their content; as a result I rarely look up during lectures.	–

APPENDIX 4.3.3 NOVEMBER 2009

Question 24	Question 25	Question 26
The hands-on [fieldtrips] made it more interesting and allowed us to apply some of the knowledge we've attained.	There needs to be more time spent on lectures as they are very detailed and deal with a lot of content most of us have never heard of, which makes us need more explanations of terms, concepts, etc.	–
The lecture handouts at the beginning of class; the style of the lecture; fieldwork assignments.	A mid-term so we know how her tests work before the final, and so the final isn't worth as much; not as many pointless articles in the seminar portion of the lab; provide clear outline of expectations on assignments.	Thanks Mary!
–	Too much group work. It was difficult to arrange a time to meet with everyone in your group every week.	–
I liked the weekly labs that helped us go move in depth with what we were learning. I also liked the [fieldtrips] that allowed us to be hands on and see everything. They were also both good ways to maintain [one's?] mark.	When you hand out your slideshow, give more space to write. There needs to be more testing opportunities than just the exam, we need to see what your exam is like.	–
	Felt the textbook was a little difficult to understand at times; we covered a lot of information in little time.	–
I liked most of the [fieldwork]. situations. It is nice to get out of the class and learn on actual situations.	I would suggest for the instructor to be very clear on what she wants for assignments and labs. Also some of labs were marked a little too hard.	–
I liked that the material was applied to our local environment, especially through the use of [fieldtrips]. As well, the use of a small group assignment each week was much easier then one large paper, and allowed to learn more because each topic was different.	Maybe a [fieldtrip] outside of [the local area], or a guest lecture.	–

(Continued)

Question 24	Question 25	Question 26
The course is applicable to everyday encounters where you can look outside and know how it had occurred. The handouts were helpful but there was too much paper used up for this.	Slower pace of info lecture and for students to have more hands on experience in the labs. Too much [stuff]? work. A mid-term would be extremely helpful and would prepare us for the final exam.	–
The handouts at beginning of class; fieldwork assignments were different.	Have a mid-term!!!; not so many articles; if articles used make them more appropriate; have CLEAR outlines and expectations on assignments; and do not change expectations [half] way through; answer questions – do not ignore them (if not sure of answer say you will let us know next class).	mid-term!!
Hand outs provided; consistent pictures in PowerPoint; topics; hands on work (labs) with? outside of course.	Less group work (make assignments worth more so there [are] less? Assignments – group work was good but there was a lot!; using one text book, or creating a course package [with] various chapters from various books, the terms were very new to me and I had a hard time understanding some concepts – could never find info because the text was not used for most of the end of course; a mid-term!!! would [be] very useful with such a course because of all the material.	Thanks Mary!
I enjoyed the outdoor labs because we were able to see the topics she was talking about in lecture in the every day environment.	Have a mid-term in this course so there will not be as much material on the final.	N/A
I liked the variety of different information that was presented during the class and examples of these concepts being presented in a handout.	I hated the labs. I viewed the labs as difficult and I hated they were all [group work]. The labs would be better if some days the activity were [individualized] instead of relying on the majority of the work replying on [group work].	Overall good course.

(Continued)

Question 24	Question 25	Question 26
Practicality of information.	I found (some) lectures sometimes confusing?	–
Human impact up(?) geological processes. It was informative without being too scientific.	The lab readings at times were hard to understand. Sometimes the group work did not work so well. The final exam is going to be massive, too much info to study.	[The author] is extremely knowledgeable in the topic and that really helped when we needed clarification on the lesson.
I learned about new topics and concepts.	Assignments were not explained well enough. The instructor seemed unapproachable when there was an issue with grading or questions regarding assignments. The textbook we purchased did not seem to be used [a lot]. The instructor used different materials, making it [seem] irrelevant to read or use the textbook. I did not enjoy the labs.	–
I like the handouts given.	I found the textbook to kind of be a waste. We used most information from other books so maybe a course package would be more [beneficial]. Also I don't like the fact that there was no mid-term, because we have covered a ton of material that will ALL be on the final... A little overwhelming.	–
That there was a variety of material taught.	Have a mid-term. There is too much material to study for just a final. Explain more of your marking scheme; explain what you are looking for more clearly.	Have a mid-term. [The author] is a nice and friendly person and I like having her as a teacher.
I liked the practicality of it. Getting out into the field, regardless if its just the port of [the city] is valuable. Helps [contextualize].	More fieldwork. More in depth labs not harder, just more "tools of the trade."	Great job [instructor].

(Continued)

Question 24	Question 25	Question 26
–	If you are going to make students purchase a required textbook it would be nice if she actually used it, nothing that the instructor [taught] was in text book. [Fieldtrips] need to be more organized. This class should not be 1 h long but the full 1.5 h for instructor to lecture slower for students to understand.	There needs to be a mid-term. It would be helpful if the instructor was approachable. Instructor would change what she said and tell people different things.
–	WAY too much group work – undirected so there were many large groups where some people were not able to contribute. I found I did not learn much other than my assigned part in the group work (labs and fieldwork). Did not effectively use text – did not refer to text after Thanksgiving – waste of $ to buy text for only ½ of the course information. Poor grading system (lack of structure and was not the same [with] all groups).	Have a mid-term to help with gauging where students are. She changes her mind about everything!! Not [consistent] with answering questions on assignments and marking. Not making PowerPoint available online – used many pictures from other sources yet could not make out photocopies. Not available for extra help. Lack of [organization] in regards to [fieldtrips] and due dates.
I liked the fact that we had weekly labs that were used to apply our knowledge that we had learned each week, rather than one or two big assignments throughout the semester. I also liked how the PowerPoints were handed out so that we could follow along and fill in info as we go.	To have a mid-term or tests so that we can become better prepared for the final exam, and to know what your testing is like so that we don't bomb the final exam since it is worth so much of our final mark. Also to leave more space on the handouts so that we can fit it all and also go a bit slower so that we can get it all down.	–
I liked the lectures and the group work.	There ought to be penalties for not pulling one's own weight in group work. More diagrams in lectures – can never have enough.	–

REFERENCES

Allen, C.D., Lukinbeal, C., 2010. Practicing physical geography: an actor-network view of physical geography exemplified by the rock art stability index. Prog. Phys. Geog. 35 (2), 227–248.

Arreola, D.D., 2001. Teaching the borderland. Geogr. Rev. 91 (1–2), 480–486.

Auer, M.R., 2008. Sensory perception, rationalism and outdoor environmental education. IRGEE 17 (1), 6–12.

Barrett, L.R, Matney, T., Park, L.E., 2004. Teaching archaeogeophysical survey and mapping any time of the year: an interdisciplinary course. JGE 52 (3), 236–244.

Bednarz, S., Witham, Chalkley, B., Fletcher, S., Hay, I., Le Heron, E., Mohan, A., Trafford, J., 2008. Community engagement for student learning in geography. J. Geogr. Higher Educ. 32 (1), 87–100.

Biggs, J., 1999. What the student does: teaching for enhanced learning. HERD 18 (1), 57–75.

Birdsall, S.S., 2003. Learning to see landscape through a flexible lens. J. Geogr. 102, 29–34.

Boyle, A., Maguire, S., Martin, A., Milsom, C., Nash, R., Rawlinson, S., Turner, A., Wurthmann, S., Conchie, S., 2007. Fieldwork is good: the student perception and the affective domain. J. Geogr. Higher Educ. 31 (2), 299–317.

Carlson, T., 2007. A field-based learning experience for introductory level GIS students. J. Geogr. 106, 193–198.

Chapman, L.S., Putnam, D.F., 1984. The Physiography of Southern Ontario. Ontario Geological Survey, Special Volume 2. Ontario Ministry of Natural Resources, Toronto.

Dadd, K.A., 2009. Using problem-based learning to bring the workplace into the classroom. JGE 57 (1), 1–10.

Dalton, R.T., 2001. What do they bring with them? The fieldwork experiences of undergraduates on entry into higher education. J. Geogr. Higher Educ. 25 (3), 379–393.

Drennan, G.R., Evans, M.Y., 2011. Introductory geological mapwork – an active learning classroom. JGE 59, 56–62.

Dyment, J.E., 2005. Green school grounds as sites for outdoor learning: barriers and opportunities. IRGEE 14 (1), 28–45.

Ellis, J.T., Rindfleisch, P.R., 2006. A coastal environmental field and laboratory activity for an undergraduate geomorphology course. J. Geogr. 105, 216–224.

Elton, L., 2001. Research and teaching: conditions for a positive link. Teach. High. Educ. 6 (1), 43–56.

Elwood, S.A., 2004. Experiential learning, spatial practice, and critical urban geographies. J. Geogr. 103, 55–63.

Eppes, M.C, 2009. Introducing field-based geologic research using soil geomorphology. JGE 57 (1), 11–22.

Fisher, J.A., 2001. The demise of fieldwork as an integral part of science education in United Kingdom schools: a victim of cultural change and political pressure? Pedagog. Cult. Soc. 9 (1), 75–96.

Fuller, I.C., 2006. What is the value of fieldwork? Answers from New Zealand using two contrasting undergraduate physical geography field trips. NZG 62, 215–220.

Fuller, I., Rawlinson, S., Bevan, R., 2000. Evaluation of student learning experiences in physical geography fieldwork: paddling or pedagogy? J. Geogr. Higher Educ. 24 (2), 199–215.

Fuller, I., Gaskin, S., Scott, I., 2003. Student perceptions of geography and environment science fieldwork in the light of restricted access to the field caused by foot and mouth disease in the UK in 2001. J. Geogr. Higher Educ. 27 (1), 79–102.

Fuller, I., Edmondson, S., France, D., Higgitt, D., Ratinen, I., 2006. International perspectives on the effectiveness of geography fieldwork for learning. J. Geogr. Higher Educ. 30 (1), 89–101.

Goudie, A.S., 2005. The Human Impact on the Natural Environment: Past, Present, and Future, 6th Edition Wiley-Blackwell, Oxford.

Harackiewicz, J.M., Durik, A.M., Barron, K.E., Linnenbrink-Garcia, L., 2008. The role of achievement goals in the development of interest: reciprocal relations between achievement goals, interest, and performance. J. Educ. Psychol. 100 (1), 105–122.

Harris, T, Tweed, F., 2010. A research-led, inquiry-based learning experiment: classic landforms of deglaciation, Glen Etive, Scottish Highlands. J. Geogr. Higher Educ. 34 (4), 511–528.

Hayes, M.A., 2009. Into the field: naturalistic education and the future of conservation. Conserv. Biol. 23 (5), 1075–1079.

Healey, M., 2005. Linking research and teaching to benefit student learning. J. Geogr. Higher Educ. 29 (2), 183–201.

Hefferan, K.P., Heywood, N.C., Ritter, M.E., 2002. Integrating field trips and classroom learning into a capstone undergraduate research experience. J. Geogr. 101, 183–190.

Herrick, C., 2010. Lost in the field: ensuring student learning in the "threatened" geography fieldtrip. Area, 42.1, 108–116.

Hirsch, P., Lloyd, K., 2005. Real and virtual experiential learning on the Mekong: field schools, e-Sims and cultural challenge. J. Geogr. Higher Educ. 29 (3), 321–337.

Hope, M, 2009. The importance of direct experience: a philosophical defence of fieldwork in human geography. J. Geogr. Higher Educ. 33 (2), 169–182.

Hubbard, B., Glasser, N.F., 2005. Field Techniques in Glaciology and Glacial Geomorphology. Wiley-Blackwell, Oxford.

Hudak, P.F., 2003. Campus field exercises for introductory geoscience courses. J. Geogr. 102, 220–225.

Hung, W., Bailey, J., Harpole, Jonassen, D.H., 2003. Exploring the tensions of problem-based learning: insights from research. NDTL 95, 13–23.

Jarvis, C., Dickie, J., 2010. Podcasts in support of experiential field learning. J. Geogr. Higher Educ. 34 (2), 173–186.

Jennings, S.A., Huber, T.P., 2003. Campus-based geographical learning: a field oriented teaching scenario. J. Geogr. 102, 185–192.

Jones, S., 2000. Enlivening" development concepts through workshops: a case study of appropriate technology and soil conservation. J. Geogr. Higher Educ. 24 (1), 76–86.

Kent, M., Gilbertson, D.D., Hunt, C.O., 1997. Fieldwork in geography teaching: a critical review of the literature and approaches. J. Geogr. Higher Educ. 21 (3), 313–332.

Lafon, C.W., 2005. Reconstructing fire history: an exercise in dendrochronology. J. Geogr. 104, 127–137.

Lai, K.C., 1999. Freedom to learn: a study of the experiences of secondary school teachers and students in a geography field trip. IRGEE 8 (3), 239–255.

Lev, S.M., 2004. A problem-based learning exercise for environmental geology. JGE 52 (2), 128–132.

Malam, L., Grundy-Warr, C., 2011. Liberating learning: thinking beyond "the grade" in field-based approaches to teaching. NZG 67, 213–221.

McKendrick, J.H., Bowden, A., 1999. Something for everyone? An evaluation of the use of audio-visual resources in geographical learning in the UK. J. Geogr. Higher Educ. 23 (1), 9–19.

McKendrick, J.H., Bowden, A., 2000. Equipped for the 21st century?: audio-visual resource standards and product demands from geography departments in the UK. J. Geogr. Higher Educ. 24 (1), 53–73.

McMorrow, J., 2005. Using a web-based resource to prepare students for fieldwork: evaluating the Dark Peak virtual tour. J. Geogr. Higher Educ. 29 (2), 223–240.

Mooney, S.J., 2006. A simple group work approach for effective field work: a soil sciences case study. JGE 54 (1), 74–79.

Moran, D., Round, J., 2010. A riddle, wrapped in mystery, inside an enigma": teaching post-socialist transformation to UK students in Moscow. J. Geogr. Higher Educ. 34 (2), 265–282.

Nairn, K, 2003. What has the geography of sleeping arrangements got to do with the geography of our teaching spaces? Gend. Place Cult. 10 (1), 67–81.

Panelli, R., Welch, R.V., 2005. Teaching research through field studies: a cumulative opportunity for teaching methodology to human geography undergraduates. J. Geogr. Higher Educ. 29 (2), 255–277.

Paradis, T.W., Dexter, L.R., 2007. Learner-centered teaching and assessment in an undergraduate field analysis course. J. Geogr. 106, 171–180.

Pawson, E., Fournier, E., Haigh, M., Muniz, O., Trafford, J., Vajoczki, S., 2006. Problem-based learning in geography: towards a critical assessment of its purposes, benefits and risks. J. Geogr. Higher Educ. 30 (1), 103–116.

Rathburn, S.L, Weinberg, A.E., 2011. Undergraduate student satisfaction and achievement at the GetWET Observatory: a fluid learning experience at Colorado State University. JGE 59, 47–55.

Resler, L.M., Kolivras, K.N., 2009. A field-based technique for teaching about habitat fragmentation and edge effects. J. Geogr. 108, 210–218.

Robson, E., 2002. An unbelievable academic and personal experience": issues around teaching undergraduate field courses in Africa. J. Geogr. Higher Educ. 26 (3), 327–344.

Salvage, K., Graney, J., Barker, J., 2004. Watershed-based integration of hydrology, geochemistry, and geophysics in an environmental geology curriculum. JGE 52 (2), 141–148.

Scott, I, Fuller, I., Gaskin, S., 2006. Life without fieldwork: some lecturers" perceptions of geography and environmental science fieldwork. J. Geogr. Higher Educ. 30 (1), 161–171.

Sidaway, J.D., 2002. Photography as geographical fieldwork. J. Geogr. Higher Educ. 26 (1), 95–103.

Simm, D.J., David, C.A., 2002. Effective teaching of research design in physical geography: a case study. J. Geogr. Higher Educ. 26 (2), 169–180.

Slaymaker, O., Spencer, T., Embleton-Hamann, C., 2009. Geomorphology and Global Environmental Change. Cambridge University Press, Cambridge.

Spronken-Smith, R., 2005. Implementing a problem-based learning approach for teaching research methods in geography. J. Geogr. Higher Educ. 29 (2), 203–221.

Spronken-Smith, R., Kingham, S., 2009. Strengthening teaching and research links: the case of a pollution exposure inquiry project. J. Geogr. Higher Educ. 33 (2), 241–253.

Spronken-Smith, R., Bullard, J., Ray, W., Roberts, C., Keiffer, A., 2008. Where might sand dunes be on Mars?. Engaging students through inquiry-based learning in geography. J. Geogr. Higher Educ. 32 (1), 71–86.

Stumpf, II, R.J., Douglass, J., Dorn, R.I., 2008. Learning desert geomorphology virtually versus in the field. J. Geogr. Higher Educ. 32 (3), 387–399.

Summerfield, M.A., 1991. Global Geomorphology: An Introduction to the Study of Landforms, 1st Edition Pearson Education Limited, London.

Thomas, N., 2000. Using independent, investigative learning to teach environmental geophysics to geology undergraduates. Teach. Earth Sci. 25 (3), 152–156.

Trenhaile, A.S, 2007. Geomorphology: A Canadian Perspective, 3rd Edition Oxford University Press, Oxford.

Trenhaile, A.S., 2010. Geomorphology: A Canadian Perspective, 4th Edition Oxford University Press, Oxford.

van der Hoeven Kraft, K.J., Srogi, L., Husman, J., Semken, S., Fuhrman, M., 2011. Engaging students to learn through the affective domain: a new framework for teaching in the geosciences. JGE 59, 71–84.

Wieman, C., 2007. Why not try a scientific approach to science education. Change 39 (5), 9–15.

Winder, G., Lewis, N., 2010. Introduction: performing a new regional geography. New Zeal. Geogr. 66, 97–104.

Chapter 5

Field Methodologies

Mary J. Thornbush*[,1], Casey D. Allen**, Faith A. Fitzpatrick†
*School of Geography, Earth and Environmental Sciences University of Birmingham, Edgbaston, Birmingham, United Kingdom; **Department of Geography and Environmental Sciences, University of Colorado Denver, Colorado, USA; †U.S. Geological Survey, Wisconsin Water Science Center, Middleton, Wisconsin, USA
[1]Corresponding author: E-mail: m.thornbush@bham.ac.uk

Although specific fieldwork methods and techniques exist for nearly every given situation, and countless articles and books have been written on each of those topics, this section focuses on *how* geomorphologists conduct fieldwork. Throughout this section, many examples of methodological and even disciplinary cross-pollination exist. Even though each chapter uses the commonly assessed fluvial landscape, they do so in a unique way. These thoughts are first echoed in Kuhn, Greenwood, and Fister's (Chapter 5.1) "Use of Field Experiments in Soil Erosion Research", where the critically important topic of soils as related to precipitation events and discussed utilizing two separate case studies. They outline the key components necessary for conducting fieldwork appropriately, noting that although experimental fieldwork has strengths and weaknesses that are different from laboratory experiments and field monitoring, these differences are often overlooked. This is especially the case when it comes to adapting an ideal situation that was possible in the laboratory, to a field-based experiment, reminding the field practitioner that interpretation is context dependent. In "A Geologic Approach to Field Methods in Fluvial Geomorphology" (Chapter 5.2), Fitzpatrick uses the topic of past river adjustments and historicity to remind readers of the important role multidisciplinarity plays in achieving greater understanding, especially when searching for linkages between landscape types. She discusses various techniques used in channel morphology and reconstruction, including the use of historic event records, historic maps, and other archival sources to help understand and prepare for the future. Whether through coring, bathymetric surveys, LiDAR, ground penetrating radar, or traditional landscape description, Chapter 5.2 reminds the readers that each technique and method has its value. Brierley and Fryirs' Chapter 5.3 "Reading the Landscape in Field-Based Fluvial Geomorphology," uses a bawdy, tongue-in-cheek acronym across three distinct field-based exercises to assess river systems in an applied setting. Their CRAPPER (Catchment-scale analysis, Research-scale investigations, Adjustment,

Developments in Earth Surface Processes, Vol. 18. http://dx.doi.org/10.1016/B978-0-444-63402-3.00010-8

Process, Pattern, Evolution, and Recovery/Rehabilitation prospects) mnemonic device aids field practitioners in developing a toolkit for interpreting and reading a landscape, fluvial or otherwise, as the CRAPPER technique can be applied to multiple locations at differing scales. Without losing sight of the important role field-based investigations play in the researcher's experience, each chapter in this section asserts fieldwork's importance by promoting the integration of multiple techniques and methods, allowing the field practitioner to gain a better understanding across methodological boundaries and enhance their own understanding related to field-based endeavors.

Chapter 5.1

Use of Field Experiments in Soil Erosion Research

Nikolaus J. Kuhn[1], Philip Greenwood, Wolfgang Fister
Physical Geography and Environmental Change Research Group, Department of Environmental Sciences, University of Basel, Basel, Switzerland
[1]*Corresponding author: E-mail: nikolaus.kuhn@unibas.ch*

5.1.1 EXPERIMENTS IN GEOMORPHOLOGY

Experiments have a long tradition in geomorphology because of the need, or desire, to reduce complexity and thus simulate otherwise unattainable conditions. *Sensu stricto*, experiments involve controlled procedures that are carried out with the aim of testing a hypothesis. The challenge for designing an experiment in order to meet the expectations of the researcher lies in identifying its location in the triangle of precision versus realism versus generalism (e.g., Kuhn, 2014). By their own virtue, experiments aim to be more precise than nature, or rather, what can be observed in nature. This necessitates a reductionist, or simplified approach, which can frequently place the representativeness of the experiment, and thus the general relevance of the results to real-world scenarios into question (e.g., Bryan 1990). For instance, selecting a suitable plot size and rainfall intensity to perform rainfall simulations illustrates this challenge and will be one of the issues discussed in the case study presented in this chapter.

In environmental sciences, the term "experiment" is commonly used in a somewhat wider context than has typically been applied in the more traditional, or "natural," sciences. The reason for this is the compromise that has to be made in order to meet a position in the precision–realism–generalism triangle that generates an answer to the question being addressed by the experiment (Slaymaker, 1991). Further limitations arise from the field situation, as these can typically limit the level of control when compared to a laboratory experiment. Therefore, three broad aims of field experiments in geomorphology can be identified:

1. Actual experiments that aim to test one or more hypotheses on the interaction of one or more components within a landscape system.
2. Process-rate measurements in the field in order to quantify a conceptual model or test the relevance of a particular process in a given landscape system.

Developments in Earth Surface Processes, Vol. 18. http://dx.doi.org/10.1016/B978-0-444-63402-3.00011-X

3. Measurements of an integrated response, such as collecting runoff or sediment from a plot or manipulated microcatchment that has been designed to generate input data for quantitative modeling.

The major difference between these three aims lies in the knowledge about the functioning of the studied system, the design of the experiment, and the quality of the data that are collected (Fig. 5.1.1). A real field experiment changes one factor of the system, largely independent of scale, and identifies the effect on the output of the system. The reaction of a system, defined by outside borders and knowledge of the of individual internal components, might be known *per se*, but the actual interaction of dominant factors and processes still represent a "grey box." Alternatively, it could be the so-called "black box" scenario, where nothing about a given system is known except the relationship between inputs and outputs. Either an input into such a system is changed and the effects studied, or more often, a property (supposedly) controlling the functioning of the system is perturbed or even altered. In this way, a hypothesis on the relevance of changing one flux or property of the system can be tested in terms of its behavior or

FIGURE 5.1.1 A rainfall simulator setup for a microscale erosion experiment (Witterswil, Switzerland).

response. The first case study on the influence of wind on erosion rates illustrates such an example of a "true" experiment.

The second type of experiment is carried out when at least a conceptual understanding of the actual internal functioning of a system exists. The experiment is often required to measure a particular process rate, for instance, when estimating its importance amongst other processes, or its influence on spatial and/or temporal parameters within a given landscape system. In this case, the interaction between the components of a landscape system has been identified and the experiment focuses on collecting information on one or a set of different processes. The second case study on infiltration experiments on terracettes falls into this category. Using information about certain soil physical properties allows inferences to be drawn regarding the spatial pattern of infiltration, and thus runoff generation. The experiment aims at measuring actual infiltration rates, which then allow the estimation of fluxes from one component of the system to another.

The final type of experiment can be considered a measurement of a predefined parameter value from a predefined (natural or artificial) landscape system, either before or after a modeling exercise. The latter case could also be considered as ground truthing or model validation/verification. One of the most prominent of these types of experiments is that used to collect erosion data for the Universal Soil Loss Equation (USLE). Originally, erosion plots were installed across a range of climatic, soil, and land-use conditions. Each plot expressed a suite of different environmental factors, such as gradient, soil type, vegetation cover, and density, which, at that time, were presumed to be most important in controlling erosion rates. The resultant data were then matched with the relevant suite of controlling environmental variables, which led to the development of an empirical model with the ability to predict erosion risks (Laflen and Moldenhauer, 2003). The stochastic nature of these relationships requires constant refinement and collection of data. In recent years, however, the model has increasingly been applied in regions or on certain types of terrain that technically lie outside of its calibrated range and beyond the original scope of erosion risk assessment (e.g., Meusburger et al., 2012). To accelerate the knowledge-generation process, rainfall simulations have been used to acquire erosion data. In this case, sediment-delivery rates for standardized rainfall are measured from standardized plots. Although comparing their values is of interest, they largely serve as input information for models rather than contributing much to an improved understanding of the actual erosion process that has generated the resultant data. One could actually argue that this type of measurement is not an experiment, but merely represents an exercise in data collection for model calibration or, if done after the modeling, aims at ground truthing or model validation. In any case, since this type of field experiment has been described very comprehensively in the scientific literature (Laflen and Moldenhauer, 2003), the examples in this chapter focus on the first two types of experiments. Before doing so, however, the use of field experiments is compared to field observations and laboratory experiments.

5.1.2 ADVANTAGES AND DISADVANTAGES OF FIELD EXPERIMENTS IN SOIL-EROSION RESEARCH

The advantages or disadvantages of field experiments, laboratory experiments, and field monitoring are controversially discussed in the surface-process community. Since all three methodological approaches have their advantages, they are used in many investigations. However, the appropriateness of their use depends on the nature of the research question and the expertise of the researchers involved in designing and executing the experiment, as well as in operating the required equipment (Greenwood et al., 2013). It is therefore prudent at this point to briefly describe and compare those methods before focusing on the applicability of field experiments. In the context of this comparison and the rest of this paper, the term "field experiments" is used to refer to rainfall simulation experiments, whereas "field monitoring" implies long-term observations made on plots receiving just natural rainfall.

5.1.2.1 Field Experiments versus Field Monitoring

The aim of field experiments is to create a controlled, or at least semicontrolled, environment that cannot be recreated in the laboratory and where monitoring is not feasible. Within this controlled environment, certain parameters can be maintained as constant as possible (Grismer, 2012), whereas others are manipulated on purpose. This is often done by conducting paired tests on plots differing in the factor of interest. One of the plots is used as a control, whereas the other ones are modified by the specific parameters of interest. Ideally, both plots are monitored similarly, so that the immediate effect of the modified parameter can be determined. However, in reality it is often difficult to select plots that have exactly the same properties, or worse, they seem to be identical, but the results show that they are not (Nearing et al., 1999). To reduce the error probability, it is therefore advisable to use more than just one set of control and experimental plots. Similar problems arise when tests cannot be carried out synchronously, as is often the case as a result of logistical constraints, and relevant factors, such as soil moisture, change in the time between tests.

Field monitoring (Fig. 5.1.2) offers the possibility to observe naturally occurring scenarios, but lacks the control over the factors of interest of a research question. Even for the unlikely case that two rainfall events are identical, other erosion-controlling factors, such as soil-surface roughness or aggregate stability, may vary (Kuhn et al. 2003), so that again no comparable results can be obtained. Therefore the lack of directly comparable results clearly limits the potential to derive an improved process understanding of a given erosion scenario solely from field-based observations. Experiments can close this gap, especially when monitoring results show a large degree of unexplained variability. Another important advantage associated with experiments is their capability to simulate conditions or scenarios that cannot be studied in current natural environments, such as varying past or future climatic conditions (e.g., Kuhn et al., 2004). Field experiments enable, for example, the simulation of

FIGURE 5.1.2 Example of a field observation station to show the possibility to monitor natural events at different scales and soil treatments (Sparacia Experimental Farm, Sicily).

events with a low frequency, but potentially a high-magnitude erosion impact (Wolman and Miller, 1960). Monitoring would require most likely a very long observation period before a reliable relation between rainfall, runoff, and erosion for a given site can be established for the upper extreme end of rainfalls (Lopes, 1996; Chaubey et al., 1999). Typical examples are the magnitude and reoccurrence interval of river floods (Molnar, 2001; Benito et al., 2003; Dettinger, 2011), but the same phenomenon has also been reported when measuring soil erosion (Boardman, 2003; González-Hidlago et al., 2007). Long-term monitoring also involves effects of trends and nonlinearities in the data sets (Kuhn et al. 2011), as well as the risk of encountering damage that should have been avoided by a timely erosion-risk assessment. Therefore field experiments offer the possibility to simulate such high-magnitude events and measurements can be accomplished more or less independent from natural events, enabling an assessment of erosion rates and the risks associated with them.

A final contribution of field experiment-based datasets is their comparability with each other, even when derived at sites that are located in a large distance, at least when collected with the same equipment and protocol. The long-term nature of field observation plots and the different research questions that often (with the exception of the USLE) motivate their dimensions, limit the comparison of results. Field experiments, on the other hand, can be carried out with a relative ease in the same way at a wide range of sites in a short period of time, which enables a comparison of a wide range of specific parameters with regard to particular research foci (e.g., Schmidt et al. 1997).

5.1.2.2 Field versus Laboratory Experiments

Laboratory experiments offer all the advantages of field experiments, but an even stronger level of control of environmental conditions. A further distinct advantage of laboratory studies is the use of rainfall simulators that generate rainfall, which is often more precise and has a greater resemblance to natural rainfall, specifically drop size and kinetic energy, than field simulators. For field equipment, portability, robustness, and quality of the rainfall always have to be balanced, often requiring compromises compared to natural rainfall. However, studying soil in the laboratory also imposes size restrictions and/or a certain level of disturbance to the soil between field sampling and lab analysis. After a certain storage time between collection in the field and experiment, the soil is prepared and placed in a soil flume prior to conducting the actual experiment. Depending on the distance of the sampling site to the laboratory, its disturbance during transport, and the time of storage, which may range from a few days to a several months, the soil has undergone a deterioration or degradation process in terms of its microbiological, as well as its structural and chemical condition (Shukla, 2014). These changes require consideration and careful interpretation of the achieved results when compared to a field context because the soil material used in the laboratory may no longer be referred to as a developed soil, but just a disturbed substrate (Kuhn et al. 2011; Hu et al., 2013). The major advantage of field experiments is therefore the undisturbed condition of the studied site, in particular the soil being tested. In addition, the general setting of the soil, for example, with regard to hill slope hydrology, remains unchanged. Consequently, to maintain this advantage, field experiments should disturb the soil as little as possible in order to preserve its natural condition. A disadvantage arising from the preservation of undisturbed conditions in the field is their variability, which requires a careful and detailed documentation. For example, infiltration differences observed during a laboratory test can be attributed entirely to the properties of the tested soil pack. In the field, the properties and arrangement of subsurface soil horizons may differ, even with similar surface soil conditions. As a consequence, a wider range of site conditions than in the lab therefore affects differences in runoff. Testing undisturbed soil samples in the laboratory would appear to offer the advantages of lab and field experiments, but the potential for this combination is limited. Only a few laboratory-based investigations have attempted the difficult task of taking intact soil monoliths into their laboratory experiments (e.g., Kinnell and McLachlan, 1989; Persson and Bergström, 1991; Laegdsmand et al., 1999; Belford, 2006). The difficulties associated with extracting an intact monolith from the surrounding soil matrix usually limit their size to a surface area of approximately 1–2 m^2.

5.1.3 GENERAL CONSIDERATIONS FOR THE PREPARATION OF FIELD EXPERIMENTS

The short comparison of field and lab experiments as well as the strengths and weaknesses of field monitoring illustrates that either approach has advantages and disadvantages. Depending on the research question, the appropriate method or combination of methods has to be identified. The following section offers some general and practical thoughts to ensure the collection of high-quality data when planning a field experiment.

The first and most important part of any study involves formulating a clearly defined research question. On the basis of this question, specific objectives can be derived and further decisions regarding the methods can be made. One of the key factors is the location of the study. The location of a study site is either the result of a site-specific research question ("Soil erosion in town A") or sites offering more general differences that are supposed to be studied, for example, along a gradient of soil properties, or, when testing the effect of texture on erosion. In either case, background information about candidate sites needs to be studied and well-documented. The latter is of particular importance when carrying out a study at geographically different sites to enable the correct interpretation of the obtained results. The identified relevant soil and environmental conditions at the selected site(s) should be measured and/or documented at the time of the experiments to place the results into an appropriate geographic context and to compare them to other results, both own and such reported in the literature.

A second complexity to consider is the actual experiments that need to be carried out at the selected site to achieve the objectives of the study. Once these key factors concerning the research question have been checked and one or more suitable locations have been identified, the logistical and other physical constraints associated with the selection need to be assessed. In principle, the objectives of a given experiment dictate the selection of the appropriate type of experiment suited best in to answer the research question. However, after the site selection, more specific questions may arise depending on the conditions found there. Transport infrastructure and the capacity to set up equipment may pose constraints on the experimental device that can be used. These compromises required to do measurements at a suitable site should constantly be checked against the aims and objectives of the study. Once a rainfall simulator has been selected, it is very important to thoroughly reflect on whether the handling characteristics of the device are good enough and understood well enough to generate reliable data at the selected site. If the answer is no, it would be very advisable to test the simulator again and, if necessary, improve the reproducibility and accuracy of the simulation conditions before moving to the field site. Alternatively, the methodological approach has to be changed to avoid reducing an experiment to a point outside the precision–realism–generalism triangle. If the potential study sites still do not offer a feasible opportunity to conduct the fieldwork, new sites have to be selected or the time, effort, or scope of the fieldwork needs to be modified.

5.1.4 CASE STUDIES: TESTING NEW AND MEASURING KNOWN PROCESSES

The more specific context for conducting good experiments beyond the general considerations mentioned above is illustrated in the following two sections by case studies. The first illustrates attempts to measure the relevance of the so far, little-studied interaction between wind and rainfall (wind-driven rain) on soil erosion. The question was triggered by the recognition that the ignoring of wind-driven rain may potentially introduce much unexplained variability into erosion model results (Fister et al., 2011). The second question focused on the effects of small-scale topography on infiltration and runoff generation on subalpine meadows. Here processes were known, but the modification introduced by the systematic patterns of the topography was unclear. Both studies are explorative. They are not presented as all-encompassing scientific research, but written and structured in a way that will enable young scientists, or those planning to conduct field experiments for the first time, to follow the process of conducting them from the beginning, when a research question is conceived, to the end, when the research question has been answered, or at least results obtained during the field experiments have been analyzed as far as possible.

5.1.5 CASE STUDY 1: DOES THE INFLUENCE OF WIND ON RAINFALL ALTER EROSION RATES ON ARABLE LAND?

5.1.5.1 Background and Research Question

The influence of wind on falling raindrops and its potential to alter soil erosion rates was already proposed during the 1960s and 1970s (Umback and Lembke, 1966; Disrud et al., 1969; Mutchler and McGregor, 1979). Considering that laboratory investigations confirmed that wind modifies the characteristics of falling raindrops, especially their impact angles (De Lima, 1989; Sharon, 1980; Umback and Lembke, 1966), impact frequencies, drop velocities (Caldwell and Elliott, 1972; Pedersen and Hasholt, 1995; Fister et al., 2011, 2012), drop sizes (Umback and Lembke, 1966; Cornelis et al., 2004a; Fister et al., 2011, 2012) and hence, their kinetic energy (Umback and Lembke, 1966; Disrud and Krauss, 1971; Helming, 2001; Erpul et al., 2005), it seems plausible that erosion rates will also change. However, apart from a few flume experiments in a laboratory (Cornelis et al., 2004; Samray et al., 2011; Erpul et al., 2011) and a few field studies that observed that splash-drift results in higher erosion rates, the subject has so far not been studied under field conditions.

Laboratory experiments cannot reflect the complexity of naturally developed soils, especially surface properties such as roughness and processes such as infiltration. Since both are essential for the interaction between wind-driven rain and the eroding surface, a field investigation is necessary to test the potential effect of wind on the erosivity of falling raindrops. The objective of the field experiment presented further was therefore to measure runoff and sediment yield during erosion events on agricultural fields, both with and without the influence of wind on rainfall.

5.1.5.2 Wind-Driven Rain Simulation

Previous studies based on field monitoring struggled to identify wind effects on erosion because linking wind direction and velocity measurements to rainfall

FIGURE 5.1.3 Setup of the push-type portable wind and rainfall simulator (length of the flow conditioning section is 5 m and 4 m for the wind tunnel; diameter of the tunnel is 0.8 by 0.8 m)

kinetic energy proved to be difficult (De Lima, 1989; van Dijk et al., 1996). Consequently, field experiments using a portable wind and rainfall simulator appeared as the best approach to study wind-driven rain, because wind velocity and direction can be controlled.

Fig. 5.1.3 shows the simulator that was built to simulate wind-driven rain in the field. A detailed description of the design and rainfall characteristics of the simulator used for this study is given in Fister et al. (2011, 2012). The process of instrument development is not subject of this study, but it is paramount to point out that whenever new field simulation equipment is developed, the characteristics of the equipment must be assessed in order to determine the driving processes they generate. This must be done systematically and under controlled conditions before conducting field experiments (Fister et al., 2011; Iserloh et al., 2012; Gartmann et al., 2011). Often this phase of the research process is judged as unimportant because it is laborious, time consuming, and with limited scientific gain. In addition, the lack of standardized test methods and standardized published results requires decisions on the extent to which the device should be calibrated to achieve the aims of the study (Grismer, 2012; Iserloh et al., 2013; Ries et al., 2009, 2013, 2014).

Rainfall simulation should replicate the natural processes as closely as possible. However, the degree to which this conformity realistically can, or must, attain depends on the research question, the available resources, time, and equipment at hand. For example, the quality of detailed process simulations depends heavily on the accuracy and known conditions of rainfall, whereas relative studies to determine erosion on different soils are strongly affected by the reproducibility of a standardized rainfall. Therefore, it is necessary to specify the boundary conditions for simulated rainfall conditions with regards to the aims and objectives of the research question. This, again, determines which type of simulator should be used. The wind-driven rain simulator employed in this study uses a pressure nozzle system for rainfall generation. A major argument for using pressure nozzles is their ability to produce rainfall that is generally considered to be more representative of natural raindrop characteristics, especially size distribution and kinetic energy if used at low drop fall heights common for field simulators (below 3 m). Furthermore, nozzles are easier to operate in a field environment compared to drip irrigators (e.g., with hypodermic needles). In addition, nozzles can also be rapidly changed and allow the application of different rainfalls. A wide range of nozzles is commercially available, each capable of producing vastly different rainfall. Although this creates many opportunities for experiments, selecting appropriate nozzles can be laborious. A detailed study of the nozzles and the rainfall they generate based on results reported in the literature is strongly advised before an experiment. To confound the situation slightly further, it is important to understand that nozzles are produced commercially for the uniform delivery of liquids, such as paint or herbicides, usually at extremely high pressures to generate a mist of liquid. When performing rainfall simulations, the nozzles are operated at much lower pressures, outside their tested design criteria, to generate large drops. The reduction of pressure requires a thorough documentation of the rainfall intensity, drop-size distribution, and kinetic energy generated at a given pressure and water temperature. Before using pressure nozzles in the field, the setup of pumps and water supply should also be tested.

The final issue that had to be addressed before this study moved to the field concerned the specific rainfall intensities and durations as well as the wind velocities that were applied. With regards to the objectives and the practicalities of the experiments, no wind and only one other wind velocity, applied for the same rainfall intensity, were sufficient to test the hypothesis. Consequently, rainfall was applied for 30 min at an intensity of ca. 95 mm/h, with a wind velocity of 7.5–8 m/s (27–28.8 km/h). This wind speed is rather low in comparison to natural storm events at the study site, but was ultimately dictated by the maximum power of the fan generating the wind. Since the focus of the experimental work was to differentiate between rainfall not affected by wind and wind-driven rain, this methodological drawback was not considered to be fundamentally important to the outcome of the investigation. By contrast, the chosen rainfall intensity was high for the study area. This error was also considered acceptable in the context of the study because the high intensity compensates for the relatively low kinetic energy of the raindrops. In addition, each rainfall simulation could be limited to 30 min, as the results from a number of pretests indicated that steady-state runoff conditions would easily be attained within this timeframe. Continuing each experiment for longer than the period stated above would be superfluous, since this would reduce the total number of experiments because water consumption would be increased accordingly.

5.1.5.3 Study Site and Date of Experiments

The study is situated in central Jutland in Denmark. It was chosen for two reasons: first, water erosion is prevalent in the area (Boardman and Poesen, 2006) and wind velocities are frequently high; second, the soil is sandy and shows little aggregation, which reduces the likelihood of a complex erosional response due to aggregate breakdown and the formation of a crust (Kuhn and Armstrong, 2012). The experiments were carried out in late summer/early autumn 2011 and 2012. Choosing two consecutive years for this project for fieldwork ensured that some interannual variability of soil conditions was captured by the experiments.

5.1.5.4 Plot Dimensions and Scale Considerations

The dominant problem associated with studying and measuring soil erosion is the fact that processes are highly dynamic and change over relatively small spatial scales (Lal, 1990; Parsons et al., 2006). For instance, the predominant erosion processes on the microsized plot presented in this case study are splash and sheet (Greenwood et al., 2013). With increasing size of the studied area, rills may form and develop into the dominant erosion process (Greenwood, 2012). On even larger spatial scales, gully erosion may become dominant over rill erosion. Ideally, the effect of wind on erosion would therefore be measured at multiple scales in order to capture the influence on all processes. However, the dimensions of the simulator were limited and thus, the size of the studied plot was only 2.2 m^2, for several reasons:

1. Size and weight: The simulator had to be small and light enough to be carried by two or three people onto a freshly tilled and sloping field;
2. Operation: Only a limited number of people are available to set up and operate the simulator, reducing the size of the plot that can be monitored; and
3. Uniformity of wind-driven rain: Wind velocity declines along the wind tunnel, which limits the length of the plot to a section with relatively uniform wind conditions.

These limitations focused the experiments on sheet wash processes. Since the kinetic energy of falling raindrops is the most relevant for the driving forces of erosion at this scale (Morgan, 1995), the scale limitation was considered to be acceptable for a first test of the hypothesized effect of wind-driven rain on erosion.

5.1.5.5 Test Sequence and of Sediment Collection

Three consecutive, 30-min rainfalls, interrupted by a 30-min break, were applied to the erosion plots. The first rainfall involved no wind and aimed at priming the plot by increasing soil moisture and leveling small roughness elements after tillage operations (Kuhn and Bryan, 2004). The second rainfall generated the no-wind erosion control, while the third one, with wind, generated soil loss by wind-driven rain. The 30-min breaks allowed the soil macro-pores to drain, generating similar antecedent soil moisture conditions, which allow erosion rates obtained for windless rainfall to be compared to wind-driven rain. This pattern is limited to soils without significant cumulative crusting and selective erosion. It would not work, for example, on well-aggregated loamy or clayey soils, which illustrates the relative position of each experiment in the precision-realism-generalism triangle. The three rainfalls formed a test sequence, after which the simulator was moved to another test plot. The combination of the three consecutive test runs is called a test sequence. Runoff and sediment were collected at a bottom weir in a 0.5 L bottle and decanted into a 10 L bucket for intervals lasting 90 s. This provided a high resolution for runoff (numbers of bottles collected in 90 seconds) and a soil erosion value for each 90-s interval. The runoff and sediment samples were transferred to a laboratory where the sediment was allowed to settle for 24 h. Afterward, the supernatant was carefully decanted from each bucket and the remaining sediment was dried and weighed.

5.1.5.6 Supplemental Parameters

A number of additional parameters were measured to support the analysis of the erosion measurements and to provide information that enables a comparison to other results, both obtained with this simulator as well as other studies. These properties include, soil profile development, soil surface roughness, bulk density, carbon content, soil-moisture content, aggregate size and stability, stone coverage, as well as cropping and tillage history, litter and vegetation density, and coverage (Ries et al., 2014). These soil and surface properties represent an essential minimum amount of information required in the context of a range of similar wind-driven rain experiments carried out through Europe. Therefore they are by no means exhaustive and should be amended according the specific research questions.

5.1.5.7 Overview of Results

The water erosion rates observed during six simulated sequences that involved 18–30-min rainfalls during the Jutland field campaign of 2011 are presented in Fig. 5.1.4. In five out of six cases, erosion during the third run, which includes the influence of wind on raindrops, was significantly increased compared to the second rainfall without wind. These results confirm the hypothesis that wind-driven rain potentially produces an increase in soil erosion rates. Although generally convincing, test sequence number three (FOU 3) shows that other factors are able to reduce erosion almost entirely. Given that the erosion driving rainfall was almost equal during each test, the low soil loss observed during FOU 3 indicates that soil resistance to erosion has a strong control over wind-driven rain on the Jutland sands. This observation is supported by the wide range of sediment yield values, ranging from less than 10 g/m^2 to over 700 g/m^2, during the experiments. They are indicative of high spatial variability of soil and surface properties

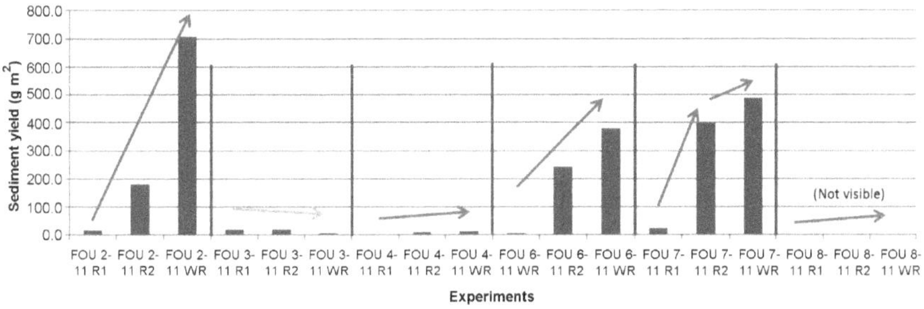

FIGURE 5.1.4 Sediment yield from wind-driven rainfall experiments in Jutland, Denmark (FOU, abbreviation for the test site near Foulum; 11, year of experiment; R1-R2-WR, one set of consecutive test runs on one plot; R1, first rainfall on dry soil; R2, second rainfall on wet soil; WR, wind-driven rainfall on wet soil. Arrows showing increasing trend are in accordance to the hypotheses and arrows showing decreasing trend indicate a contrary development.)

between the test plots (Fister et al., 2013). The high variability of results caused by the soil conditions presents a strong contrast to the controlled rainfall applied to the plots. This emphasizes the need for careful site selection, a thorough documentation of soil and site conditions as well as a sufficient number of experiments to achieve statistically significant and spatially representative results.

5.1.5.8 Discussion and Conclusions

The results of the Jutland field campaign illustrate that the field experiments used to test the wind-driven rain hypothesis worked very well, despite the limitations of the setup, for example, lack of natural rainfall and small plot size. Still, the obtained data must be interpreted and extrapolated with caution. Despite the fact that plots are located next to each other, the high variability of soil conditions generates such a wide range of erosional responses that even averaging the obtained results will not produce a good mean value for wind-driven rain erosion effects on the tested soil. Although this may appear disappointing and to limit the result of the Jutland campaign to a largely qualitative result, it actually illustrates the value of field experiments very well. In the laboratory, with disturbed, but much more uniform soil conditions, the results would almost certainly show much lower variability. This supposedly better result would, however, give a false impression of the relevance of the tested phenomenon, in this case wind-driven rain, in the natural environment where more factors interact with erosion processes. The variability also shows that the next step to move toward a quantitative assessment of the relevance of wind-driven rain has to include the spatial pattern of soil resistance to erosion. A false impression of a fairly uniform effect from a laboratory-based study would have carried the risk of masking this factor. A similar argument can be made when comparing the results to long-term field observations. The latter do provide a much wider range of rainfall and soil conditions, but these are often difficult to entangle into simple cause-and-effect chains. In the case of wind-driven rain, for example, it is highly unlikely that two rainfalls, differing just in wind speed, would fall onto soil in similar conditions, that is, vegetation cover, state of tillage, preceding erosion history and soil moisture. The perceived drawbacks of field experiments, such as the remaining variability of the studied surface and the limited scope of the simulated rainfall compared to the range of naturally occurring conditions, are thus actually strengths, because the relevance of one factor can be studied in a systematic manner by applying controlled rainfalls onto soil surfaces in natural conditions.

5.1.6 CASE STUDY 2: RUNOFF AND INFILTRATION EXPERIMENTS ON TERRACETTES USING SIMULATED RAINFALL

For this particular example of a field-based experiment, a series of simulated rainfall events were conducted in order to test the hypothesis that terracettes on subalpine or steep-sloping environments influence surface hydrology during erosion events.

5.1.6.1 Introduction

Before entering into the technicalities of the experiment, it is firstly prudent to provide the reader with a little background information about terracettes, and of the overall motivation for undertaking this series of experiments. In countries dominated by mountainous environments, subalpine and steeply-sloping terrain provides a vital source of fodder for grazing livestock. In addition, however, terrain such as this also has the potential to act as noncomplex flow-routes, with the capability of linking high altitude grazing areas to meadow and rangelands located at lower elevations (Kuhn and Zhu, 2008; Michna et al., 2013; Hoffman et al., 2014). Given the often acute gradients encountered within terrain of this nature (Konz et al., 2012), their presence represents a convenient and potentially highly efficient runoff conveyance route, with the capacity to induce instability into such environments (Schaub et al., 2010) by facilitating the rapid and largely unimpeded downslope transfer of runoff and associated substances during high-intensity rainfall events. However, many such slopes often support a series of small uniform steps known as terracettes. With regard to their characteristics and appearance, terracettes look like a wide staircase with distinct runners and risers (Fig. 5.1.5), with individual steps usually running parallel across slope and spaced more or less at equidistant intervals from each other. Although much has been written about their origin and the processes that are thought to have led to their formation and development (e.g., Ødum, 1922; Rahm, 1962; Buckhouse and Krueger, 1981; Auzet and Ambrose, 1996; Henck et al., 2010), two theories usually prevail. The first, concerning their formation, revolves around the fact that the upper layer of soil gradually moves downslope under the influence of gravity by soil creep (Carson and Kirkby, 1972; Anderson and Cox, 1978; White, 2000; Heimsath et al., 2002). The gravitational influence is thought to be further facilitated by repeated wetting and drying cycles (Vincent and Clarke, 1980; Auzet and Ambrose, 1996), and possibly exacerbated during seasonal thaw (Jaesche et al., 2002), both factors of which are believed to be responsible

FIGURE 5.1.5 Terracettes formed across acute and often noncomplex slopes. Their juxtaposition against the natural downslope flow-path may provide them with the ability to impede or even capture surface runoff during erosion events; thereby acting as preferential flow-paths or even temporary storage areas.

for the generally uniform downslope displacement of surface soil and to the subsequent formation of a series of almost uniform individual steps, or terracettes. Although terracettes may have initially been formed by a combination of hillslope processes and gravity, many people have argued, however, that these systems are accentuated and become more pronounced by the trampling action of livestock, as they regularly use them as convenient pathways to gain access across-slope to different areas of grazing (Buckhouse and Krueger, 1981).

5.1.6.2 Expectations of a Field Experiment

For this particular example of a field-based experiment, a lack of supporting literature about the role of terracettes, certainly in the context of their influence on surface hydrology, meant that the hypothesis was theory-based and largely conceptual. Despite this apparent limitation, however, it was deemed possible to at least obtain proof of concept of the role of terracettes on surface hydrology by undertaking an *in situ*, or field experiment. The investigation was thus driven by conceiving a succinct research question, namely: do terracettes potentially influence the surface hydrology of steep-sloping environments? The key objective was to measure infiltration and soil moisture change using a rainfall simulator.

Time and a certain amount of prior knowledge about geomorphology, surface hydrology, and soil physics are advantageous as they provide the investigative team with an ability to predetermine which particular analyses are most likely to yield relevant information needed to answer to the research question (see Greenwood et al., 2014). Existing knowledge of the first two disciplines, combined with selected information on the physical properties of the soil, allowed us to make an educated guess of the most relevant analytical methods that could be applied in this experimental scenario. These included *in situ* soil bulk density and soil-pore measurements (Shukla, 2014). Both tests, although conducted retrospectively in a laboratory, are fundamental to investigations of this nature, inasmuch as they both provide crucial information on the density and porosity of a given soil, which collectively give an indication of how rapidly water will percolate down or through the vadose zone, or is already saturated, over the soil surface (Morgan, 1995). Results published by Greenwood et al. (2014) indicated a systematic difference in bulk density between runners and risers, which would point to reduced infiltration on the former.

An arguably most fundamental analytical method to check the effect of bulk density on infiltration is to calculate surface runoff coefficients based on rainfall simulation tests. These measurements integrate a multitude of variables that potentially control the surface hydrology. As such, they include fundamental parameters such as water infiltration, retention, and interception (Ries et al., 2013). This particular analytical measure is considered to be invaluable in aiding researchers in geomorphology or hydrology, as it can provide them with a means of precisely quantifying the proportion of water leaving the study area as surface runoff relative to the proportion of cumulative water initially applied as rainfall. Finally, in-situ volumetric soil moisture values can complement the infiltration data, principally because the technique is rapid and data generation is immediate. Consequently, taking measurements before and after rainfall application provided a very preliminary yet insightful indication of net changes in soil moisture content (Imeson, 2011).

5.1.6.3 Key Resources and Logistical Constraints

Once a suite of relevant analytical methods have been selected, it is then possible to identify the necessary resources needed to undertake those analyses, as well as gauge the varying logistical constraints that dictate where the experiment can actually most effectively be undertaken and rendered operational. The main resources and logistical constraints under consideration for this particular experiment included:

1. Labor-force required (i.e., how many people are needed to undertake the experiment?)
2. Time availability (how much time can workforce devote to the experiment?)
3. Local knowledge (i.e., how do you identify a potentially suitable experimental site?)
4. Mode of transport and distance to the site (i.e., what are the travel arrangements and distance to the site)
5. Equipment (exactly what is required and how much?)
6. Site accessibility (i.e., can the equipment be transported and unloaded close to the site, or does it have to be carried over long distances?)

Although the above list is not supposed to be comprehensive, devoting sufficient time to consider those and other necessary factors prior to undertaking experimental work is paramount. The main reason for this is that these practical yet fundamental issues ultimately dictate whether the work can physically be achieved or not. After thoroughly considering all of the above factors, it was decided that 2 days of experimental work, plus one additional day of organizational time, would be sufficient to generate a large enough data-set, whereupon the hypothesis could be sufficiently tested. Within this timeframe, it was felt that eight separate simulated rainfall events could realistically be undertaken at eight different locations within the same terracette system over the 2-day period. As such, 3 days in total were considered to be sufficient time to select and initially load the equipment into a trailer, transport it to the site, unload and initially configure the equipment, conduct each rainfall simulation and take measurements simultaneously and then resite the equipment at the next location, before finally dismantling and reloading it all back into the trailer and returning from the field site. Given the steep and difficult terrain, individual sites were selected based on how developed each terracette section was, as well as its position on the slope relative to the source of water, which needed to be pumped some distance upslope in order to reach the rainfall simulator.

5.1.6.4 Study Site

The chosen study site on which the investigations was undertaken is located in Canton Jura, approximately 10km northwest of the town of Delémont, in northwest Switzerland (47°25.00'N; 7° 18.30'W). A more detailed description of the study area can be found in Greenwood et al. (2014) and in Greenwood and Kuhn (2014). In brief, however, the geology of upland areas within the Jura region is composed of Jurassic limestone. Elevations range from ca. 500–1,300m asl and the study site is at ca. 710m asl. Annual precipitation is around 1,100mm, with maximum monthly rainfall, at ca. 91mm, usually falling between May to September. The area can receive as much as 80% of the maximum monthly total during November to March period, but much of this frequently falls as snow. Sub-zero temperatures usually commence from around December onward and can persist until early March. This means that snowfall frequently accumulates, particularly on east and north-facing slopes and in shady valley bottoms, where it can remain until mid-April.

The well-developed terracette system within the chosen study area (Fig. 5.1.5) is located on a west-facing slope with gradients ranging from ca. 25–40° (46–84%). The acute nature of the terrain precludes any form of intensive agriculture taking place, but the site is extensively grazed by cattle from around April to September. The depth of the rendzina soil on the hillslopes is limited to around 15–20cm, and with a pH around 8–8.5, is highly calcareous and hence, relatively infertile. The west-facing exposure of sites such as this means that soils are regularly subjected to aspect-related physical influences that include desiccation, particularly during the warmer months (Rorison, 1990). In addition to the above, bare, partially vegetated, or recently cultivated soils on valley bottoms are also prone to surface crusting, even after relatively light rainfall. Vegetation associated with pastoral areas within the Jura region consist predominantly of a relatively dense cover of fine-leaved grasses, and these are frequently interspersed with a wide variety of flowering alpine and sub-alpine plant species. Despite the relative infertility and limited nutrient availability associated with these soils, areas within the Jura generally renowned for providing rich, high-quality grazing (Rodwell, 1998).

5.1.6.5 Rainfall Simulation Experiments

A detailed description of the rainfall simulator used in this study is given in Hikel et al. (2013). Four rainfall simulations were conducted per day and each plot was subjected to rainfall equivalent to 100 mm h. Although this represents an exceptionally high intensity for this particular region, it was felt that this was justified as the volume of water applied represented the best chance of creating runoff within a relatively short period of time. Secondly, each rainfall event could realistically be limited to 30 min duration. This also meant that overall water consumption, and hence, the logistical demands associated with importing larger volumes of water to a remote area, was reduced. These logistical constraints to the rainfall appeared acceptable for this initial phase of the terracette research, which basically just aimed at testing their potential role in surface hydrology. Artificial rain was created by pumping water, via a portable generator and electric pump, from 150 L barrels to a lightweight single-spray rainfall simulator. The rainfall simulator is, in turn, supported by a strong, yet lightweight aluminum frame that, despite the steep and narrow nature of the terrain, can be adjusted to provide a level and stable frame. A selection of the field equipment and part of a bounded plot can be seen in Fig. 5.1.6. The plots ranged in size between 0.5 and 1 m^2, with shapes varying from square to triangular. This variation of size and shape was required to encompass the natural topography of the terracettes into a micro-catchment that generated runoff and routed it to the lowest point, where a weir could be installed. The differences were considered acceptable due to the small size of the plots and the scale-independent measurement of soil moisture across the plot surfaces after the experiment.

5.1.6.6 Antecedent Soil Moisture Conditions

Antecedent soil moisture, or the amount of water held within soil pores at a given location prior to a given point in time, such as a high-intensity rainfall event, can represent a major control on the timing and rate of runoff generation (Anderson and Anderson, 2011). With this in mind, a climate station was installed close to the experimental site in order to record some of the more obvious weather parameters, such as precipitation and air temperature, but also the less commonly used, soil matric potential. Soil matric potential measures how strongly water is adsorbed to soil particles. As such, it can represent a proxy

FIGURE 5.1.6 A selection of some of the equipment used when undertaking field-based rainfall simulation experiments.

measure of the moisture content within a particular soil matrix. Precipitation ultimately controls matric potential and matric potential is, in turn, predominantly controlled by air temperature (Shukla, 2014). Collectively, therefore, these three parameters were deemed to provide a holistic indication of how quickly, if at all, selected surface runoff might be generated.

When calculating runoff coefficients, the target area over which rainfall is applied (hereafter referred to as a plot) must be enclosed, or bounded, as this makes it possible to calculate the equivalent cumulative rainfall (mm/h) and hence, the cumulative volume of surface runoff (mL) leaving the area (m^2) over a given time period (minutes). By collecting total runoff, which in this instance can be defined as a steady and continuous flow of water leaving the plot outlet, the proportion of water leaving the plot as surface runoff can be calculated and reported as a percentage of the total rainfall applied. For these experiments, runoff collection was facilitated by embedding a metal channel flush with the soil surface at the downslope end of each plot so that any runoff was readily funneled and collected in a series of containers, which are marked, for instance, in chronological order in which each have been filled. They should then be sealed with a watertight stopper, and the time in which runoff is collected recorded separately.

5.1.6.7 Surface Hydrology and Volumetric Moisture Content Related to Runoff Generation

The runoff coefficients observed during the tests conducted on the Jura terracettes ranged from 0.1 to 23.5% and showed no systematic change between plots with different proportions of runners and risers. Lower values indicate less surface runoff, whereas higher values indicate more runoff. This result is surprising because based on the increased bulk densities, a positive relationship between the proportion of plot area covered by a runner and runoff coefficient had been expected. As the infiltration capacity of a soil generally tends to control how easy runoff can or cannot be generated (Imeson, 2011; Ries et al., 2013; Shukla, 2014), the wide-ranging values provided a tentative indication that the ease with which runoff was generated is more variable than bulk densities suggest.

Volumetric soil-moisture conditions were measured at pairs of runners and risers within each plot before applying rainfall, and then remeasured approximately 30 s after rainfall was terminated. Moisture values from each runner and riser were averaged for each plot and post-rainfall values were subtracted from prerainfall values in order to provide an indication of the net change in soil-moisture content. The underlying rationale for this was that a net increase in volumetric soil-moisture content after applying rainfall would indicate the ingression of water into soil pore spaces, and hence indicate infiltration. In contrast, no net soil-moisture gain indicates that those plots shed water more readily, due to their lower infiltration. The results of these measurements also varied widely, but generally correspond to the observed runoff coefficients (Greenwood et al., 2014).

5.1.6.8 Potential Relevance of Observed Infiltration on Terracettes for Hillslope Hydrology

Even from the preliminary information presented in this example of a field experiment, it is possible to draw some very provisional conclusions that first help to establish proof of concept, and second, contribute to answering the original research question. Importantly, and as is commonly the case, these findings also potentially raise additional questions to which follow-up experiments need to be specifically tailored in order that these can be fully addressed. In terms of what can be concluded so far, however, is that higher runoff-coefficient values suggest that certain areas within this particular terracette system have reduced infiltration rates. The interpretation from this is that those areas have probably experienced some degree of compaction. Whether this is by livestock or not remains out of the analytical scope of this current investigation, but this question could represent a focus for future work. Taking this primary conclusion a stage further, the ability of those areas to readily generate runoff suggests, by default, that these areas could intercept and convey seepage and surface runoff downslope and thus exert a notable influence on surface hydrology. In contrast, however, low runoff-coefficient values suggest that some areas within the same terracette system maintain high infiltration. The reason for this counterintuitive result is unknown, but one can speculate that for example individual macropores introduce a runoff sink (e.g., Munyankusi et al. 1995). Despite being subjected to high-intensity rainfall for a prolonged period of time, these areas neither generate surface runoff nor convey surface runoff downslope. Interestingly, as this break in the natural downslope conveyance process can be construed as a "hiatus," or disconnection in the natural downslope conveyance process, this finding has implications in terms of determining the fate of any runoff-associated material and the stability of the slope experiencing increased infiltration and thus water saturation (Caviezel et al. 2014).

5.1.6.9 Field Experiments: Dealing with Unexpected Findings, or a Case of Serendipity?

Despite being subjected to high-intensity rainfall for a prolonged period of time, low runoff at certain locations suggested that infiltration continued to exceed precipitation. It is entirely feasible, therefore, that areas such as those could act as infiltration and even sediment-depositional sites during a real runoff and erosion event. In the light of the compaction of the runners, this finding was unexpected. This result is surprising since it represents the almost exact antithesis of what was originally anticipated. As such, it could thus be argued that it represents a case of serendipity, or in other words, an informative, even beneficial, yet unexpected discovery was made purely by accident. Unexpected findings such as the one described above open up exciting opportunities and provide adequate justification for undertaking further work. However, any new experiments would need to be tailored to take account of the possibility that terracettes may actually reduce runoff and in doing so, act as temporary stores of runoff-associated substances. Obviously any new experiments would need to assess this likelihood, perhaps by including relevant analyses capable of elucidating this possibility. Such analyses could focus on, for instance, determining the relative runoff-trapping efficiency of terracettes in depositional areas, or even seek ways in which the residence time of stored sediment on depositional areas could possibly be measured (Greenwood et al., 2012). What can be stated from the results generated by this investigation so far, however, is that they represent a basis upon which future experimental work incorporating additional or different analytical techniques could be undertaken. At the same time, repeating the experiment, but this time with the benefit of hindsight, would provide an opportunity to evaluate what did and what did not work, and thus incorporate any small changes or refinements into those previously employed methods. From the perspective of this particular example of a field experiment, any changes and amendments to future experiments would capitalize on prior knowledge gained from the first suite of tests and likely focus on creating greater uniformity between erosion-plot characteristics. Such an approach would allow runoff data to be more readily compared between plots.

5.1.7 CONCLUSIONS: FIELD EXPERIMENTS AS A TOOL TO EXPLORE SURFACE PROCESSES

In the first part of this chapter of the volume, different types of experiments were presented that are often used in environmental sciences depending on the research question and the degree of knowledge about this question. The two case studies that were presented above fall largely in the first category of experiments aiming at gaining a process understanding. Reflecting on the results of two case studies, it also becomes clear that scientists sometimes have to deviate from the ideal experimental configuration in order to attain any objectives of their research. Basic knowledge and literature offer straightforward criteria to design and ideal experiment, use the ideal type of simulator and draw a list of all other relevant environmental and technical factors that should be met to answer a research question. The difficult task for an experimental scientist is to adjust the "ideal" configuration to the best adaptation possible in field conditions. To do so, they have to decide where and when it is acceptable to make compromises between scientific relevance, portability, and functionality of the experiment, but still stay inside the precision–realism–generalism triangle.

It can also be seen that despite the best possible preparation the obtained data show unexpected results. Possible reasons include natural spatial variability, unknown artifacts in the simulator, errors in the setup or with the supplemental measurements that are not sufficiently comprehensive to explain the results of the experiment. If despite all efforts the data are incomprehensible, three options for remedial action can be considered:

1. If the results are not accepted as valid, in spite of everything apparently working well, then the theory about the process has to be re-evaluated and a new plan must be formulated in order to find a solution. The results of the terracette experiments fall partly into this category.
2. If something is thought to be erroneous, the instrumentation has to be checked for any problems in order to understand the equipment and its interaction with the studied environment more comprehensively.
3. Research questions may have to be adjusted to the answer delivered by the data. The high variability of the erosion data generated in the Jutland campaign fall to a certain degree into this category: the potential influence of wind-driven rain is visible, but the actual relevance remains unclear.

In conclusion, field experiments represent an important tool that enables soil erosion researchers to investigate processes through measuring rates of processes under controlled driving conditions, but variable natural surface properties. This basic principle enables a detailed insight into the functioning of specific process domains, that is, plots, but is also limited, mostly in spatial scale and the number of scenarios of interaction between driving rainfall and runoff and the surface they act upon. The interpretation of field experiments therefore requires a careful consideration of the context they were collected in.

REFERENCES

Anderson, E.W., Cox, N.J., 1978. A comparison of different instruments for measuring soil creep. Catena 5, 81–93.

Anderson, R.S., Anderson, S.P., 2011. Geomorphology: The Mechanics and Chemistry of Landscapes. Cambridge University Press, New York.

Auzet, A.V., Ambrose, B., 1996. Soil creep dynamics, soil moisture and temperature conditions on a forested slope in the granitic Vosges Mountains, France. Earth Surf. Proc. Land. 21, 531–542.

Belford, R.K., 2006. Collection and evaluation of large soil monoliths for soil and crop studies. J. Soil Sci. 30 (2), 363–373.

Benito, G., Diez-Herrero, A., De Villalta, M.F., 2003. Magnitude and frequency of flooding in the Tagus Basin (central Spain) over the last millennium. Clim. Change 58, 171–192.

Boardman, J., 2003. Soil erosion and flooding on the eastern South Downs, Southern England, 1976–2001. T. I. Brit. Geogr. 28 (2), 176–196.

Boardman, J., Poesen, J. (Eds.), 2006. Soil erosion in Europe. Wiley, Ghent, Belgium.

Bryan, R. (Ed.), 1990. Soil Erosion: Experiments and models. Catena Supplement 17. Catena Sciences Ltd, p. 208.

Buckhouse, J.C., Krueger, W.C., 1981. What caused those terracettes? Rangelands 3 (2), 72–73.

Caldwell, D.R., Elliott, W.P., 1972. The effect of rainfall on the wind in the surface layer. Bound.-Layer Meteor. 3, 146–151.

Carson, M.A., Kirkby, M.J., 1972. Hillslope Form and Process. Cambridge Geographical Studies No. 3. Cambridge University Press, Cambridge, UK.

Caviezel, C., Hunziker, M., Schaffner, M., Kuhn, N.J., 2014. Soil–vegetation interaction on slopes with bush encroachment in the central Alps – adapting slope stability measurements to shifting process domains. Earth Surf. Proc. Land. 39 (4), 509–521.

Cornelis, W.M., Erpul, G., Gabriels, D., 2004. The I.C.E. wind tunnel for wind and water interaction research. Visser, S., Cornelis, W.M. (Eds.), Wind and Rain Interactions in Erosion, 50. Tropical Resource Management Papers, Ghent, Belgium, pp. 197–224.

Cornelis, W.M., Oltenfreiter, G., Gabriels, D., Hartmann, R., 2004a. Splash-saltation of sand due to wind-driven rain: horizontal flux and sediment transport rate. Soil Sci. Soc. Am. J. 68 (1), 41–46.

Chaubey, I., Haan, C.T., Grunwald, S., Salisbury, J.M., 1999. Uncertainty in the model parameters due to spatial variability of rainfall. J. Hydrol. 220, 48–61.

De Lima, J.L.M.P., 1989. Raindrop splash anisotropy: slope, wind and overland flow velocity effects. Soil Technol. 2 (1), 71–78.

De Lima, J.L.M.P., van Dijk, P.M., Spaan, W.P., 1992. Splash-saltation transport under wind-driven rain. Soil Technol. 5, 151–166.

Dettinger, M., 2011. Climate change, atmospheric rivers, and floods in California – a multimodel analysis of storm frequency and magnitude changes. J. Am. Water Resour. Assoc. 47 (3), 514–523.

van Dijk, P.M., Stroosnijder, L., de Lima, J.L.M.P., 1996. The influence of rainfall on transport of beach sand by wind. Earth Surf. Proc. Land. 21, 341–352.

Disrud, L.A., Lyles, L., Skidmore, E.L., 1969. How wind affects the size and shape of raindrops. Agr. Eng. 50 (10), 617.

Disrud, L.A., Krauss, R.K., 1971. Examining the process of soil detachment from clods exposed to wind-driven simulated rainfall. Transactions of the ASAE 14, 90–92.

Erpul, G., Gabriels, D., Cornelis, W.M., Samray, H.N., Guzelordu, T., 2008. Sand detachment under rains with varying angle of incidence. Catena 72, 413–422.

Erpul, G., Norton, L.D., Gabriels, D., 2003. Sediment transport from interrill areas under wind-driven rain. J. Hydrol. 276, 184–197.

Erpul, G., Gabriels, D., Norton, D.L., 2005. Sand detachment by wind-driven raindrops. Earth Surf. Proc. Land. 30, 241–250.

Erpul, G., Gabriels, D., Norton, L.D., Flanagan, D., Huang, C., Visser, S.M., 2011. Interactions between raindrop impact and shallow interrill flow under wind-driven rain. In: Vermang, J., Gabriels, D., Cornelis, W.M., De Boever, M. (Eds.), Land Degradation Processes and Assessment. UNESCO Chair of Eremology, Ghent, Belgium, pp. 9–18.

Fister, W., Heckrath, G., Greenwood, P., Kuhn, N.J., 2013. Erodibility of biochar from a sandy soil in Denmark. In: Ammann, W.J. (Ed.), United Nations Convention to Combat Desertification (UNCCD). 2nd Scientific Conference Publication, Davos, Switzerland, pp. 1–5.

Fister, W., Iserloh, T., Ries, J.B., Schmidt, R.-G., 2011. Comparison of rainfall characteristics of a small portable rainfall simulator and a combined portable wind and rainfall simulator. Zeitschrift für Geomorphologie 55 (Suppl. 3), 109–126.

Fister, W., Iserloh, T., Ries, J.B., Schmidt, R.-G., 2012. A portable wind and rainfall simulator for in situ soil erosion measurements. Catena 91, 72–84.

Gartmann, A., Fister, W., Schwanghart, W., Müller, M.D., 2011. CFD modelling and validation of measured wind field data in a portable wind tunnel. Aeolian Res. 3, 315–325.

González-Hidlago, J.C., Pena-Monné, J.L., de Luis, M., 2007. A review of daily soil erosion in Western Mediterranean areas. Catena 71, 193–199.

Greenwood, P., 2012. Tracing fine-sediment using artificial radionuclides. British Society for Geomorphology (BSG), Geomorphological Techniques, Chapter 3, Section 5.2, Oxford, UK, pp. 1–10.

Greenwood, P., Fister, W., Kinnell, P.I.A., Rüegg, H.R., Kuhn, N.J., 2013. Developing and testing a precision erosion measurement facility for elucidating mobilization mechanisms in shallow-flow conditions. In: De Boever, M., Khlosi, M., Delbeque, N., De Pue, J., Ryken, N., Verdoodt, A., Cornelis, W.M., Gabriels, D. (Eds.), Desertification and Land Degradation: Processes and Mitigation, 4th United Nations Educational, Scientific & Cultural Organization (UNESCO) Conference on Desertification and Land Degradation. UNESCO Chair of Eremology, Gent University, Belgium, pp. 105–111.

Greenwood, P., Kuhn, N.J., 2014. Does the invasive plant, *Impatiens glandulifera*, promote soil erosion along the riparian zone? An investigation on a small watercourse in northwest Switzerland. J. Soils Sediments 14 (3), 637–650.

Greenwood, P. Kuonen, S. Fister, W. Kuhn, N.J., 2014. The possible influence of terracettes on the surface hydrology of steep-sloping and sub-alpine environments. Geographica Helvetica Special Edition, "Mapping, Measuring and Modeling in Geomorphology."

Grismer, M., 2012. Standards vary in studies using rainfall simulators to evaluate erosion. California Agr. 66 (3), 102–107.

Heimsath, A., Chappell, J., Spooner, N.A., Questiaux, D.G., 2002. Creeping soil. Geology 30, 111–114.

Helming, K., 2001. Wind speed effects on rain erosivity. In: Stott, D.E., Mohtar, R.H., Steinhardt, G.C. (Eds.), Sustaining the Global Farm. Selected papers from the 10th International Soil Conservation Organization Meeting, May 24–29, 1999, Purdue.

Henck, A., Taylor, J., Lu, H., Li, Y., Yang, Q., Grub, B., Breslow, S.J., Robbins, A., Elliott, A., Hinckley, T., Combs, J., Urgenson, L., Widder, S., Hu, X., Ma, Z., Yuan, Y., Jian, D., Liao, X., Tang, Y., 2010. Anthropogenic hillslope terraces and swidden agriculture in Jiuzhaigou National Park, northern Sichuan, China. Quat. Res. 73, 201–207.

Hikel, H., Yair, A., Schwanghart, W., Kuhn, N.J., 2013. Experimental investigation of soil ecohydrology on rocky desert slopes in the Negev Highlands, Israel. Zeitschrift für Geomorphologie 57 (1), 39–58.

Hoffman, U., Hoffman, T., Juraniski, G., Glatzel, S., Kuhn, N.J., 2014. Assessing the spatial variability of soil organic carbon stocks in an alpine setting (Grindelwald, Swiss Alps). Geoderma 232-234, 270–283.

Hu, Y., Fister, W., Kuhn, N., 2013. Temporal variation of enrichment ratio of SOC in eroded sediment from interrill erosion over prolonged rainfall events, in: EGU General Assembly Conference Abstracts, p. 9592.

Imeson, A., 2011. Desertification, Land Degradation and Sustainability. John Wiley and Sons Ltd, p. 344.

Iserloh, T., Fister, W., Marzen, M., Seeger, M., Kuhn, N.J., Ries, J.B., 2013. The role of wind-driven rain for soil erosion – an experimental approach. Zeitschrift für Geomorphologie 57 (1), 193–201.

Iserloh, T., Fister, W., Seeger, M., Willger, H., Ries, J.B., 2012. A small portable rainfall simulator for reproducible experiments on soil erosion. Soil Till. Res. 124, 131–137.

Iserloh, T., Ries, J.B., Arnaez, J., Boix Fayos, C., Butzen, V., Cerdá, A., Echeverría, M., Fernández-Gálvez, J., Fister, W., Gómez, J.A., Gómez-Macpherson, H., Lázaro, R., León, F.J., Martínez-Mena, M., Marzen, M., Martínez-Murillo, J.F., Ortigosa, L., Peters, P., Regüés, D., Ruiz-Sinoga, J.D., Seeger, M., Solé-Benet, A., Wengel, R., Wirtz, S., 2013. European small portable rainfall simulators: a comparison of rainfall characteristics. Catena 110, 100–112.

Jaesche, P., Huew, B., Stingl, H., Veit, H., 2002. Temporal variability of Alpine solifluction: a modelling approach. Geographica Helvetica 3, 157–169.

Kinnell, P.I.A., McLachlan, C., 1989. Shallow soil monoliths for laboratory studies on soil erosion from undisturbed soil surfaces. Aust. J. Soil Res. 27, 227–231.

Konz, N., Prasuhn, V., Alewell, C., 2012. On the measurement of alpine soil erosion. Catena 91, 63–71.

Kuhn, N.J., 2014. Experiments in Reduced Gravity Sediment Settling on Mars. Elsevier Science Ltd.

Kuhn, N.J., Armstrong, E.K., 2012. Erosion of organic matter from sandy soils: Solving the mass balance. Catena 98, 87–95.

Kuhn, N.J., Schütt, B., Baumhauer, R., 2011. Managing the impact of Climate Change on the Hydrology of the Gallocanta Basin, NE-Spain. J. Environ. Manage. 92, 275–283.

Kuhn, N.J., Zhu, H., 2008. GIS-based modeling of runoff source areas and pathways. Geographica Helvetica 63, 48–57.

Kuhn, N.J., Bryan, R.B., 2004. Drying, soil surface condition and interrill erosion on two Ontario soils. Catena 57 (2), 113–133.

Kuhn, N.J., Yair, A., Grubin, M.K., 2004. Spatial distribution of surface properties, runoff generation and landscape development in the Zin Valley Badlands, northern Negev, Israel. Earth Surface Processes and Landforms 29, 1417–1430.

Kuhn, N.J., Bryan, R.B., Navar, J., 2003. Seal formation and interrill erosion on a smectite-rich Kastanozem from NE-Mexico. Catena 52, 149–169.

Laegdsmand, M., Villholth, K.G., Ullum, M., Jensen, K.H., 1999. Process of colloid mobilization and transport in macroporous soil monoliths. Geoderma 93, 33–59.

Laflen, J.M., Moldenhauer, W.C., 2003. Pioneering Soil Erosion Prediction: The USLE Story. Special Publication No. 1. World Association of Soil and Water Conservation (WASWC), Bejing, China.

Lal, R., 1990. Soil Erosion in the Tropics: Principles and Management. McGraw-Hill, New York.

Lopes, V.L., 1996. On the effect of uncertainty in spatial distribution of rainfall on catchment modelling. Catena 28, 107–119.

Meusburger, K., Steel, A., Panagos, P., Montanarella, L., Alewell, C., 2012. Spatial and temporal variability of rainfall erosivity factor for Switzerland. Hydrology and Earth System Sciences, 10.5194/hess-16-1-2012.

Michna, P., Eugster, W., Hiller, R.V., Zeeman, M.J., Wanner, H., 2013. Topoclimatological case-study of Alpine pastures near the Albula Pass in the eastern Alps. Geographica Helvetica 68, 249–263.

Molnar, P., 2001. Climate change, flooding in arid environments, and erosion rates. Geology 29, 1071–1074.

Morgan, R.P.C., 1995. Soil Erosion and Conservation. Longman Scientific & Technical, New York.

Munyankusi, E., Gupta, S.C., Moncrief, J.F., Berry, E.C., 1994. Earthworm macropores and preferential transport in a long-term manure applied typic Hapludalf. J. Environ. Qual. 23, 773–784.

Mutchler, C.K., McGregor, K.C., 1979. Geographical differences in rainfall. Proceedings of the rainfall simulator workshop, March 7–9, 1979, Tucson (Arizona).

Nearing, M.A., Govers, G., Norton, L.D., 1999. Variability in soil erosion data from replicated plots. Soil Sci. Soc. Am. J. 63, 1829–1835.

Ødum, H., 1922. Om "Faarestiernes" Natur ("On the nature of the so-called sheep tracks"). Dansk. Geol. Fören Medd. 6, 1–29.

Parsons, A.J., Wainwright, J., Brazier, R.E., Powell, D.M., 2006. In sediment delivery a fallacy? Earth Surface Process and Landforms 31 (10), 1325–1328.

Pedersen, H.S., Hasholt, B., 1995. Influence of wind speed on rainsplash erosion. Catena 24, 39–54.

Persson, L., Bergström, L., 1991. Drilling method for collection of undisturbed soil monoliths. Soil Sci. Soc. Am. J. 55 (1), 285–287.

Rahm, D.A., 1962. The terracette problem. Northwest Science 36 (3), 65–80.

Ries, J.B., Iserloh, T., Seeger, M., Gabriels, D., 2013. Rainfall simulations – constraints and challenges for a future use in soil erosion research. Zeitshrift für Geomorphologie 57 (1), 1–10.

Ries, J.B., Marzen, M., Iserloh, T., Fister, W., 2014. Soil erosion in Mediterranean landscapes – Experimental investigation on crusted surfaces by means of the Portable Wind and Rainfall Simulator. J. Arid Environ. 100-101, 42–51.

Ries, J.B., Seeger, M., Iserloh, T., Wistorf, S., Fister, W., 2009. Calibration of simulated rainfall characteristics for the study of soil erosion on agricultural land. Soil Till. Res. 106, 109–116.

Rodwell, J. (Ed.), 1998. British Plant Communities, Volume 3 Grasslands and montane communities. Cambridge University Press, Cambridge.

Rorison, I.H., 1990. Soils, mineral nutrition and climate. In: Hillier, S.H., Walton, D.W.H., Wells, D.A. (Eds.), Calcareous Grassland Ecology and Management. Proceedings of a Joint British Ecological Society/Nature Conservancy Council Symposium. Bluntisham Books, UK, pp. 21–28.

Samray, H., Erpul, G., Gabriels, D., 2011. Raindrop detachment and raindrop-induced flow transport (RD-RIFT) under different fall trajectories of winddriven raindrops. In: Vermang, J., Gabriels, D., Cornelis, W.M., De Boever, M. (Eds.), Land Degradation Processes and Assessment. UNESCO Chair of Eremology, Ghent, Belgium, pp. 9–18.

Schaub, M., Konz, N., Meusberger, K., Alewell, C., 2010. Application of in-situ measurements to determine ^{137}Cs in the Swiss Alps. J. Appl. Radioact. 101, 369–376.

Schmidt, W., Kramer, M., Michael, A., 1997. Beregnungsversuche zur Abschätzung von Aodenabtrag, Aberflächenabfluss, Nährstoff und Pestizidausträgen. Arch. Agron. Soil 42 (2), 107–117.

Sharon, D., 1980. The distribution of hydrologically effective rainfall incident on sloping ground. J. Hydrol. 46 (1-2), 165–188.

Shukla, M.K., 2014. Soil Physics: An Introduction. CRC Press, Taylor Francis Group, London, UK.

Slaymaker, O. (Ed.), 1991. Field Experiments and Measurement Program in Geomorphology. Balkema and UBC Press, Vancouver, p. 224.

Umback, C.R., Lembke, W.D., 1966. Effect of wind on falling water drops. T. Am. Soc. Agr. Eng. 9, 805–808.

Vincent, P., Clarke, V., 1980. Terracette morphology and soil properties: a note on a canonical correlation study. Earth Surf. Proc. Land. 5 (3), 291–295.

White, R.E., 2000. Principles and Practice of Soil Science: The Soil as a Natural Resource, third ed. Blackwell Science Publishers, London, UK.

Wolman, M.G., Miller, J.P., 1960. Magnitude and frequency of forces in geomorphic processes. J. Geol. 68 (1), 54–74.

Chapter 5.2

A Geologic Approach to Field Methods in Fluvial Geomorphology

Faith A. Fitzpatrick[1]
U.S. Geological Survey, Wisconsin Water Science Center, Middleton, Wisconsin, USA
[1]Corresponding author: E-mail: fafitzpa@usgs.gov

5.2.1 INTRODUCTION

Managing rivers and river processes is aided by geomorphic knowledge of past river behavior and how rivers change over time and space (Gregory et al., 2008). A widespread management issue for fluvial geomorphologists is how past human disturbances affect modern fluvial processes (Gregory et al., 2008). Fluvial processes and landforms are affected by both changes in runoff characteristics as well as sediment dynamics.

Human-accelerated erosion and sedimentation has plagued rivers across the world for millennia (Montgomery, 2007). Soil erosion and gullying associated with clearing of upland vegetation for agriculture has resulted in thick accumulations of sediment in river bottomlands, and the agricultural regions of North America are no exception (Lowdermilk, 1934; Happ et al., 1940; Leopold, 1956; Knox, 1972, 1987, 2006; Trimble, 1974; Beach, 1994; Faulkner, 1998; Montgomery, 2007) (Fig. 5.2.1). Much of the sediment eroded during the eighteenth, nineteenth, and twentieth centuries in North America remains stored in floodplains, along channel margins, or in former mill ponds (Costa, 1975; Knox, 1972, 1977, 1987; Magilligan, 1985; Trimble, 1974; Fitzpatrick et al., 2009; Walling and Owens, 2002; Allmendinger et al., 2007; Walter and Merritts, 2008) (Fig. 5.2.2).

The US Soil Conservation Service used topographic surveying and geologic and soils field techniques to quantify valley filling as a consequence of accelerated soil and gully erosion in the early twentieth century across the United States (Happ et al., 1940). Knox, (1972, 1977, 1987, 1999, 2001, 2006); Knox and Daniels (2002); Faulkner (1998), and Fitzpatrick et al. (2009) used similar methods in the Upper Mississippi River Basin to describe stream behavior patterns in erosion and sedimentation from Holocene climate change and agricultural development following Euro-American settlement. Belby (2009) related

Developments in Earth Surface Processes, Vol. 18. http://dx.doi.org/10.1016/B978-0-444-63402-3.00012-1

FIGURE 5.2.1 Photos of gully erosion in the Buffalo River, Wis.: (A) 1922 (*photo by O. Zeasman*) and, (B) 1992 (*photo by D.J. Faulkner*). Widespread gully erosion happened during intense rains and quickly expanded drainage networks. Over time the gullies became overgrown with trees but still remain efficient conduits of runoff, sediment, and nutrients from agricultural fields. See Faulkner (1998) for a detailed description of how data from valley transects and coring were used to interpret spatial variations in historical alluviation.

historical sedimentation processes to nutrient sequestration in Upper Mississippi River bottomlands. Trimble (1999) carried on the work of Happ in construction of historical sediment budgets for Coon Creek. Trimble's budgets included longitudinal river responses and lag times in sediment dynamics after rates of soil erosion had decreased from upland sources because of implementation of conservation practices. Finally, historical fluvial responses in channel behavior and sediment dynamics from clear-cut logging and log drives in the Upper Great Lakes Region were reconstructed using similar techniques (Fitzpatrick et al., 1999; Fitzpatrick and Knox, 2000).

Along with increased soil erosion were other human-caused adjustments that elevated erosion and sedimentation – ditching and channelization; headward expansion of drainage systems; construction of impoundments, and dams; and interbasin transfers. Widespread soil conservation efforts in the mid to late

FIGURE 5.2.2 Photos of bottomland aggradation from agriculture-related erosion in uplands (A) Stafford Happ in the foreground with a foot on buried fencepost in the Kickapoo River valley, Wis. (Happ et al., 1940), and (B) bank cut along the Little Platte River, southwest Wisconsin (150 mi^2 drainage area, photo by James Knox). Post Euro-American settlement overbank sediment was usually light brown in color, layered, and more sandy than its pre Euro-American counterpart. The pre-Euro-American settlement bottomland soil was generally dark, organic rich, and composed of more silty material representing in many places wet prairies, meadows, and forested wetlands.

twentieth century reduced runoff and sediment from uplands, but valley bottoms remain filled with legacy sediment and lowland rivers remain clogged with fine-grained sediment (Knox, 1977, 2001) (Fig. 5.2.2). Many river systems are still adjusting to historical gullying and headward drainage extension and concomitantly large pulses of sediment to downstream overbank areas, resulting in a basin-wide disconnection of channels from their floodplains (Trimble, 1999).

Many fluvial geomorphology studies are done as part of a multiple-lines-of-evidence approach on projects that concern fisheries management, conservation, or water quality issues. Example questions might be "Is the burial of spawning gravels by fine sediment caused by natural or human factors?" or "Have elevated sediment loads in the past affected modern channel hydraulics and morphology?" Fish habitat improvements in the 1960s and 1970s and stream restoration in the 1990s brought on a new set of management questions for fluvial geomorphology studies. Fisheries and water-resource managers frequently ask fluvial geomorphologists to determine "natural" rates of erosion, sedimentation, bank erosion, and lateral migration of rivers (Fitzpatrick et al., 1999). In North America "natural" usually means rates equivalent to before Euro-American settlement, or 100–200 years ago, compared to "modern" over the last 20 years or so.

Putting patterns of erosion and deposition into a context of watershed and landscape position (Brierley and Fryirs, this volume; Fryirs and Brierley, 2013) provide a picture of how rivers respond as a function of drainage network position, geologic setting, and glacial geomorphic history which can then be linked to system wide changes in hydrology, sediment production, and base level (Knox, 1977; Fitzpatrick and Knox, 2000). This can be done in an historical way for many different sized watersheds. Methods used to develop modern sediment budgets (Reid and Dunne, 1996; Reid and Trustrum, 2002) can be applied to expand watershed-wide erosion, transport, and deposition into historical and Holocene contexts (Notebaert et al., 2014).

This chapter focuses on how sedimentary records in valley bottoms, when combined with mapping surficial and buried landforms, can be used to interpret historical river behavior over extended time and spatial scales beyond modern record keeping and site specific data collection activities (Malanson et al., 2014).These methods largely come from the field of geology integrated with physical geography, soils, hydrology, and hydraulic engineering – in particular paleohydrology, stratigraphy, sedimentology, and soil geomorphology. Selected results from two studies of channel change and floodplain dynamics from Wisconsin's Northern Lakes and Forest (Great Lakes Region) and the Southwest Driftless Area (Upper Mississippi River Basin) are included as examples. Sedimentary records in fluvial landforms contain a window into how rivers respond to changes in hydrology and sediment production. These records contain a wealth of information about past denudation rates and river conditions (incision, aggradation, lateral migration, floodplain/channel vertical connections, adjustments in channel size and planform, channel types, and textural changes) and help extend relatively short records (few years to maybe multiple decades) of field measurements and data collection.

A more comprehensive description of a range of surficial geology approaches to fluvial geomorphology can be found in the book "Tools in Fluvial Geomorphology" Chapter 2 "Surficial Geologic Tools in Fluvial Geomorphology" (Jacobson et al., 2003). Bridge (2003) contains detailed descriptions and diagrams of fluvial sedimentary environments and expected sedimentary sequences. For aquatic environments, methods for reservoir sedimentation are helpful (USDA and SCS, 1983). Mapping fluvial landforms, past and present, through a combination of topographic surveying and interpretation of sedimentary records, is one of the most powerful tools for getting a glimpse of past and possibly future river behavior. The period of interest may be just the last 150 years or so for understanding changes since Euro-American settlement in the United States or over the last 10,000 years to understand Holocene responses to climate change. Understanding the past responses to climate and land use change adds to the predictive power of understanding future behavior to large watershed-wide drivers such as climate and land use.

Outcomes and products from surveying and identifying sedimentary sequences in fluvial landforms may include topographic maps, profile sections, and geologic cross-sections specific to valleys (Lahe, 1961, p. 409; Leopold and Miller, 1954), 3-dimensional geological maps (Thorleifson et al., 2010), soil maps and descriptions (Schaetzl and Anderson, 2005), and paleolimnology profiles (Oldfield, 1981; Smol, 1992). Study by Faulkner (1998) of alluviation in the Buffalo River, Wis., using these methods, provides examples of how historical channel incision, as well as valley bottom width and watershed size, helped to explain patterns of variability in rates of floodplain sedimentation.

5.2.2 PREFIELD ACTIVITIES

As with any fieldwork, prefield activities involve the thought process and design of the framework for the fieldwork, including questions of when, where, and what is done (Kondolf and Piegay, 2003). The first step is to clearly identify management questions that can be answered with geologic approaches to fluvial geomorphology. A list of objectives and overall hypotheses should be compiled for the fieldwork.

Fluvial landforms and related floodplain deposits may extend a few feet or many miles beyond the present-day location of a river channel. Therefore, it is especially advantageous and time saving to learn everything possible about the watershed and the valley before driving away from the office and stepping out in the field and setting up instruments. The coring and surveying methods described in this chapter are time consuming and selection of the locations for doing this work needs to be planned out in advance. In addition, the work usually requires landowner permission to gain access to study sites and calling the local diggers hotline to avoid buried utility lines.

Prefield activities include identifying where to go for the first visit to the river and its valley, which is usually called a reconnaissance tour. Maps, historical documents, photos, and other data pertinent to the study such as photographs should be collected as much as possible prior to the reconnaissance trip. These range in difficulty to obtain, except that many have been scanned and are available to the public on the internet. Thematic maps, aerial photographs and historical data are loaded into a geographic information system before the reconnaissance trip. Hydrography maps show overall patterns and variability in drainage patterns and sinuosity. Especially important is to obtain watershed digital elevation data or contour maps at the smallest scale available, similar to or higher resolution than the US Geological Survey's 1:24,000 quadrangle maps and delineate the surface-water drainage boundaries of the watershed. It is helpful to identify areas of a watershed that vary in topography, surficial geology, structure, and valley development.

Without getting into too much detail, there are a few basic maps that are informative for most studies. These include soil survey maps (county scale, soil mapping units, SURRGO soils data base (http://www.nrcs.usda.gov/wps/portal/nrcs/soilsurvey/soils/survey/state/), bedrock geology and structure, glacial landforms, Quaternary deposits, depth to bedrock, and land cover. Especially important is an understanding of the bedrock type and structure that underlies the surficial deposits. Sometime river patterns and structures are altered by intersection with an erosion resistant bedrock layer. In the United States, a map of ecoregions describes an integration of physiography, natural vegetation, and land cover (Omernik, 1987). Detailed county soil surveys can be used in larger watersheds to delineate river-related soils, including terraces of different ages and origins and floodplains. Especially useful in the soil survey descriptions is the identification of underlying layers that the soils are developed into.

Historical records from USGS streamflow-gaging stations contain a detailed record of channel morphology changes, including bed elevation shifts (Juracek and Fitzpatrick, 2009). Biological data collections usually include reach-scale aquatic habitat and possibly riparian-vegetation assessments that include measurements of geomorphic conditions, including stream morphology, streambed and bank particle size, geomorphic units, fine sediment deposition, and bank heights and erosion (e.g., Fitzpatrick et al., 1998; Kaufmann et al., 1999).

Especially useful are records of past flood events such as those accounted in newspapers. Historical county almanacs have accounts of major rainstorms and flood damages that then can provide dates for newspapers to look dates and photos of flooded rivers. Cities and villages that are near rivers also have historical accounts of floods and historical river use. Past and present dam files at state regulatory dam safety departments also contain information on floods.

Constructing a longitudinal profile of the main stem and any tributaries of interest is done before the reconnaissance trip. The longitudinal profile provides a glimpse at how the slope of a stream changes longitudinally throughout the watershed. There are already useful published methods for constructing

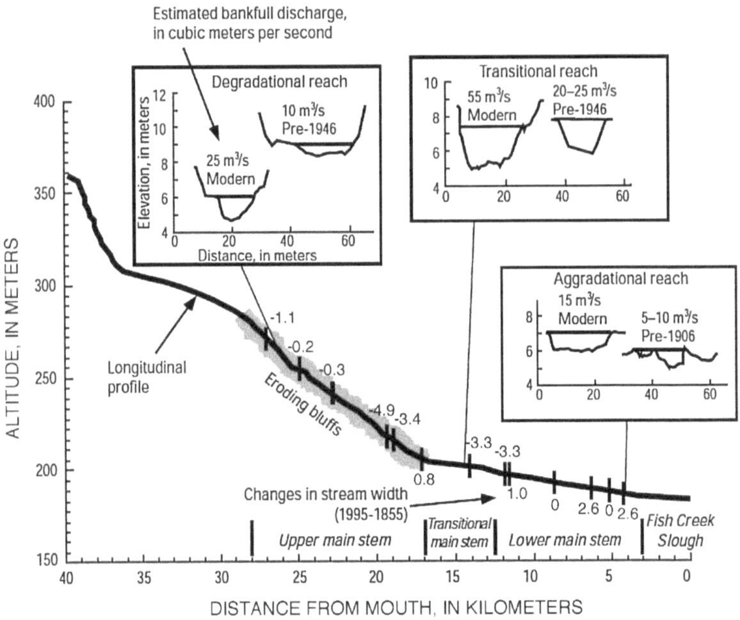

FIGURE 5.2.3 Example of summarizing transect and coring data of paleochannels along a longitudinal profile along North Fish Creek, Wis. (Fitzpatrick et al., 1999).

longitudinal profiles from contour maps and describing their characteristics so those details are not described here (Hack, 1957, 1965; Leopold et al., 1964; Knighton, 1998). Slope changes represent large changes in the energy gradient and commonly signify distinct channel, floodplain, and valley conditions, including geologic setting, hydraulics, channel bed sizes, and ultimately and sediment sources, delivery, and storage (Hack, 1965; Montgomery and Buffington, 1998). Such changes in the profile, combined with valley characteristic form the backbone of dividing a stream network into segments or landscape units of similar fluvial geomorphology (Fitzpatrick et al., 1999; Fitzpatrick et al., 2006). It can also help to define valley to stream lateral connections for sediment supply and storage as well as longitudinal connections (Fig. 5.2.3). Thus the longitudinal profile provides a backbone of sediment production, transfer, and deposition along a longitudinal continuum from headwater to river mouths (Gregory, 2006). For streams in older unglaciated landscapes that intersect uniform geology (surficial and underlying formations), the profile will usually be concave up (steep in the headwaters and gentle near its mouth). For geologically young streams in recently glaciated landscapes, such as the Midwest United States or Great Lakes Region, where streams are generally less than 15,000 years old, or for streams that intersect varying geologic units or are in structurally active areas, the profile may be concave or have abrupt changes and inflections in the profile that reflect other landforms or changes in the geologic setting (Fig. 5.2.3).

Features that may change the longitudinal continuum in sediment transport should be mapped relative to the longitudinal profile. These features include dams (historic and present human and beaver constructed), lakes, diversions, major tributary confluences, bridges and culverts crossings, log jams, and any channel alterations. It is important to extend the coverage of looking for these features to downstream areas, as downstream changes can affect upstream river behavior, such as sediment accumulation upstream of a dam, channelization, or gravel mining.

Maps and notes that have been particularly helpful for the Upper Mississippi River and Great Lakes region are the public land surveys conducted by the US Bureau of Land Management (Federal General Land Office (GLO)) in the early to mid-1800s just prior to European settlement (e.g., notes for Wisconsin http://digicoll.library.wisc.edu/SurveyNotes/). These notes are for the United States Public Land Survey System which established township, range, and section grids that formed the basis for land ownership plots across the United States except for the thirteen original colonies, Kentucky, part of Louisiana, Texas, and Hawaii (Mozola and Huebner, 2007). The notes and maps contain records of vegetation and river location and channel width, and sometimes have notes on stream bottom sediment size, water clarity, and flow conditions (swift, tranquil), and so forth. Important features noted were river crossings, fords, and milldams. On larger rivers meanders along both banks were measured. For smaller rivers, the general pattern inside of sections (between boundary lines) is less detailed and usually involved more guesswork or rough knowledge as to direction compared to the very quantitative measurements along the section lines. Detailed maps of the Mississippi River system were done by G.K. Warren (1867). Another useful set of maps that predates aerial photographs in Wisconsin are the "Bordner Surveys" which were land economic inventories done during the Depression Era (late 1920s–30s) to inventory land resources in the State that could be utilized more productively (WDAWGNHS, 1938). These maps have very detailed land use and land cover, and some predate widespread ditching and channelizing activities (http://uwdc.library.wisc.edu/collections/EcoNatRes/WILandInv).

Taken altogether, historical documents, aerial photos, maps, and longitudinal profiles provide an idea of which sections of a river have been prone to lateral migration and avulsions, where the river has been, and where floodplain sedimentation might be the thickest (e.g., Fitzpatrick et al., 2007, 2009) (Fig. 5.2.4). Previous channel locations can be plotted in a GIS using the previously described sources of information. Overlain on the overall longitudinal profile, these maps can help define and zone in on areas that have erosion and deposition issues. The maps also help to date cutoff meanders that can be included in field studies of incision, aggradation, and channel shape and size and can indicate more dynamic sections of rivers where older cutoff channels also may be located (O'Connor et al., 2003).

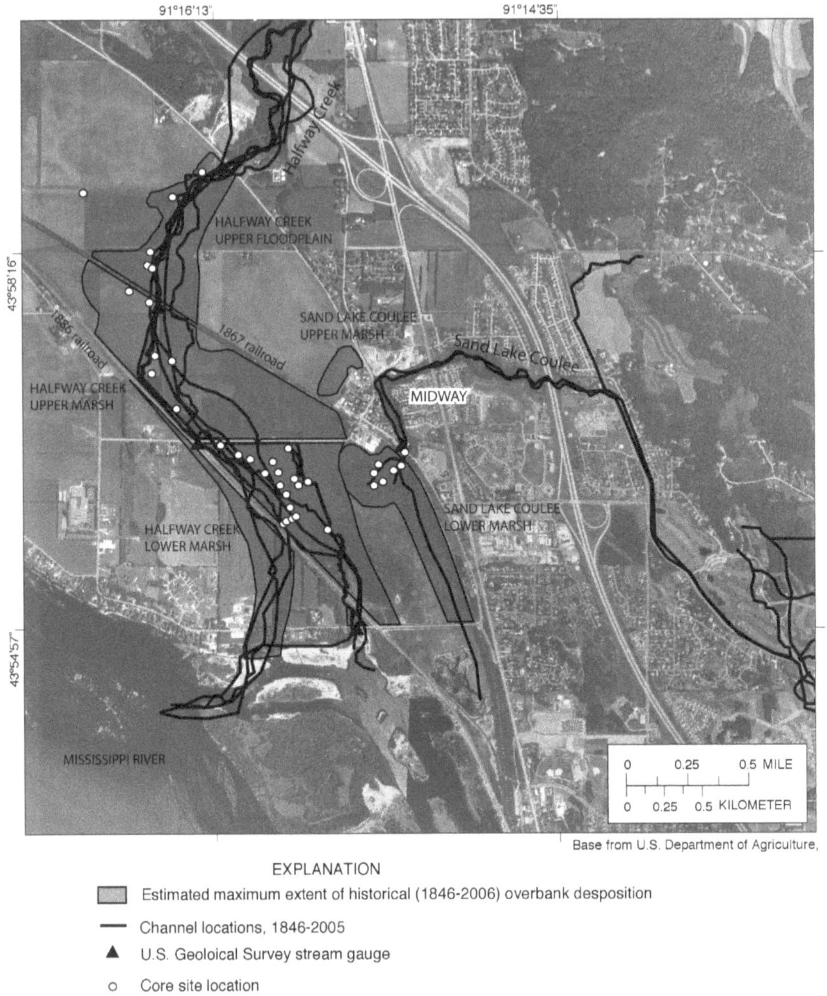

FIGURE 5.2.4 Mapping historical channel locations relative to overbank sedimentation in the Halfway Creek Basin, Upper Mississippi River system, Wis. (1846–2005) (*modified from Fitzpatrick et al., 2007*).

5.2.3 RECONNAISSANCE TRIP

A reconnaissance trip involves an overall look of the river relative to its valley along a considerable distance and at many locations in the watershed to get an idea of the major longitudinal shifts in valley development, river patterns, channel characteristics, and sediment dynamics. This is usually done by combination of several modes of transportation – a windshield drive-by, helicopter overflight, paddling in a canoe, or perhaps motoring in a larger vessel, or even walking depending on the size of the river. It is important to see as much of the stream network as possible, including ephemeral channels and gullies. Remnants of dams, even small, barely noticeable milldams should be noted because they may still affect sediment storage and delivery (Walter and Merritts, 2008). Any other features that don't show up on maps or are too small to see on aerial photographs (Fitzpatrick, 2005; Fitzpatrick et al., 2009). Other features such as bedrock outcrops, falls, knickpoints, unexpected changes in valley width, changes in geologic setting and physiography, exposed infrastructure, large sections of erosion, natural levees, and locations of abandoned channels are located and recorded with a hand-held Global Positioning System (GPS) or higher resolution instrument.

Part of the reconnaissance trip involves talking with landowners and visiting local historical societies and libraries to learn more about the human alterations in the watershed. These include dams as well as major historical ditching and drainage alterations, diversions, mining and stream restoration or habitat improvements. Speaking with local landowners is necessary to get permission to access private lands. Sometimes large tracts of riparian land are in public ownership or easements, which requires alerting a local representative that access is needed.

Preliminary locations for valley transects are usually established during the reconnaissance trip, including the best access to the location, whether it be by land or water. Depending on the size of the river and its valley bottom, sometimes exploratory coring is done to get an idea of the thickness of recent sedimentation and the types of deposits in the river and on the adjacent riparian land.

5.2.4 GEOLOGIC APPROACH TO FLUVIAL GEOMORPHOLOGY FIELD METHODS

The following section provides a summary of a geologic approach that was adapted for the specific purpose of interpretation of historical river behavior and floodplain dynamics. The field methods include topographic mapping and surveying of fluvial landforms and describing fluvial sedimentary sequences. These components are done mainly along transect lines that run longitudinally and laterally along the river and its valley, generally bisecting or following the general flow direction of the river (Fig. 5.2.5) (Fitzpatrick, 1998). Sometimes transects are set up along alluvial fans as well along larger valley margins (Fig. 5.2.6) (Fitzpatrick et al., 2007, 2009).

5.2.4.1 Topographic Mapping and Surveying

Topographic mapping and surveying of fluvial landforms requires expertise in levels, transits, and theodolites and measuring distances with tapes or laser diffraction (McCormac, 1991). Surveying skills are usually taught in engineering departments as a field-based class. In the 1990s, total station instruments became very popular because they measure slope distances and horizontal and

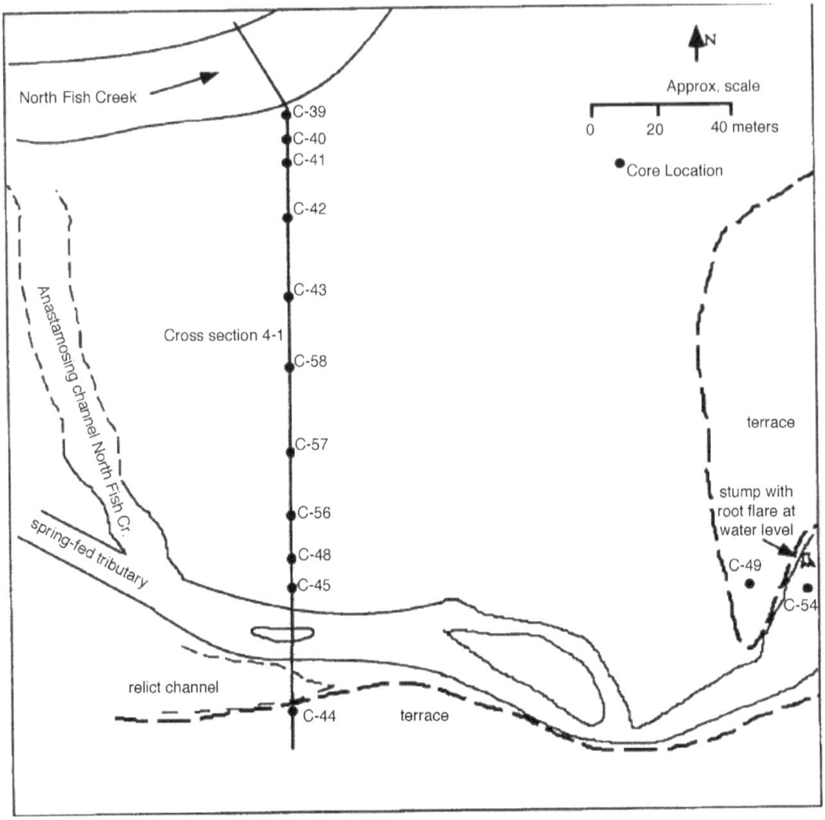

FIGURE 5.2.5 Example diagram of valley transect and cores collected for the aggradational mouth of North Fish Creek, Wis. (Fitzpatrick, 1998). A topographic survey for this site included a channel cross section of the modern North Fish Creek channel (top) with an overland transect to an abandoned channel with a springfed tributary (bottom). Location of this transect along the longitudinal profile and relative elevations of the two channels is shown in the "aggradational reach" box at River 2.6 km in Fig. 5.2.3.

vertical angles all in one step and data are easily uploaded with data loggers and storage devices. Regardless of the equipment, all surveying requires knowledge of the basic fundamentals of topographic surveying. In addition, familiarity with each brand and model of surveying equipment is needed before fieldwork is conducted in terms of data repeatability and automation, map projections and units, accuracy and precision and environmental conditions that might decrease the resolution of the data collected, such as rain, snow, and wind.

Permanent markers are needed on transect endpoints as well as benchmarks to establish elevation and survey controls. Markers and benchmarks can be made out of many materials, including brass plates mounted in cement posts, pipes, or capped concrete reinforcement bars (rebar) secured in the ground (Fig. 5.2.7); large nails or spikes in tree trunks along wooded sections, or washers or markers

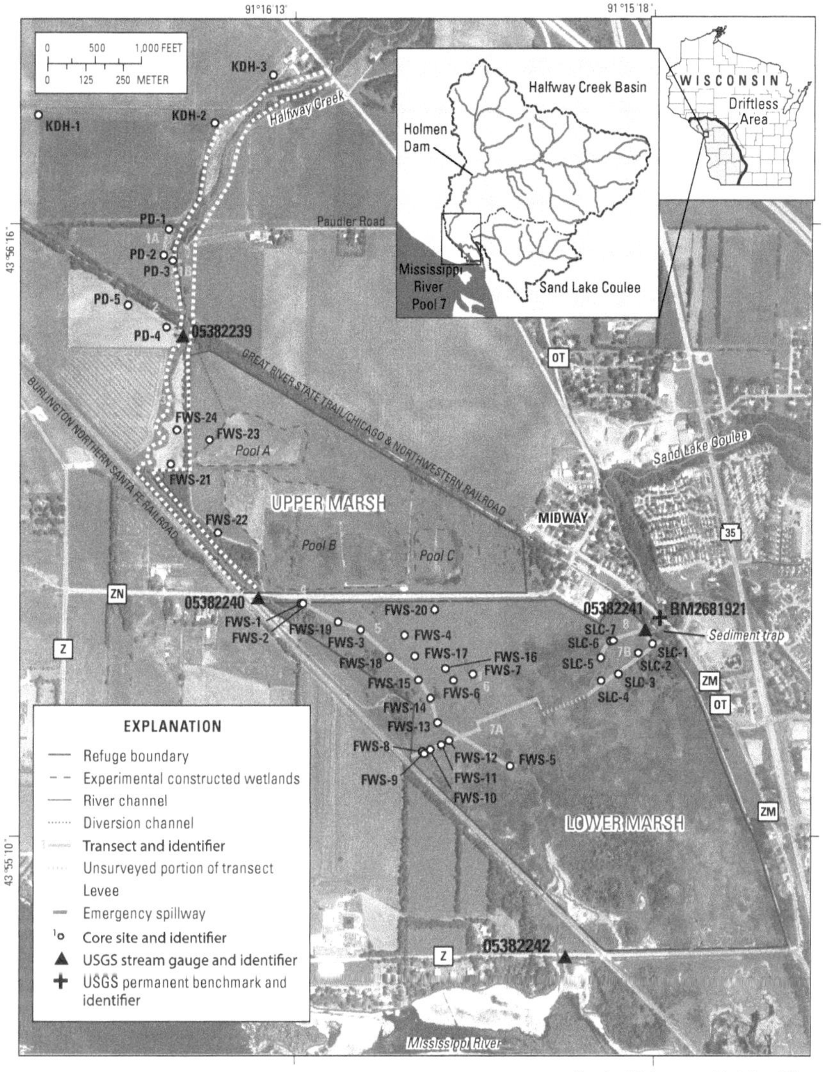

FIGURE 5.2.6 Transect layout for core site locations for the Halfway Creek, Wis. historical sedimentation study (Fitzpatrick et al., 2006). All of the transects were placed perpendicular to the general flow direction of the studied geomorphic units, except for Transect 5, which longitudinally follows the mid-length of an post-Euro-American settlement alluvial fan.

glued to bedrock (Harrelson et al., 1994). Benchmarks are described in detail and mapped and measured to other known objects so that they could be found again and surveys repeated over time (Peppler and Fitzpatrick, 2005; Moody and Meade, 2013). The endpoint markers are placed inland away from nearbank channel flows, but sometimes they might be completely buried or pulled out by

FIGURE 5.2.7 Surveying channel cross sections along the Powder River with an automatic level: (A) Robert Mead records information about a rebar pin set along a channel cross section (B) John Moody holds a surveying rod along a metal tagline with brass bead markings which was strung tightly without any sag in the line perpendicular to the channel, (C) John Moody setting up an auto level, (D) a typical cross-section endpoint set up with a rebar marked with orange flagging with the tag line reel and a cable clamp for keeping the tag line taught over long distances. A separate rebar is set to hold the end of the tagline (Moody and Meade, 2013)

debris or humans. So it is helpful to have enough information to measure to an exact point where the benchmark should be. Their height above ground surface is recorded to track burial or erosion over time, and also to help find it during the next round of surveying.

Since the early 1990s GPSs have become more common, with resolution now possible to the nearest 1 cm in vertical and horizontal planes. These systems use several satellites and cell phone service to determine x, y, and z coordinates on the earth's surface. They are sometimes faster than levels and total stations because there is no need to have to traverse a long distance between the study site and a control point or known benchmark with a known altitude. In the United States, the National Geodetic Survey has many control monuments of varying resolutions for calibration and validating high resolution GPS data (http://www.ngs.noaa.gov/cgi-bin/datasheet.prl). Most recently, the GPS use cell-phone lines to determine their relative position and are called Real-Time-Network (GPS-RTN) units (http://water.usgs.gov/osw/gps/real-time_network.html). These units require a number of reference stations and are only available in areas that have cell-phone coverage. The high resolution GPS systems have multiple electronic units, cables, and batteries that need to be well cared-for; otherwise the unit will fail at the most inopportune time, usually remote locations during freezing, drizzly weather. They also need to be programmed before fieldwork for specific areas and require control point coordinates to be loaded prior to fieldwork.

Widespread collection of remotely sensed digital elevation data with airborne or terrestrial Light Detection And Ranging (LiDAR) sensors for topographic mapping has exploded in the last decade. Like all other geodetic data, it is important to document the accuracy not only of the LiDAR data but also the digital elevation model data that are constructed from the raw data (Hodgson and Bresnahan, 2004). Experimental Advanced Airborne Research LiDAR (EAARL) has capabilities of collecting bathymetric data through relatively clear water (http://coastal.er.usgs.gov/lsrm/tech/eaarl/).

Bathymetric surveys of small river channels can be done with levels, total stations, and high resolution GPSs (about 1-cm vertical accuracy is required) with transect lines that extend from land into water. For large unwadeable rivers, land surfaces above water are surveyed in the same fashion, but because the water is so deep, elevation data for the river bottom are usually collected with multibeam sonar (e.g., Lee, 2013) or some kind of acoustic device. LiDAR data may be used for above water surfaces if available. Large rivers require some "stitching" of topographic and bathymetric data in a geographic information system (GIS). Sometimes this is done with surveys of the riverbanks with high-resolution GPS or terrestrial LiDAR.

Channel cross-section surveys, typically done for many fluvial studies (Harrelson et al., 1994), usually make up part of a longer transect that extends onto the floodplain and across valley bottoms. Surveyed points along channel cross sections are done at unevenly spaced points to be able to capture changes in the bed topography. A rough estimate for small streams is 15–20 points, including the banks. Channel cross sections should be extended at least two channel widths inland to incorporate depositional features from overbank flows and floods, such as natural levees.

Harrelson et al. (1994) also contains some quick basics on river surveying with a differential or auto level for field-based longitudinal profiles of individual reaches of the river of interest. Survey points for longitudinal profiles are usually spaced about one channel width apart or more often to capture any undulations in the channel bed such as knickpoints. Typical surfaces measured for reach-scale longitudinal profiles are the edges of water, thalweg, channel midpoint, bankfull depth, and top of banks (channel capacity flows). It is very important to measure the edge of the water surface, top of bank, or thalweg with as much accuracy as possible because the longitudinal elevation changes associated with these locations is used to estimate the slope of the channel and for calculating bankfull discharge based on Manning's Equation and for comparing modern channel morphology and discharge to that in older abandoned channels.

Many lowland rivers have accumulations of muck or soft sediment (silt, clay, and maybe some sand with high water content). Traditionally during channel cross-section surveying, the surveying rod was placed at the hard bottom, easily penetrating the soft muck, because it was thought that the most important elevation was the hard channel bottom for purposes of hydraulic geometry studies (Leopold and Maddock, 1953; Rosgen, 1994). However, for many studies concerning sediment budgets or clean sediment total daily maximum loads, it is important to measure the thickness and extent of the soft sediment because it may amount to multiple years of annual loading for a river. It is best to measure two elevations for each point along a channel cross-section or longitudinal profile when soft sediment is present in a channel – the top of the soft sediment and the top of the hard bottom.

For transects that span valley bottomland, surveyed points tend to be spaced further apart than those within channels. Spacing of points along the transect line also varies depending on the amount of variation in landform elevations across the valley. The direction of the transect line may be shifted at an intersecting point to keep it perpendicular to a channel and its valley. Transects are best extended to the valley sides, including terraces to upland surfaces, depending on the study objectives.

5.2.4.2 Characterizing Fluvial Sedimentary Sequences

This section describes how to characterizing fluvial sedimentary sequences and sedimentation history through coring, trenches and pits, and bank cuts. Geophysical techniques, such as ground penetrating radar (GPR), can augment and extend data and interpretations from cores and are gaining popularity (Bristow and Jol, 2003; Vandenberghe and van Overmeeren, 1999). GPR can be paired with acoustic-wave based sub-bottom profilers in underwater freshwater environments to distinguish coarse grained and fine grained deposits (Lin et al., 2009).

Similar to mapping and surveying surficial fluvial landforms, collecting cores requires different devices on land compared to under water. There are many different styles of coring devices. Their suitability varies depending on the desired depth, particle sizes, water and organic material content, and sampling and preservation requirements. It is a good idea to have multiple coring devices available because it usually takes a few tries to get the best device to work since every site usually has different circumstances. Elevations and water depths of coring locations are included as points in the topographic surveys in order to be able to plot the alluvial stratigraphy relative to the land surface and the fluvial landforms. For underwater cores, the edge of the water is surveyed with a high-resolution GPS (within 1 cm vertical accuracy) before and after coring.

Underwater coring devices usually have a closed barrel and a valve or piston that creates a vacuum when a core is collected and helps hold the core sample in the barrel (Fig. 5.2.7) (see http://lrc.geo.umn.edu/laccore/assets/pdf/sops/coringdevices.pdf for more examples). The coring devices are usually pushed by hand in shallow water (less than about 5 m), or attached to a cable and dropped in deep water (greater than 5 m). Gravity corers are a common type of corer used in deep water settings, and the weight of the corer provides the force to push the corer through the sediment. The diameter of the core barrels range from about 4–10 cm and are usually about 1 m long. Box corers are sometimes used to obtain a large sample over a small depth interval (Edgington and Robbins, 1991). Some cores have a vibration device that helps penetration through coarse-grained layers. Some corers have an acrylic liner that helps keep the core sample clean if there are plans for chemical analyses of the sample.

Most coring does not result in complete sampling and recovery of the penetrated material on land or under water, requiring the depth of penetration and thickness of recovered material to be recorded from which a recovery ratio is calculated. Recovery ratio is the length of the core retrieved divided by the penetration depth. Even though the sediment is considered fully saturated for underwater cores, recovery ratios are commonly less than 1 because of gas bubbles, rodding (sediment is moved to the side of the core tip as it is pushed through the sediment), blasting (top layer is blown away by the impact of the sampler meeting the bottom), compaction, or core lost out of the bottom of the barrel as the corer is lifted to the surface (Wright, 1991). A core catcher is not recommended in most instances for underwater cores because it can impede sediment from entering the core tip. A recovery ratio greater than 1 results if the technician pulls too hard on the line connected to the piston or surges the valve in a core tube. If the recovery ratio is low or high, another core should be collected until a satisfactory ratio is achieved. It is also important not to oversample the barrel length. The most likely reason for core shortening should be recorded in the field while the experience and surroundings are fresh in one's mind. A clear sided barrel helps to document that the sediment–water interface remains intact (water in the core barrel is clear above the sediment).

The majority of cores collected in water are single drive only, meaning that an open-tipped core barrel is pushed as far down as one can go once and then retrieved (Figs. 5.2.8A, B). A second interval or more in underwater environments requires a casing to be able to find the core hole again. In addition, the core barrel needs a tip that can be held tightly in place until the desired interval is reached. A modified Livingston push corer has these features (Wright, 1967). The precise fit of the piston, however, makes this coring device more suitable for lakes with fine-grained deposits or peat than for sandy channels but it may work in impoundments, backwaters, and oxbows. A hydraulic tapping-type coring device mounted on a four-wheel all-terrain vehicle can be used over thick ice during the winter in cold climates and multiple drives can be completed to the depth of interest using an incremental core tube, which like the modified Livingston has a tip that is held in place until the desired depth is reached (Fig. 5.2.8C). For impoundment coring, a good combination to use is a large 7.5-cm diameter piston corer or box corer for the surficial deposits down to about 2 m, followed by the smaller 4-cm diameter hydraulic probe corer to penetrate deeper to below the pre-dam surface. It is difficult to near impossible

FIGURE 5.2.8 Examples of typical coring devices for inundated sediment: (A) small valve push corer, (B) piston push corer, and (C) hydraulic tapping-style probe mounted on a 4-wheel drive mule on a frozen impoundment in northern Wisconsin.

in an impoundment to capture all of the soft sediment along with the more consolidated pre-dam floodplain or channel deposits in one hand-pushed core. Wright (1991) contains some useful coring tips for piston coring.

Terrestrial corers tend to be constructed differently than underwater corers because they are more suited to collect consolidated and dense deposits where core holes remain relatively open. A 2.5-cm diameter, 30-cm long soil probe or soil sampler works well for exploratory coring and can reach depths of 10 m or so if the hole can be kept open (Figs. 5.2.9, 5.2.10A, B). This style

FIGURE 5.2.9 Example of a soil probe in use by J.C. Knox in the Halfway Creek valley.

FIGURE 5.2.10 Examples of vertical accretion deposits on floodplains: (A) exploratory soil probe of riparian wetland soil (histosol) representing slow sedimentation rates and high organic matter accumulation in an offchannel wetland and (B) exploratory soil probe showing a crevasse splay of light colored sand over dark, fine-grained buried prairie soil (mollisol) near a levee break. Gravel deposits are representative of many channel bed materials: (C) modern gravel bar, and (D) Geoprobe core sample of a buried gravel layer indicating the location and elevation of the bottom of a buried abandoned channel (*photos by the author*).

of open-barrel probe is typically used by soil scientists to efficiently traverse large areas when constructing county soil survey maps at depths penetrating about 1.5 m. The probe can also be used in channels to roughly estimate major changes in sediment texture from silt/clay to sand to gravel or bedrock. The probe yields fast results for identifying major changes in texture and soil development (Figs. 5.2.10A, B). For sand and gravel units, a hand auger can also be used. Usually the maximum depth reached by these coring devices depends on the depth to the water table and saturated sand. A vibracorer with a long tube can sometimes overcome this but the vibracorer requires much experience and finesse to get a representative core with an acceptable recovery ratio. A hydraulic push-style coring device (Fig. 5.2.8C) can be used in selected locations identified with the exploratory coring for collecting larger diameter and deeper cores for physical or chemical laboratory analyses.

Trenches or pits, either machine or hand dug, can also be placed along surveyed transect lines similar to what is done for describing sedimentary units, stratigraphy studies, and soil delineations (Krumbein and Sloss, 1951; Compton, 1962; Harms and Fahnestock, 1965; Bridge, 2003; Schaetzl and Anderson, 2005). Pits and trenches are useful to document detailed layering, facies changes, and erosion surfaces for fluvial deposits over short distances (Moody et al., 1999). Trenches work best in landscapes with minimal tree cover. They can also be extended longitudinally and perpendicular to flow direction to help identify the fluvial landform and lateral or longitudinal sequencing of landforms such as bar formation (Bridge, 2003).

Bank cuts along rivers (eroding banks) are very common and can be used in tandem with cores to produce stratigraphic diagrams of floodplain, terrace and valley-side deposits (Fig. 5.2.11). Methods are similar to those used to describe road cuts and rock outcrops (Compton, 1962). A clean face of the bank must

FIGURE 5.2.11 Example of a bank cut with post-settlement Euro-American sandy vertical accretion underlain by peat (wetland) in lower main stem of the Bad River, Wis. Photo by M. Peppler, U.S. Geological Survey.

be cleared prior to describing the section with a hoe, spade, or a pickaxe. The chiseled flat end of the pickaxe works well for cleaning off hard, dry sedimentary units such as a glacial diamicton. A small hand trowel can be used for fine cleaning for delicate thin layering. A vertically held folding ruler is used to measure depth below ground surface, and the section is usually described down to the edge of water. The cuts may expose floodplain fluvial deposits or occasionally the river course may cut into terrace deposits or upland (valley side). It is important to notice where the river abuts into nonfluvial deposits as these can be major sources of fine- or coarse-grained sediment and gravel to downstream areas (Fitzpatrick et al., 1999). This may be viewed as a desirable or undesirable source of sediment, depending on if the current sediment load and river adjustments are accelerated or starved by human intervention.

5.2.4.3 Describing Cores and Exposures in the Field or Laboratory

Field descriptions of sedimentology, alluvial stratigraphy, and soils are required for exploratory cores, augering, trenches, pits, and bank cuts. Description of the sedimentary units should be objective, explicit, and unambiguous (Bridge, 2003) using terminology for logging cores and sedimentary units in from geology and soils manuals (Hudson-Edwards, 2007; Schoeneberger et al., 2012). Helpful guides for collecting and describing terrestrial cores and cuts are the "Soil Survey Field and Laboratory Methods Manual" (Burt, 2009) and the "Field Book for Describing and Sampling Soils" (Schoeneberger et al., 2012). The US Department of Agriculture textural triangle and color chart (USDA et al., 1951; Munsell Color, 1975) are used to classify all cores for texture and color. Particle size texturing in the field is done using the soil texture classes arranged on a texture triangle (Soil Survey Division Staff, 1993) and grading of texture by rubbing soil between the fingers (Milfred et al., 1967). The field descriptions should be as similar as possible to methods used to describe cores collected in liners and brought back to the lab.

All descriptions contain the following characteristics: texture, color, amount of soil development, type and amount of organic matter, structure, grading, sorting, density, transitions between layers (abrupt or gradual), gravel shape (angular or round), degree of wetness, recovery ratios, smell (many fluvial environments downstream of urban areas or industries contain buried organic and other contaminants), cultural artifacts, and any other sediment or biological features of interest. Even with the small 2-cm diameter of the soil probe (Figs. 5.2.10A, B), samples of wood and sediment, and any other material can be collected and stored in foil, plastic bags, jars, or vials for subsequent laboratory analyses.

Buried soils (paleosols) are commonly found in flood-plain deposits and are indicators of past stability in floodplain surfaces (Happ et al., 1940; Birkeland, 1984; Retallick, 1985; Fenwick, 1985; Bettis, 1992) (Figs. 5.2.2B,

Fig. 5.2.10A, B, Fig. 5.2.11). In general, modern floodplain soils are usually poorly developed because most watersheds have land cover disturbance and elevated sedimentation rates. Many buried floodplain surfaces show incipient A horizon development of a thin dark zone at the surface of the soil caused by an accumulation of decomposing organic matter. The A horizon is a dark zone at the surface of the soil caused by accumulation of decomposing organic matter. The buried floodplain surfaces can be recognized by one or more of the following characteristics: presence of an A horizon, organic matter accumulation from flood-plain or wetland vegetation, lateral extent of a buried surface that is parallel to the land surface and that may truncate geologic bedding, root traces, and soil structure (Birkland, 1984; Retallick, 1985). Sometimes part of a buried soil may be removed by discontinuous and episodic scour activity associated with flooding, channel migration, or channel avulsion (Moody et al., 2002).

In humid temperate climates, a stable floodplain surface with a relatively slow or negligible sedimentation rate will have a dark organic rich deposit at the surface, similar to what is found on presettlement floodplain surfaces (Knox, 1987). Episodic sedimentation events are noted by deposits of lighter colored sand or sandy loam deposits over dark fine deposits, indicating a change in the rate of fluvial deposition (Figs. 5.2.2, Fig. 5.2.10B, Fig. 5.2.11). Sometimes these may be isolated features such as a crevasse splay from a levee breach (Fig. 5.2.10B), or continuously cover the entire valley bottom, such as the bottomlands of the Driftless Area in Wisconsin (Fig. 5.2.2 (Knox, 1977). Besides being lighter in color and more sandy, postsettlement alluvium tends to be less compacted compared its presettlement counterpart. Recent fluvial deposits tend to have prominent stratified bedding representative of vertical accretion (settling of suspended sediment) and lack soil development on the floodplain surface compared to presettlement deposits (Knox, 1987) (Fig. 5.2.11). Other landforms that contain historical alluvium are levees, alluvial fans and crevasse splays (Fig. 5.2.10B) (Vanoni, 2006). Lateral accretion deposits, associated with lateral migration in meandering rivers, make up a much smaller portion of the post-settlement sediment record than overbank vertical accretion (Bridge, 2003).

Buried channels are notable in cores by the presence of very coarse sand, gravel, cobbles, or boulders that show up in bank exposures as lenses of coarse grained material similar to the width of the modern channel (Figs. 5.2.10C–D). In general, gravels and larger sized materials can only be transported as bedload (salting or sliding along the streambed) and therefore are a good indicator of past locations and elevations of buried channels. However, gravel-sized deposits and larger may end up in near-channel areas during large floods and can be distinguished from channel bed deposits by their vertical and horizontal positioning.

Further quantification of cores and exposures typically includes collecting samples from exposures and subsampling cores for particle size analyses and organic matter content. These methods are discussed elsewhere in detail (Guy, 1969; Burt, 2009).

5.2.4.4 Age Determinations of Field Sediment Samples

To answer questions concerning natural versus human-influenced sedimentations rates, age determinations need to be done on sediment samples or wood or organic material found in field sediment samples or core samples. A brief review of some of the most common methods for fluvial sediment is included below. Berglund (1986) is a standard reference of many of the terrestrial methods for Quaternary studies. For aquatic environments, methods employed by paleolimnology and lake sedimentation studies are helpful (Hakanson and Jansson, 2002).

Radiometric techniques based on the uranium decay series, such as cesium-137 and lead-210 (Schelske et al., 1994; Olsson, 1986; Cohen, 2003) and are best done with profiles of fine-grained sediment that have had no erosion, chemical or physical remobilization, bioturbation, pedogenesis, decomposition, or diagenesis. This technique is most often used for aquatic sediment profiles but can also be used carefully for fine-grained vertical accretion deposits in terrestrial floodplain settings.

Radiocarbon (carbon-14) dating (Libby et al., 1949) has a resolution back to about 50,000 years and can be done wood and organic matter as well as carbonate precipitates. This technique has been especially useful for dating wood in buried abandoned channels and can be used in both terrestrial and aquatic environments where organic matter is present.

For floodplain sediment optically stimulated luminescence dating has been used which has a resolution back to about 800,000 years (Lundstrom et al., 2008). This technique calculates the time since the sediment was last exposed to sunlight or intense heat and uses quartz or potassium feldspar that are commonly found in sand. This technique is helpful for determining the age of coarse-grained fluvial deposits.

5.2.5 CALCULATIONS AND INTERPRETING FIELD DATA

Stratigraphic and sedimentary records combined with topographic surveys make for a powerful set of data for interpreting the magnitude, timing, and extent of channel changes, floodplain dynamics, and sedimentation history, especially if the transect bisects relict channels from meander cutoffs or avulsions (Knox, 1977, 1999; Knox and Daniels, 2002). For studies of natural versus human accelerated sedimentation rates, these data can be used to map the vertical and spatial extent of floodplain deposition and channel changes (Fitzpatrick and Knox, 2000). When sedimentary layers have age determinations from multiple transects, historical sediment budgets can be estimated (Fitzpatrick and Knox, 2000; Fitzpatrick et al., 2009). The following section shows examples of constructing stratigraphic diagrams and historical sediment budgets.

5.2.5.1 Interpreting River Behavior and Floodplain Dynamics

Stratigraphic diagrams are constructed on the basis of mapping and surveying landforms across a floodplain combined with an understanding of fluvial processes responsible for the underlying deposition. In an example from North Fish Creek, a tributary to Lake Superior along the south shore of Lake Superior, a stratigraphic diagram near the lower main stem at river kilometer 2.6 shows how overbank sedimentation rates are higher and floods were larger than they were a century or more ago before Euro-American settlement (Fig. 5.2.12). In this watershed, the channels are less than 10,000 years old following the last deglaciation, and the river system is naturally eroding in the upper reach and aggrading in the lower reach at its mouth with Lake Superior. However, watershed clearing and agriculture accelerated the longitudinally connected behaviors. Near the mouth, in the aggradational reach of North Fish Creek, the elevation of the channel bed in 1996 was approximately 1 m higher than its 1906 counterpart (Fig. 5.2.12). Through comparison with pre-settlement sedimentation rates of nearby floodplain cores, the 1990s sedimentation rate in this aggradational reach of the river was 4–6 times higher than its pre-settlement rate (Fitzpatrick and Knox, 2000). The timing of increased sedimentation in the aggradational

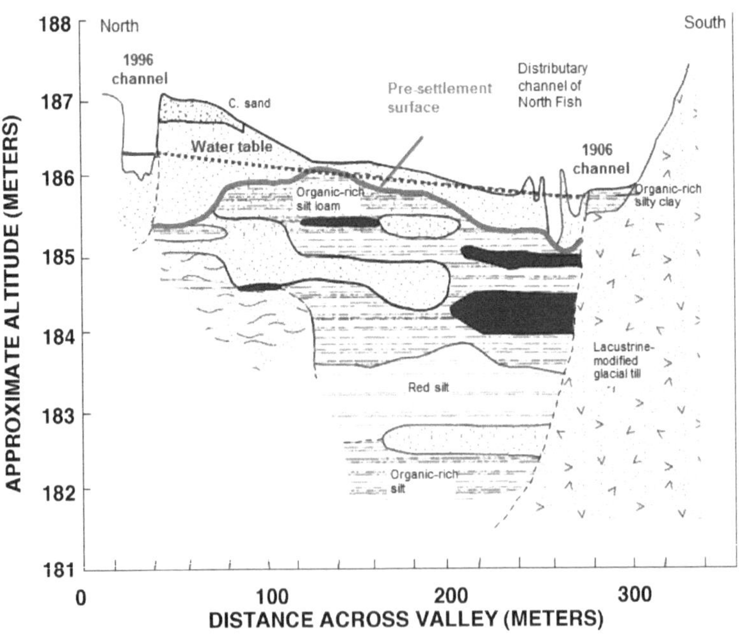

FIGURE 5.2.12 Partial valley transect of the aggradational reach of North Fish Creek, WI (Fitzpatrick, 1998; Fitzpatrick et al., 1999). Location of this transect along the longitudinal profile and relative elevations of the two channels are shown in the "aggradational reach" box at river 2.6 km in Fig. 5.2.3. Transect and core locations are shown in Fig. 5.2.5.

reach coincided with increased erosion in the upper main stem (Fig. 5.2.3). The upper main stem channels are more than 2 m deeper than their pre-settlement counterparts and are cutoff from their floodplains (Fitzpatrick et al., 1999; Fitzpatrick and Knox, 2000).

Survey and sedimentation data can be used to compare modern channel to paleochannel characteristics. The 1996 channel capacity discharge of North Fish Creek in the aggradational reach was about 15 m^3/s compared with 5–10 m^3/s for the 1906 channel (Fig. 5.2.3) (Fitzpatrick et al., 1999; Fitzpatrick and Knox, 2000). This gives an indication that channel-forming flows were larger in 1996 than in 1906. Upstream, channel-capacity flows are about twice as large. The larger size of the channels reflects that the hydrology of the watershed has not returned to its presettlement state, even though most of the watershed has a regrowth of forest.

In the aggradational reach of North Fish Creek the 1996 channel is surrounded by natural levees of coarse sand, and in general almost the full width of the valley has had some historical sandy sedimentation, much more than indicated for presettlement times (Fig. 5.2.12). This has affected the type of vegetation along channel margins and changed the water-table depth, sediment texture and organic content.

5.2.5.2 Historical Floodplain Sedimentation Depths and Volumes and Rates

Data on the degree of soil development and texture variations determined from sediment cores, in combination with age determinations, reconstruction of historical channel locations, and known dates for human activities (such as artificial levee construction), can be used to develop overbank sedimentation patterns, rates, thicknesses, and volumes (Fig. 5.2.13) (e.g., Fitzpatrick et al., 2009). Rumsby (2000) discusses the use of volumetric and depth-based estimates and some important limitations to consider. Volume of overbank sediment (Vanoni, 2006) can be converted to tons by a conversion factor based on the sediment dry bulk density analyses. Overbank sediment thickness from individual cores is usually averaged for similar geomorphic settings and history of human disturbance. Relative comparisons can be made for order-of-magnitude differences in overbank sedimentation volumes, rates, and loads.

Historical sedimentation calculations were important for the marsh at the mouth of Halfway Creek, where it flows into ecologically important backwater areas along the upper Mississippi River near Onalaska, Wis. (Fitzpatrick et al., 2009). Overbank sedimentation rates from 1846 to 2006 in the marsh associated with an alluvial fan at the mouth of Halfway Creek could be grouped into six time periods. Cultural and natural features that helped define the breaks in time periods included a buried well developed presettlement wetland or prairie soil, construction of railroads and levees, channel planform shifts and drainage network extensions, and upstream dam failures.

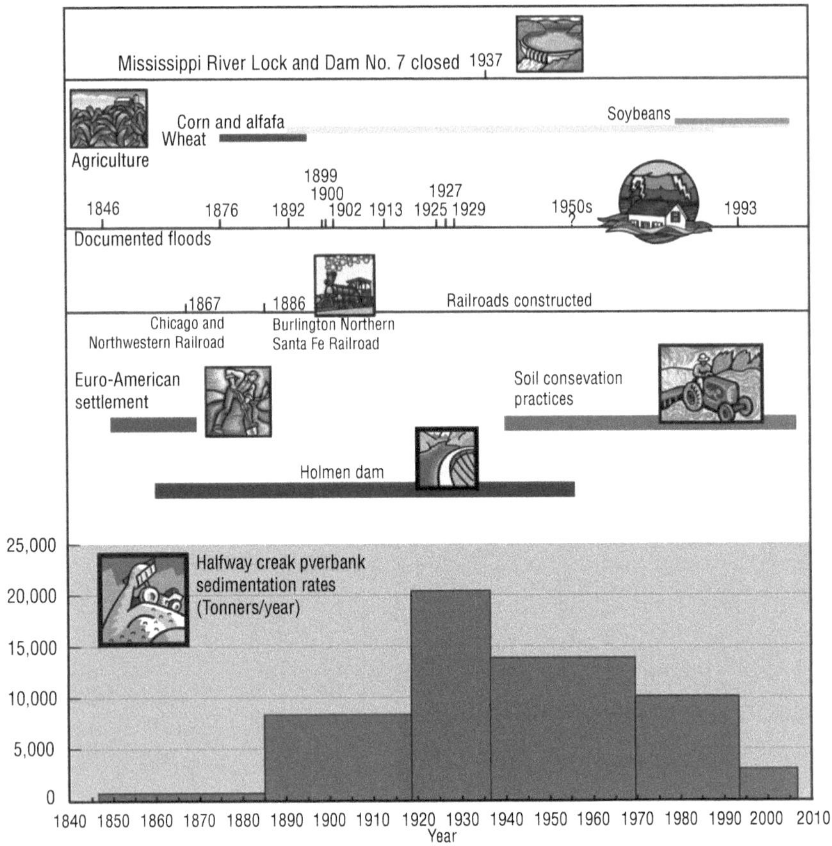

FIGURE 5.2.13 Example of plot of historical overbank sedimentation rates, in tonnes per year, for Halfway Creek, Wis., based on sedimentation rates from cores and transects and mapping of overbank sedimentation (modified from Fitzpatrick et al., 2009).

Linear sedimentation rates ranged from 0.1 cm/year before settlement and agricultural development in the watershed to a high of 2.9 cm/year from 1919–1936 (Fitzpatrick et al., 2009). This study helped to pinpoint areas of the marsh that were experiencing different sedimentation rates, depending on their proximity to channel flood flows. From 1994–2006, the lower marsh still had sedimentation rates of 1.0 cm/year, an order higher in magnitude that presettlement rates but about one third less than during peak agricultural activity before soil conservation efforts were in place. The historical sedimentation rates were important in defining the success of a restored marsh upstream specifically designed to trap sediment and nutrients to prevent them from reaching downstream areas that needed aquatic habitat protection (Kreiling et al., 2013).

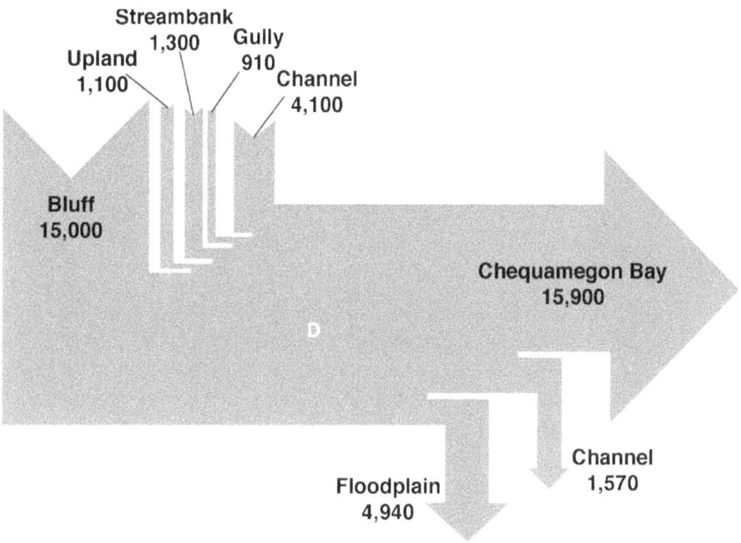

FIGURE 5.2.14 Example of a historical watershed annual sediment budget, in tonnes per year, for the North Fish Creek, Wisconsin watershed (Fitzpatrick et al., 1999).

5.2.5.3 Historical Watershed Sediment Budgets

A watershed sediment budget is one that identifies the magnitude of sediment sources and sinks in a watershed relative to watershed output (Reid and Trustrum, 2002). These budgets can be done for any time period of interest, including rivers that have had a major perturbation, such as conversion of forest to agriculture, in their history. For North Fish Creek the major historical sediment sources were upland soil erosion, gullying, bank erosion, valley bluff erosion, and incision. Sediment sinks included the channel and floodplain (Fig. 5.2.14) (Fitzpatrick et al., 1999; Fitzpatrick and Knox, 2000). Modeling simulations combined with modern sediment loads can be used to estimate loads transported out of a watershed (Fitzpatrick, 1998). In the example of North Fish Creek (Fig. 5.2.14) the aorta-diagram style of portraying sediment budget is helpful to visualize that bluff erosion contributes the majority of sediment to Chequamegon Bay at its mouth, but that there also is important sediment storage happening in the floodplain and river channel.

5.2.6 CONCLUSION

Using a geologic approach is especially useful to expand the time frame beyond modern instrument records and be able to tell how river behavior and floodplain dynamics are linked along a longitudinal river continuum from headwaters to rivermouths. Evidence for erosion in the upper parts of watersheds can be tied with sedimentation in lower reaches. These techniques are helpful for answering questions about natural (climate) versus human-altered river morphology and sedimentation issue. Other applications include expanding the extreme flood record, fisheries and habitat management, nutrient cycling and storage in floodplains, reservoir storage, causes for channel instability, and stream restoration reference conditions.

REFERENCES

Allmendinger, N.E., Pizzuto, J.E., Moglen, G.E., Lewicki, M., 2007. A sediment budget for an urbanizing watershed, 1951–1996, Montgomery County, Maryland, USA. J. Am. Water Resour. Assoc. 43 (6), 1483–1498.

Beach, T., 1994. The fate of eroded soil: sediment sinks and sediment budgets of agrarian landscapes in southern Minnesota, 1851–1988. Ann. Assoc. Am. Geogr. 84, 5–28.

Belby, C.S., 2009. Human impacts on sedimentation and nutrient sequestration in the Upper Mississippi River floodplain. University of Wisconsin-Madison Dissertation, 331 p.

Berglund, B.E., 1986. Handbook of Holocene palaeoecology and palaeohydrology. John Wiley and Sons, Ltd., Chichester, U.K., 869 p.

Bettis, E.A., III, 1992. Soil morphologic properties and weathering zone characteristics as age indicators in Holocene alluvium in the Upper Midwest. Iowa Quaternary Studies Group Contribution 30, pp. 119-144.

Birkland, P.W., 1984. Soils and geomorphology. Oxford University Press, New York, 372 p.

Bridge, J.S., 2003. Rivers and Floodplains, Forms, Processes, and Sedimentary Record. Blackwell, Oxford, UK, 404-416.

Bristow, C.S., Jol, H.M., 2003. An introduction to ground penetrating radar (GPR) in sediments-211Special Publications, Geological Society, London, 1–7.

Burt, R. (editor/compiler), 2009. Soil survey field and laboratory methods manual. Lincoln, NE, US Department of Agriculture, Natural Resources Conservation Service National Soil Survey Center, Soil Survey Investigations Report No. 51, v. 1.0, 435 p.

Cohen, A.S., 2003. Paleolimnology—the history and evolution of lake systems. Oxford University Press, Oxford, UK, 528 p.

Compton, R.R., 1962. Manual of Field Geology. John Wiley & Sons, New York.

Costa, J.E., 1975. Effects of agriculture on erosion and sedimentation in the Piedmont Province. Geol. Soc. Am. Bull., Maryland, 86, 1281-1286.

Edgington, D.N., Robbins, J.A., 1991. Standard operating procedure for collection of sediment samples. Lake Michigan Mass Balance Compendium Document, http://www.epa.gov/grtlakes/lmmb/methods/sediment.pdf.

Faulkner, D.J., 1998. Spatially variable historical alluviation and channel incision in west-central Wisconsin. Ann. Assoc. Am. Geogr. 88 (4), 666–685.

Fenwick, I., 1985. Paleosols: Problems of recognition and interpretation. In: Boardman, J. (Ed.), Soils and Quaternary landscape evolution. John Wiley & Sons, Chichester, pp. 3–21.

Fitzpatrick, F.A., 1998. Geomorphic and hydrologic responses to vegetation, climate, and base level changes, North Fish Creek, Wisconsin. University of Wisconsin-Madison, Madison, Wis, Ph.D. dissertation, p. 275.

Fitzpatrick, F.A., 2005. Trends in streamflow, sedimentation, and sediment chemistry for the Wolf River, Menominee Indian Reservation. US Geological Survey Science Investigations Report, Wisconsin, 1850-1999, 2005–5030, p. 47.

Fitzpatrick, F.A., Knox, J.C., 2000. Spatial and temporal sensitivity of hydrogeomorphic response and recover to deforestation, agriculture, and floods. Phys. Geogr. 21 (2), 89–108.

Fitzpatrick, F.A., Knox, J.C., Whitman, H.E., 1999. Effects of historical land-cover changes on flooding and sedimentation, North Fish Creek. US Geological Survey Water-Resources Investigations Report, Wisconsin, 99–4083, p. 12.

Fitzpatrick, F.A., Knox, J.C., Schubauer-Berigan, J.P., 2009. Channel, floodplain, and wetland responses to floods and overbank sedimentation, 1846-2006, Halfway Creek Marsh, Upper Mississippi Valley, Wisconsin. In: James, L.A., Rathburn, S.L., Whittecar, G.R. (Eds.), Management and Restoration of Fluvial Systems with Broad Historical Changes and Human Impacts. Geological Society of America Special Paper 451, pp. 23–42.

Fitzpatrick, F.A., Peppler, M.C., DePhilip, M., Lee, K.E., 2006. Geomorphic characteristics and classification of Duluth-Area streams. US Geological Survey Scientific Investigations Report, 2006–5029, p. 54.

Fitzpatrick, F.A., Waite, I.R., D'Arconte, P.J., Meador, M.R., Maupin, M.A., Gurtz, M.E., 1998. Revised methods for characterization stream habitat in the National Water-Quality Assessment Program. US Geological Survey Water–Resources Investigations Report, 98–4052, p. 68.

Fryirs, K.A., Brierley, G.J., 2013. Geomorphic Analysis of River Systems: An Approach to Reading the Landscape. Blackwell Publishing, Oxford, UK, 345 p.

Gregory, K.J., 2006. The human role in changing river channels. Geomorphology 79, 172–191.

Gregory, K.J., Benito, G., Downs, P.W., 2008. Applying fluvial geomorphology to river channel management: Background for progress towards a palaeohydrology protocol. Geomorphology 98, 153-172.

Guy, H.P., 1969. Laboratory theory and methods for sediment analysis. Chapter C1, Book 5. Techniques of Water-Resources Investigations of the United States Geological Survey, 58 p.

Hack, J.T., 1957. Studies of longitudinal stream profiles in Virginia and Maryland. US Geological Survey Professional Paper 294B.

Hack, J.T., 1965. Postglacial drainage evolution and stream geometry in the Ontonagon area, Michigan. US Geological Survey Professional Paper 504–B, 40 p.

Hakanson, L., Jansson, M., 2002. Principles of lake sedimentology. The Blackburn Press, Caldwell, New Jersey, 316 p.

Happ, S.C., Rittenhouse, G., Dobson, G.C., 1940. Some Principles of Accelerated Stream and Valley Sedimentation. US Department of Agriculture Technical Bulletin 695. US Government Printing Office, Washington, DC, 695.

Harms, J.C., Fahnestock, R.K., 1965. Stratification, bed forms, and flow phenomena (with an example from the Rio Grande). Spec. Pub. Soc. Econ. Pal. Min. 12, 84–115.

Harrelson, Cheryl, C., Rawlins, C.L., Potyondy, John, P., 1994. Stream channel reference sites: an illustrated guide to field technique. Gen. Tech. Re. RM-245. Fort Collins, CO. USDA For. Serv. R. M. Res., 61.

Hodgson, M.E., Bresnahan, P., 2004. Accuracy of airborne LiDAR-derived elevation: empirical assessment and error budget. Photogramm. Eng. Remote Sens. 70 (3), 331–339.

Hudson-Edwards, K., 2007. Fluvial environments. In: Perry, C., Taylor, K. (Eds.), Environmental Sedimentology. Blackwell, Malden, MA, pp. 75–108, Chapter 3.

Jacobson, R.B., O'Connor, J.E., Oguchi, T., 2003. Surficial geologic tools in fluvial geomorphology, Chapter 2. In: Kondolf, M., Piegay H. (Eds.), Tools in Fluvial Geomorphology. John Wiley & Sons, Ltd, Chichester, UK, pp. 26–57.

Juracek, K.E., Fitzpatrick, F.A., 2009. Geomorphic applications of streamgage information. Riv. Res. Appl. 25, 329–347.

Kaufmann, P.R., Levin, P., Robison, E.G., Seeliger, C., Peck, D.V., 1999. Quantifying physical habitat in streams. US Environmental Protection Agency, EPA/620/R-99/003 Washington, D.C., http://www.epa.gov/emap/html/pubs/docs/groupdocs/surfwatr/field/phyhab.html.

Knighton, D., 1998. Fluvial Forms and Processes. John Wiley & Sons, New York.

Knox, J.C., 1972. Valley alluviation in southwestern Wisconsin. Ann. Assoc. Am. Geogr. 62 (3), 401–410.

Knox, J.C., 1977. Human impacts on Wisconsin stream channels. Ann. Assoc. Am. Geogr. 67 (3), 323–342.

Knox, J.C., 1987. Historical valley floor sedimentation in the Upper Mississippi River Valley. Ann. Assoc. Am. Geogr. 77 (2), 224–244.

Knox, J.C., 1999. Long-term episodic changes in magnitudes and frequencies of floods in the Upper Mississippi River Valley. In: Brown, A.G., Quine, T.A. (Eds.), Fluvial processes and environmental change.. John Wiley & Sons, Chichester, U.K, pp. 255–282, Chapter 14.

Knox, J.C., 2001. Agricultural influence on landscape sensitivity in the Upper Mississippi River Valley. Catena 42, 193–224.

Knox, J.C., 2006. Floodplain sedimentation in the Upper Mississippi Valley – natural versus human accelerated. Geomorphology 79, 286–310.

Knox, J.C., Daniels, M., 2002. Watershed scale and the stratigraphic record of large floods. In: House, P.K., Levish, D.R., Webb, R.H., Baker, V.R. (Eds.), Ancient Floods, Modern Hazards. Principles and Applications of Paleoflood Hydrology:. American Geophysical Union, Washington, DC, pp. 237–255.

Kondolf, G.M., Peigay, H., 2003. Tools in fluvial geomorphology: problem statement and recent practice, Ch. 1. In: Kondolf, M., Piegay, H. (Eds.), Tools in Fluvial Geomorphology. Wiley, pp. 3–22.

Kreiling, R.M., Schubauer-Berigan, J.P., Richardson, W.B., Bartsch, L.A., Hughes, P.E., Cavanaugh, J.C., Strauss, E.A., 2013. Wetland management reduces sediment and nutrient loading to the upper Mississippi River. J. Environ. Qual. 42 (2), 573–583.

Krumbein, W.C., Sloss, L.L., 1951. Stratigraphy and Sedimentation, second ed. W.H. Freeman and Company, San Francisco.

Lahe, F.H., 1961. Field Geology. McGraw-Hill Book Co., New Yorkp. 926.

Lee, K.G., 2013. Estimation of reservoir storage capacity using multibeam sonar and terrestrial LiDAR, Randy Poynter Lake, Rockdale County, Georgia, 2012. US Geological Survey Scientific Investigations Ma, 3265.

Leopold, L.B., 1956. Land use and sediment yield. In: Williams, Jr., T.L. (Ed.), Man's Role in Changing the Face of the Earth.. University of Chicago Press, Chicago, pp. 639–647.

Leopold, L.B., Maddock, T., 1953. The hydraulic geometry of stream channels and some physiographic implications. US Geol. Surv. Prof. Pap., 252.

Leopold, L.B., Miller, J.P., 1954. A postglacial chronology for some alluvial valleys in Wyoming. US Geol. Surv. Water Supply Pap. 1261, 89.

Leopold, L.B., Wolman, M.G., Miller, J.P., 1964. Fluvial processes in geomorphology. W.H. Freeman and Company, San Francisco, 522 p.

Libby, W.F., Anderson, E.C., Arnold, J.R., 1949, Age determination by radiocarbon content: worldwide assay of natural radiocarbon. Science 109 (2827): 227-228.

Lin, Y.T., Schuettpelz, C.C., Wu, C.H., Fratta, D., 2009. A combined acoustic and electromagnetic wave-based techniques for bathymetry and subbottom profiling in shallow waters. J. Appl. Geophys. 68, 203–218.

Lowdermilk, W.C., 1934. Acceleration of erosion above geologic norms. American Geophysical Union, 15th Transactions, Pt. 2pp. 505–509.

Lundstrom, S.C., Mahan, S.A., Paces, J.B., Hudson, M.R., House, P.K., Malmon, D.V., Blair, J.L., Howard, K.A., 2008. Late Pleistocene aggradation and degradation of the lower Colorado River: Perspectives from the Cottonwood area and other reconnaissance below Boulder Canyon. Geol. Soc. Am. Spec. Pap. 439, 411–432.

Magilligan, F.J., 1985. Historical floodplain sedimentation in the Galena River Basin, Wisconsin and Illinois. Ann. Assoc. Am. Geogr. 75, 583–594.

Malanson, G., Scuderi, L., Moser, K.A., Willmott, C.J., Resler, L.M., Warner, T.A., Mearns, L.O., 2014. The composite nature of physical geography: moving from linkages to integration. Prog. Phys. Geogr. 38 (3), 3–18.

McCormac, J.C., 1991. Surveying Fundamentals, second ed. Prentice Hall, Englewood Cliffs, New Jersey, 567 p.

Milfred, C.J., Olson, G.W., Hole, F.D., 1967. Soil resources and forest ecology of Menominee County, Wisconsin. University of Wisconsin, Geological and Natural History Survey, Soil Survey Division, Bulletin 85, Soil Series No. 60: Madison, Wisconsin, p. 203, 3 plates.

Moody, J.A., Meade, R.H., 2013. Powder River: Data for cross-channel profiles at 22 sites in southeastern Montana from 1975 through 2013. US Geol. Surv. Data Set, Doi:10.5066/F70Z719C.

Moody, J.A., Meade, R.H., Martinson, H.A., 2002. Erosion and deposition of sediment at channel cross sections on Powder River between Moorhead and Broadus Montana, 1980-98. US Geol. Surv. Water-Res. Invest. Rep., 2002–4219.

Moody, J.A., Pizzuto, J.E., Meade, R.H., 1999. Ontogeny of a flood plain. GSA Bull. 111 (2), 291–303.

Montgomery, D.R., 2007. Dirt: The Erosion of Civilizations. University of California Press, Berkeleyp. 295.

Montgomery, D.R., Buffington, J.M., 1998. Channel processes, classification, and response. In: Naiman, R.J., Bilby, R.E. (Eds.), River Ecology and Management. Springer-Verlag, Inc., New York, pp. 13–42.

Mozola, A.J., Huebner, D.J., 2007. US Public Land Survey Grid. In: Walker, J.D., Cohen, H.A. (Eds.), The Geoscience Handbook.. American Geological Institute, Section 2.4, pp. 45–50.

Munsell Color, 1975. Munsell soil color charts. Baltimore. Munsell color Division of Kollmorgen Corp., variable pagination.

Notebaert, B., Berger, J.F., Brochier, J.L., 2014. Characterization and quantification of Holocene colluvial and alluvial sediments in the Valdaine Region (southern France). The Holocene 24 (10), 1320–1335.

O'Connor, J.E., Grant, G.E., Haluska, T.L., 2003. Overview of geology, hydrology, geomorphology, and sediment budget of the Deschutes River Basin, Oregon: Geology and Geomorphology of the Deschutes River, Oregon. Water Science and Application 7, American Geophysical Union, 9–31.

Oldfield, F., 1981. Peats and lake sediments: formation, stratigraphy, description and nomenclature. In: Goudie, A., Anderson, M., Burt, T., Lewin, J., Richards, K., Whalley, B., Worsley, P. (Eds.), Geomorphological Techniques. edited for the British Geomorphological Research Group. George Allen, Unwin, London, pp. 306–326.

Olsson, I.U., 1986, Radiometric dating, *in* Berglund, B.E., ed., Handbook of Holocene palaeoecology and palaeohydrology: Chichester, U.K., John Wiley and Sons, p. 273–312.

Omernik, J.M., 1987. Ecoregions of the conterminous United States Map (scale 1:7,500,000). Ann. Assoc. Am. Geogr. 77 (1), 118–125.

Peppler, M.C., Fitzpatrick, F.A., 2005. Methods for Monitoring the Effects of Grazing Management on Bank Erosion and Channel Morphology, Fever River, Pioneer Farm. US Geological Survey Fact Sheet 2005-3134, Wisconsin, 2004.

Reid, L.M., Dunne, T., 1996. Rapid evaluation of sediment budgets. Catena Verl, Reiskirchen, Germanyp. 164.

Reid, L.M., Trustrum, N.A., 2002. Facilitating sediment budget construction for land management applications. J. Environ. Plan. Manage. 45 (6), 865–887.

Retallick, G., 1985. Laboratory exercises in paleopedology. Course Handout, University of Oregon, Eugene, 74 p.

Rosgen, D.L., 1994. A Classification of Natural Rivers. Catena 22:169-199.

Rumsby, B., 2000. Vertical accretion rates in fluvial systems: a comparison of volumetric and depth-based estimates. Earth Surf. Proc. Land. 25, 617–631.

Schaetzl, R., Anderson, S., 2005. Soils: Genesis and Geomorphology. Cambridge University Press, New York, 817 p.

Schelske, C.L., Peplow, A., Brenner, M., Spencer, C.N., 1994. Low-background gamma counting – Applications for ^{210}Pb dating of sediments. J. Pal. 10, 115–128.

Schoeneberger, P.J., Wysocki, D.A., Benham, E.C., Staff, Soil Survey, 2012. Field book for describing and sampling soils, Version 3. 0. Natural Resources Conservation Service. National Soil Survey Center, Lincoln, NE.

Smol, J.P., 1992. Paleolimnology: an important tool for effective ecosystem management. J. Aquat. Ecosyst. Health 1, 49–58.

Soil Survey Division Staff. 1993. Soil survey manual. Soil Conservation Service. U.S. Department of Agriculture Handbook 18.

Thorleifson, H., Berg, R.C., Russell, H.A.J., 2010. Geological mapping goes 3-D in response to societal needs. GSA Today 20 (8), 27–29.

Trimble, S.W., 1974. Man-induced soil erosion on the southern Piedmont, 1700–1970. Soil Conservation Society of America, Ankeny, IAp. 180.

Trimble, S.W., 1999. Decreased rates of alluvial sediment storage in the Coon Creek Basin Wisconsin, 1975–93. Science 285, 1244–1246, (5431).

US Department of Agriculture. Soil Conservation Service, 1983. National Engineering Handbook, Section 3, Sedimentation. US Department of Agriculture, Soil Conservation Service 210-VI-NEH-03, variable pagination.

US Department of Agriculture, Soil Conservation Service, Soil Survey Staff, 1951. Soil survey manual. US Department of Agriculture. Agricultural Handbook No. 18, Washington, D.C, 503 p.

Vandenberghe, J., van Overmeeren, R.A., 1999. Ground penetrating radar images of selected fluvial deposits in the Netherlands. Sediment. Geol. 128 (3), 245–270.

Vanoni, V.A., ed., 2006. Sedimentation engineering—Chapter III, Sediment measurement techniques. American Society of Civil Engineers Manuals and Reports on Engineering Practice no. 54, pp. 227-233.

Walling, D.E. and Owens, P.N., 2002. The role of flood plain sedimentation in catchment sediment and contaminant budgets. Proceedings from the International Symposium Structure, Function, and Management Implications of Fluvial Sedimentary Systems, IAHS Publication No. 276, pp. 407-416.

Walter, R.C., Merritts, D.J., 2008. Natural streams and the legacy of water-powered mills. Science 319, 299–304.

Warren, G.K., 1867. Survey of the Upper Mississippi River. US House of Representatives Executive Document 58, 39th Congress, 2nd Session, Washington, D.Cpp. 5–38.

Wisconsin Department of Agriculture and Wisconsin Geological and Natural History Survey, 1938. Land economic inventory. La Crosse County, Wis., scale 1:63,360, accessed March 6, 2007, at http://digicoll.library.wisc.edu/SurveyNotes/.

Wright, H.E., 1967. A square-rod piston sampler for lake sediments. J. Sediment. Petrol. 37, 975–976.

Wright, H.E., 1991. Coring tips. J. Pal. 6, 37–49.

Chapter 5.3

Reading the Landscape in Field-Based Fluvial Geomorphology

Gary Brierley*[,1], Kirstie Fryirs**
*School of Environment, University of Auckland, Auckland, New Zealand; **Department of Environmental Sciences, Macquarie University, North Ryde, New South Wales, Australia
[1]Corresponding author: E-mail: g.brierley@auckland.ac.nz

5.3.1 INTRODUCTION

> *"The most engaging and interesting intellectual work on geomorphic forms such as river channels has not come from computer specialists and theoretical models but from field measurements and observations."*
>
> Luna B. Leopold (2004, p. 10)

Geomorphology is an observationally framed, interpretative science that analyzes landscape forms, processes, patterns and evolution (Baker, 1999; Church, 2011; Frodeman, 1995; Schumm, 1991; Stoddart and Adams, 2004). Effective environmental management builds upon place-based understandings of landscapes generated through geomorphic applications. Unlike various other branches of scientific inquiry, reasoning in field-based geomorphology is based largely on abduction (inferential associations) rather than induction or deduction (Magnani, 2001). As highlighted in Chorley's oft-cited quote (Chorley, 1978, p. 1): "Whenever anyone mentions theory to a geomorphologist, he (*sic*) instinctively reaches for his (*sic*) soil auger." In most instances the interplay between theory and practice tends to emphasize what is seen/inferred, then relating findings to theoretical principles, rather than vice versa. Observations inform descriptions, from which interpretations engender explanation. Detective-style investigations build upon selective evidence (the preserved record), from which many key elements may be missing, and remaining facets may initially appear to be contradictory. Fascinating and intriguing processes of inquiry entail supposition and debate. This is often far from a linear process, as a particular piece of evidence may reframe the whole argument. Each landscape has its own story (Phillips, 2012). As Wohl (2013, p. 51) comments: "Geomorphology is neither solely physics (universal) nor stamp-collecting (site-specific studies) because of the influence of contingency on geomorphic form and process. Contingency and patterns can be most effectively elucidated using field-based data."

Developments in Earth Surface Processes, Vol. 18. http://dx.doi.org/10.1016/B978-0-444-63402-3.00013-3

Encounters with contingency can be framed as "unraveling causality" – striving to work out sequences of events and their drivers/controls that fashioned particular process interactions that generated specific landscape forms and patterns. Importantly, these analyses are most effectively performed using a "whole of landscape" perspective (Baker and Twidale, 1991), appraising the changing nature of process relationships over time. Nested hierarchical frameworks provide a pedagogical toolkit to support scaffolded learning of river systems. Such analyses are typically performed as the scale of physiographic zones (ecoregions), catchments, reaches, landforms (geomorphic units), hydraulic units, and bed material size (see Frissell et al., 1986; Brierley and Fryirs, 2005). Temporal analyses frame the contemporary pattern and behavior of landforms and their configuration at the landscape scale. Assessments of evolutionary history frame prediction of likely future evolutionary trajectories. What is seen today is related to assessments of what has gone before, unraveling the memory of past events (Brierley, 2010).

Training, experience, and instinct (intuition) fashion the key concepts, principles, and conceptual models that influence how practitioners observe and hence what is seen (Church, 2011; Church, 2013; Grant et al., 2013; Sauer, 1956; Wohl, 2013). Various approaches to geomorphic inquiry have been evident over time, but it is now widely recognized that diversity of approaches is a disciplinary asset (i.e., geomorphic pluralism is healthy; Jennings, 1973; Preston et al., 2011; Slaymaker, 1997). Recent developments in geomorphology apply sophisticated technological applications that support increasingly detailed and accurate endeavors in field mapping, measurement, and dating (e.g., Carbonneau et al., 2012; Roering et al., 2013; Seijmonsbergen, 2013; Tooth, 2013). Resulting data explosions must be carefully appraised through informed sets of questions and analytical techniques, ensuring that applications meaningfully relate site-specific understandings to general theoretical principles (Brierley et al., 2013; Phillips, 2007; Rhoads and Thorn, 2011). The art of field-based teaching and interpretation cannot be replaced in this process.

The approach to teaching interpretative landscape analysis skills that is outlined in this chapter builds upon a scaffolded set of cognitive exercises that promote masterly learning. Three longstanding fieldwork exercises developed by the coauthors for undergraduate/postgraduate teaching and for professional short courses are documented. These exercises generate contextualized place-based understandings in fluvial geomorphology. They entail catchment-scale analysis of river character and behavior (river reach analysis), interpretation of river patterns along longitudinal profiles, and analysis of river evolution. Key aims (learning outcomes) of these exercises include:

1. To develop a practitioner's ability to "read the landscape."
2. To analyze relationships between river processes and landforms.
3. To interpret controls upon catchment-scale patterns of rivers.

4. To interpret river evolution, placing human disturbance in context of "natural" variability.
5. To assess likely river futures based upon appraisal of their capacity to adjust in response to disturbance events, and their linkages/connectivity at the catchment scale.

Rather than providing a "how to" overview of prescriptive step-by-step instructions, this chapter outlines how each exercise uses foundation principles in fluvial geomorphology to support field-based investigations that "ask the right questions" in making interpretations of landscape forms, processes, patterns, and evolution. Working under the mantra "Know your Catchment," these exercises can be used to generate and apply systematic geomorphic understandings of river systems as a component of river-management applications.

The first two exercises address spatial interpretations of fluvial landscapes, building directly upon principles of river reach analysis (Kellerhals et al., 1976) and the River Styles framework (Brierley and Fryirs, 2005). The first exercise prompts practitioners to assess river diversity, asking questions about process-form linkages that determine how rivers look and behave. The second exercise asks students to appraise how river reaches "fit together" in their catchment context (analysis of river patterns). The third (temporal) exercise frames interpretations of contemporary river character and behavior in an evolutionary context. Geomorphic and sedimentologic attributes of contemporary landforms are analyzed to make inferences about formative processes, how they have changed over time, and prospective future adjustments. Collectively, these exercises are embedded in an approach to "Reading the Landscape" that is briefly presented in the first part of this chapter (Fryirs and Brierley, 2013). This open-ended set of procedures can be used to interpret rivers across the range of environmental settings. Each stage of the approach is scaffolded to produce increasingly sophisticated knowledge of river form, process and evolution. Field-based analyses are framed alongside remotely sensed interpretations in a "multiple lines of evidence" approach that develops the critical thinking skills of practitioners.

Inevitably, this chapter reflects the personal perspective, training, and outlook of its authors. The authors are field scientists who are passionate about the health of river systems. Among their primary professional motivations is the desire to use field investigations as a platform for informed management actions that work with nature. Their approach to teaching promotes an ethos of learning and enjoyment. This chapter outlines teaching exercises that the authors have run collectively or individually for over 30 field classes in Australia and New Zealand. Related professional short courses have been taught in over 10 countries. The exercises are generic, and can be readily adapted and applied elsewhere. Emphasis is placed upon development of independent skill sets that promote mastery learning, rather than rote-learning applications. Such endeavors move beyond check-list, tick-box applications by asking the right questions (Brierley et al., 2013; Fryirs and Brierley, 2013).

The personal teaching outlook of the authors promotes diverse encounters with river systems. Differing perspectives and meanings of rivers are highlighted in an exercise that builds upon teaching experience entitled "Draw me a River" by (1997). In an opening postgraduate class (indeed, before any introductions or any work have been completed), student groups are asked to "Act me a River," wherein other groups have to interpret the type of river that is being performed through analysis of its behavioral regime (i.e., emphasizing river adjustment, rather than what the river looks like). As yet, these exercises have not been extended to a "river dance" but you never know …

5.3.2 AN APPROACH TO "READING THE LANDSCAPE"

"Geography and natural history are indeed related by their manner of observation. Much of what both identify and compare lies outside of quantitative analysis. … An innate aptitude to register on differences and similarities is joined to a ready curiosity and reflection on the meaning of likeness and unlikeness. There is, I am confident, such a thing as the "morphologic eye," a spontaneous and critical attention to form and pattern (and implicitly process-based understandings). … (T)he principal training of the geographer should come, wherever possible, by doing field work."

Carl O. Sauer (1956, p. 290, 296) (our comment in brackets)

Inevitably, fieldwork practices reflect the issues to be addressed and associated questions that are asked (Fig. 5.3.1). Practitioners do not simply head out into the field without carefully thinking through what they are trying to achieve (i.e., the motivation and purpose for the work) and how they intend to go about it. Fieldwork is often expensive, so targeted interventions are required. Having said this, individual pursuits become a way of life for many geomorphologists, and may cost very little as they are piggy-backed onto everyday ventures (often to the bewilderment and frustration of family members or travelling companions).

Once a clear sense of aims and goals has been established, a big picture (top−down) overview is required to clearly contextualize the situation. In geomorphic terms, site-specific attributes must be framed in their landscape context. Concerns for representativeness relate local landforms to landscape-scale patterns of features. Appreciation of temporal context is also important – has the landscape been subjected to major disturbance events (e.g., recent earthquakes, floods, fires, etc.); are prevailing fluxes that influence biophysical relationships in the study area "typical" of longer-term trends? In "reading the landscape," bottom-up practices at a given locality are framed in relation to these bigger picture considerations.

Reading the landscape is a constructivist (building-block) framework for analysis and interpretation of river forms and processes over various spatial and temporal scales (Fryirs and Brierley, 2013; Fig. 5.3.2). It entails four steps. This "bottom-up" approach synthesizes the behavior and evolution of river landscapes through systematic analysis of valley-floor landforms (termed geomorphic units).

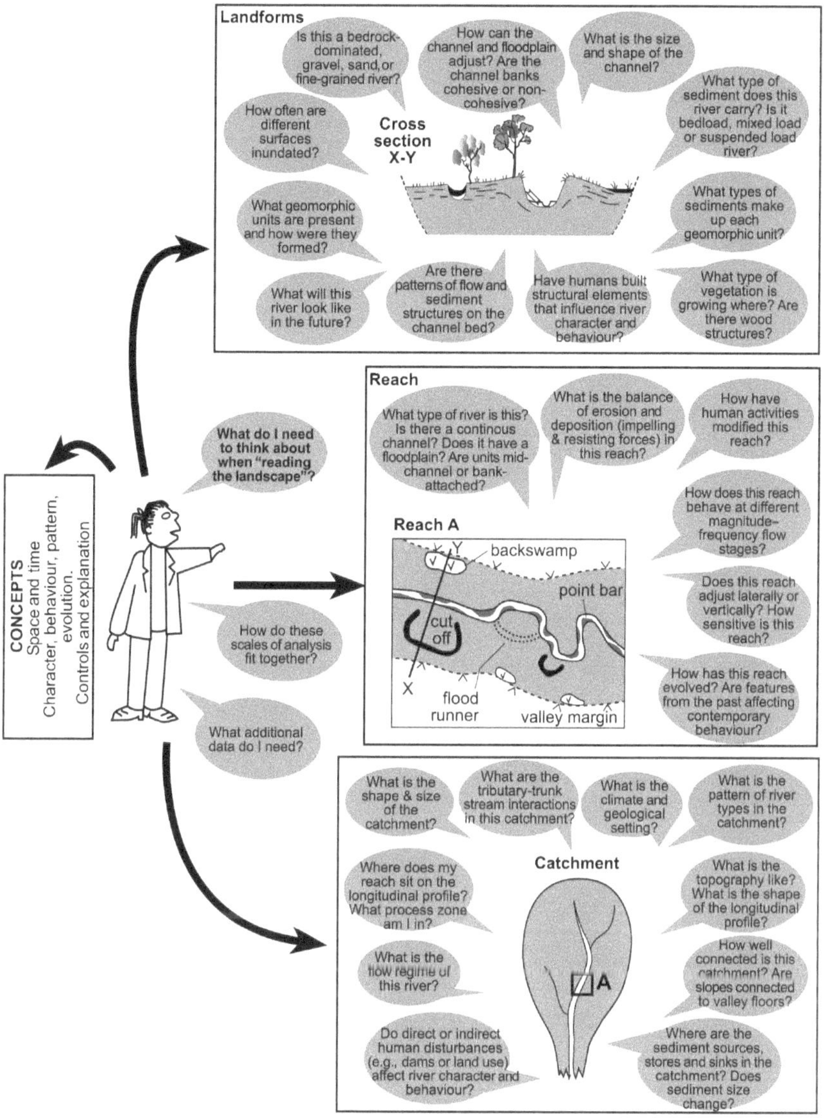

FIGURE 5.3.1 Asking the right questions when making field-based interpretations and reading the landscape (amended from Fryirs and Brierley, 2013)

Particular sets of process-form interactions generate these features at particular positions in landscapes. Each geomorphic unit is characterized by distinctive material (sedimentologic) properties. Features are influenced to differing degrees by flow-sediment-vegetation interactions. At a broader scale, river reaches are comprised of differing landforms that are formed and reworked within a particular behavioral regime. Downstream patterns of reaches, in turn, make up the catchment.

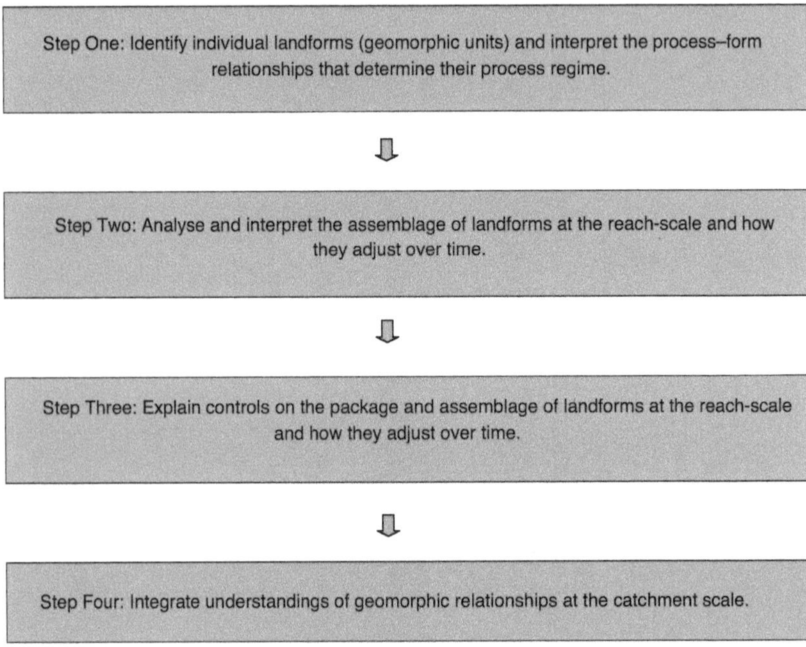

FIGURE 5.3.2 An approach to reading the landscape (modified from Fryirs and Brierley, 2013)

Reach-to-reach and tributary-to-trunk stream linkages determine fluxes of water, sediment, and vegetation, which drive river behavior, evolution, and responses to human disturbance.

Each stage of the approach is scaffolded to produce increasingly sophisticated knowledge of river form, process, and evolution (Fig. 5.3.2). Baseline information generated in Step 1 is required before "moving on" to Steps 2, 3, and 4. As the approach builds, increasing levels of experience and training are required. In applying the principles within this framework, it is recommended that levels of "confidence" are assigned to the interpretations made, and that users identify limitations or gaps in their own knowledge base and skills set. It is important to note that in this chapter, little attention is given to preparatory analyses that are undertaken prior to fieldwork. Literature searches, historical analyses, and remote sensing work are integral stepping stones for effective, contextualized field inquiry, but our focus here lies solely with the field component itself.

5.3.2.1 Step One: Identify Individual Landforms (Geomorphic Units) and the Process-Form Relationships That Determine Their Behavior

Landforms (or geomorphic units) are the component parts of a landscape. In general terms, they form under a given set of energy conditions at particular locations in a landscape. Particular set of processes produce characteristic forms with distinctive shape and size. Although philosophical concerns can be raised as to whether identified features are "real" (i.e., natural kinds; see Rhoads and Thorn, 2011) or simply convenient mental constructs to impose some degree of regularity on a continuum of surface forms (Richards and Clifford, 2011), a constructivist approach to landform identification allows the same procedures and analytical framings to be applied across the spectrum of river diversity (Brierley, 1996; Brierley and Fryirs, 2005).

Fluvial geomorphologists have established good understanding of the process–form (morphodynamic) relationships that create and rework landforms, whereby the process affects the form, and vice versa. Individual landforms have certain material and sedimentologic properties with a characteristic geometry and bounding surfaces (i.e., erosional or depositional contacts). Geomorphic units commonly have distinct vegetation associations reflecting hydrologic and substrate conditions (among many considerations; see Yu et al., 2014). Combinations of erosive and depositional processes that sculpt, create, and rework the feature define the range of behavior of each unit. Analytical procedures can be applied to assess magnitude–frequency relations of formation and reworking. This allows interpretation of the sensitivity/resilience of that feature when it is subjected to disturbance events (i.e., whether the feature will simply have additional deposits added to it, whether it will be partially reworked, or whether it will be destroyed (eroded and removed)). Importantly, analysis of formative flows includes appraisal of the balance of impelling and resisting forces, emphasizing the variable importance of vegetative and biotic interactions at any given locality.

5.3.2.2 Step Two: Analyze and Interpret the Package and Assemblage of Landforms at the Reach-Scale and How They Adjust Over Time

Sections of river with a distinct assemblage and pattern of geomorphic units that reflect particular combinations of erosional and depositional processes are referred to as reaches. By definition, reaches upstream and downstream of a reach boundary are characterized by different packages of landforms (and associated process interactions). Reading the landscape at the reach scale entails assessment of which types of geomorphic units are present (or absent), what types of sediments they are made of, and whether the units are formed and reworked by genetically linked contemporary processes or they reflect former conditions. Interpretation of the array of process–form relationships for the range of geomorphic units along a reach, and associated channel–floodplain interactions (if present), is used to analyze river character and behavior. Adjustments around a characteristic state over geomorphic timeframes determine the *range of behavior* of a river reach, as systems respond to disturbance events. For example, an active meandering river adjusts laterally, whereas a braided river is subject to thalweg shift. Inevitably, the magnitude–frequency domains with which features are generated and interact may vary from system to system.

Significant insights into landscape history can be gained through analysis of whether adjacent features in a landscape are genetically linked or not (i.e., whether they formed contemporaneously, or whether they formed over differing periods of time). This provides guidance into the evolutionary history of a landscape, highlighting erosional events that rework landscape features (i.e., temporal discontinuities). For example, terraces are older than adjacent floodplain and channel features, and they were often formed by quite different processes under differing environmental conditions. Analysis of the pattern of features and the nature of boundaries between them (erosional or depositional contacts) can provide insight into the likely sequence of events that fashioned the evolutionary history (Steps 3 and 4). This includes assessment of what landforms/sediment have been removed (i.e., what is missing due to erosion and limited preservation potential).

5.3.2.3 Step Three: Explain Controls on the Package and Assemblage of Landforms at the Reach-Scale and How They Adjust Over Time

All landscapes adjust and evolve. Determination of the timeframe over which differing features are created and/or reworked is one of the many inherent complexities in analysis of landscapes. Analysis of landscape evolution enables determination of whether the contemporary system adjusts around a characteristic state, adjusts among differing states, or is adjusting along an emergent evolutionary pathway. These interpretations can be used to relate landscape responses to human disturbance to the *natural range of variability* of a system (see Fryirs et al., 2012). Challenges are faced in unraveling how adjustments to one part of a system affect responses elsewhere in that system (e.g., upstream influences such as a sediment slug, or downstream influences such as a head cut). The true value of geomorphic understanding lies in being able to explain the controls that drive process interactions and how they have changed/adjusted over time, and interpreting what has triggered these changes/adjustments.

Differing controls upon landscape behavior operate over variable spatial and temporal scales (Fig. 5.3.3). By definition, the package of geomorphic units at the reach scale is fashioned by a consistent set of controlling factors. Valley setting (slope and width) is the primary determinant of *imposed boundary conditions* that are typically set over timeframes of thousands of years or longer (recognizing that hillslope processes may dramatically alter these relationships in some instances; see Korup, 2013; May et al., 2013). In contrast, flow and sediment transfer relationships that recurrently adjust over shorter timeframes set the *flux boundary conditions*. Primary differences in geomorphic setting (and associated behavioral regime) can be attributed to these geologic (imposed) and climatic (flux) controls. Geologic factors such as tectonic setting, lithology, and resulting topography affect the erodibility and erosivity of a landscape. Climatic factors influence the nature and rate of process activity, such as the geomorphic effectiveness of flood events and associated magnitude-frequency relationships.

Imposed boundary conditions set by:

- Valley confinement
- Valley slope (base level at bedrock riffles and gorges)
- Low relief topography with terraces, rolling, hills and rounded ridges

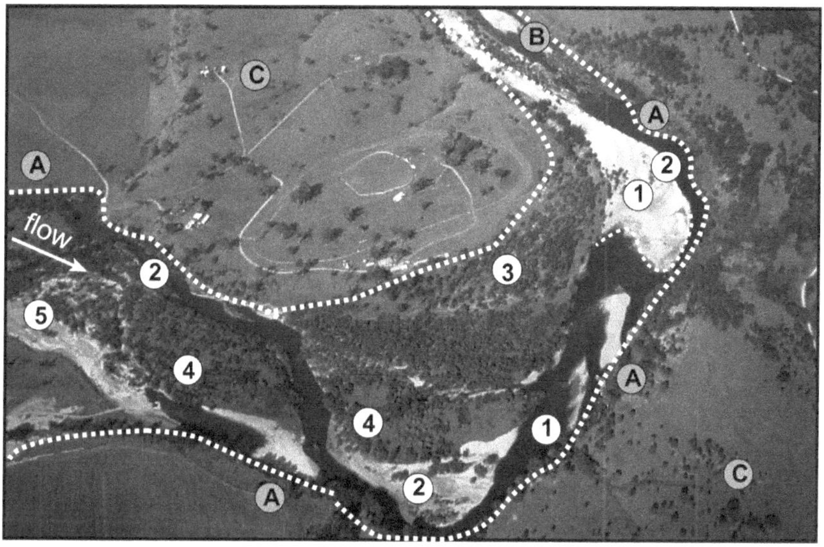

Flux boundary conditions set by interactions between water, sediment, and vegetation:

1. Flow and sediment interaction dictate patterns of erosion, sediment transfer and deposition
2. Erosion and deposition processes produce different types of instream geomorphic units (e.g., bars, riffles, pools)
3. Deposition and reworking processes generate floodplain features
4. Vegetation interactions influence morphodynamics of different geomorphic units (e.g., ridges)
5. Flow and vegetation interactions reflect and induce variable roughness

FIGURE 5.3.3 Imposed and flux-boundary conditions fashion the behavioral regime of a river. In this example, controls upon the behavioral regime of a partly-confined valley with bedrock-controlled discontinuous- floodplain river style are shown (modified from Brierley and Fryirs, 2008).

5.3.2.4 Step Four: Integrate Understandings of Geomorphic Relationships at the Catchment Scale

Drainage basins are comprised of relatively self-contained, gravitationally induced sets of biophysical relationships. The balance of erosional and depositional processes varies markedly in source, transfer, and accumulation zones (Schumm, 1977). Analysis of source to sink relationships at the catchment scale provides the most logical basis to consider the linked nature of spatial and temporal adjustments in landscapes, enabling meaningful interpretation of lagged and offsite responses to disturbance events. The unique configuration and temporal sequence of drivers, disturbances, and responses of each landscape, along with the historical imprint, result in system-specific behavioral and evolutionary traits. It is at this scale that contextual considerations come to the fore, as each reach is viewed in relation to its position within a catchment and associated process relationships. In the second exercise outlined below, these relationships are interpreted through analysis of downstream patterns and controls along longitudinal profiles.

5.3.3 THREE FIELDWORK EXERCISES TO INTERPRET RIVER FORMS, PROCESSES, PATTERNS, AND EVOLUTION

Field-based interpretative skills outlined in this chapter build upon prefieldwork preparation that incorporates analyses of the regional setting (geology, climate, hydrology, flood history, land use change, etc.). Preliminary geomorphic mapping and derivation of longitudinal profiles is typically completed prior to field visits, applying remotely sensed analyses within a GIS framework. This is supported by analysis of environmental history (e.g., flood events), historical air photographs, and associated imagery. In this way, all field sites are framed in their catchment (and landscape) context.

5.3.3.1 Exercise 1: River Reach Analysis

The primary aim of the river-reach analysis exercise (Kellerhals et al., 1976) is to analyze catchment-scale variability and patterns (connectivity) of river character and behavior. In the approach outlined here, river character and behavior are assessed at the reach scale as "River Styles" (Brierley and Fryirs, 2005). River Styles are identified on the basis of three key parameters (channel planform, the assemblage of geomorphic units that make up a reach (both channel and floodplain components), and bed-material texture). Although each River Style has a discrete set of distinguishing attributes, overlap may occur in the range of attributes. Particular emphasis is placed upon analysis of geomorphic units, as these are the key interpretative agent with which to assess process-form interactions (morphodynamics). Four key groupings of instream geomorphic units are differentiated (Fryirs and Brierley, 2013): (1) sculpted erosional bedrock; (2) depositional midchannel; (3) depositional bank attached; and (4) sculpted erosional fine grained. These analyses are framed alongside interpretations of floodplain geomorphic units. In this process-based analytical framing, concern is given to both formative (erosion/deposition) and reworking processes.

Before heading to the field, a preliminary analysis of river reaches is undertaken from remotely sensed data (typically Google Earth© or aerial photographs). From this, reach boundaries are designated based on valley confinement (the distribution of floodplain features) and downstream changes to the nature, distribution, and assemblage of channel and floodplain landforms. Representative (type) locations are designated for each type of river (River Style), for which detailed geomorphic maps are compiled.

Many attributes that are used to distinguish and characterize reaches require field investigation. In particular, the range of geomorphic units, their material properties, bed-material texture, and vegetation associations must be analyzed. Field procedures are used to verify boundaries between River Styles. These boundaries are subsequently plotted atop longitudinal profiles (see Exercise 2). Distinguishing attributes of a reach are differentiated, indicating what has changed upstream and downstream of a reach boundary. These morphologic changes are related to process associations, forming the basis to interpret downstream controls upon types and patterns of rivers. Attention is given to the nature of the reach boundary – is it abrupt, diffuse, or gradual, and why?

By interpreting the form-process associations of geomorphic units, reach-based interpretations of river behavior are made in the field. On the basis of the distribution and presence or absence of geomorphic units, practitioners interpret how the river behaves at low flow, bankfull, and overbank stages. In most cases, low-flow conditions can be directly observed. Flow distribution within the channel is considered, assessing flow interactions with bed materials. To interpret bankfull-stage behavior, students are asked to consider flow alignment within the channel, interpreting how each instream geomorphic unit is formed and reworked and how channel geometry may (or may not) adjust under these conditions. This entails assessment of midchannel and bank-attached features, and erosional and depositional forms. Sedimentologic properties of landforms are analyzed. For example, elongate step-shaped features at channel margins are differentiated into benches if they are depositional (accretionary) forms, or ledges if they are erosional forms that have been excised from the adjacent floodplain. Analyses of overbank-stage behavior require interpretations of the presence/absence and character of floodplain features (e.g., flat-topped floodplains, the presence or absence of levees, ridge and swale topography, palaeochannels, etc.). Formation and reworking processes are considered. Collectively, these insights are used to assess the behavioral regime of a given reach. This incorporates assessment of the form, capacity, ease, and recurrence of geomorphic adjustments (i.e., its sensitivity to adjustment). Analysis of the contemporary behavioral regime underpins subsequent assessment of river evolution (Exercise 3).

Data on each River Style are recorded in a systematic manner to aid the appraisal and communication of river character and behavior (e.g., Brierley and Fryirs, 2005). River Styles proformas are finalized in the field, filling in details on geomorphic units, bed-material texture, and river behavior. Annotated photographs, cross-sections, and planform maps accompany these proformas in the final presentation (see Table 5.3.1, Fig. 5.3.4). In more advanced versions of this exercise, valley-scale cross-sections are surveyed in the field. These data are then combined with discharge data for the 2, 5, 10, 50, and 100 year recurrence

TABLE 5.3.1 Procedures Used to Interpret River Styles in the Field

Task	Procedure
Rectify the planform map	• Walk the reach, amending the geomorphic unit base map as required. • Record the position, assemblage, pattern and boundaries of instream and floodplain geomorphic units on the map. • Note within reach and downstream changes in the number of channels, lateral stability, sinuosity, channel geometry, bank height and character, sediment storage, channel obstruction and vegetation structure and character on the map.
Produce a valley-wide cross-section and estimate Manning's *n* for each surface	• Based on the rectified geomorphic unit map, select a "representative" site to complete a valley-scale cross-section, assessing various attributes of each geomorphic unit. More than one cross-section may be required to ensure that attributes of all geomorphic units are recorded for that reach. • Working from left to right (looking downstream), record all breaks in slope across the valley cross-section, recording the dimensions and geometry of each geomorphic unit on the proforma. • Visually assess Manning's *n* roughness coefficient for each geomorphic surface across the cross-section.
Assess channel geometry and shape	• Determine channel width, depth and width-depth ratio, noting variability along the reach. • Characterize channel shape (symmetrical, asymmetrical, compound or irregular) • Analyze bed and bank sedimentology
Assess the shape, size, and sediment character of each geomorphic unit	• Measure the width, length and depth of each geomorphic unit • Assess the sediment character of each instream and floodplain geomorphic unit (size, sorting and sediment mix (i.e., loose or cohesive)).
Assess bed material texture	For a gravel-bed river: • Differentiate among the "active" gravel load (recently moved fraction) and the coarser "fabric" of the bed. Perform separate "b" axis bed material size analyses for each fraction (ideally 100 clast samples or more). For a sand bed river: • Analyze the grain size distribution using a 0.5Ø particle size analysis card.
Interpret vegetation associations	• Note the distribution and type of vegetation (and any wood) on the planform map. • Assess riparian vegetation associations (type, size, density, percent cover) for each instream and floodplain geomorphic unit.
Photograph the River Style	• Annotate representative photographs for each River Style, clearly differentiating key (distinguishing) geomorphic units. Include the date and location of each photograph.

From Brierley and Fryirs (2005).

Reach number and name: **Type of river (River Style):**
Grid reference for upstream boundary: **Grid reference for downstream boundary:**
What defines the upstream and downstream boundary of the reach?

DETAILS OF ANALYSIS
Map sheet(s) air photographs used:
Analysts: *Date:*

RIVER CHARACTER	
Valley-setting	
River planform • Sinuosity • # of channels • forms of channel adjustment	
Channel geometry (size and shape)	
Bed material texture	
Geomorphic units (geometry, sedimentology)	*Instream*
	Floodplain
Vegetation associations and presence/absence and character of wood	*Instream geomorphic units*
	Floodplain geomorphic units

RIVER BEHAVIOUR
Low flow stage
Bankfull stage
Overbank stage

CONTROLS	
Within-catchment position	
Sediment process zone	
Valley Morphology (size and shape)	
Valley slope	
Stream power	

Schematic valley-scale cross-section, planform map, and annotated photograph

FIGURE 5.3.4 Proforma used to summarize attributes of river character and behavior as a part of field-based assessments of River Styles (from Brierley and Fryirs, 2005). Factors that influence the process regime and pattern and distribution of River Styles are also assessed.

flows, as well as bed material texture and field-based assessments of Manning's n, to model the range of unit stream power conditions and sediment transport capacity relationships for any given reach. These data are postprocessed in the classroom and added to the River Styles proformas. Collectively, this information can be used to explain the influence and balance of erosion and deposition processes in a reach and resulting adjustments to channel shape and size. These insights feed into analyses of flux boundary condition controls on the patterns of river types within any given river system (see Exercise 2).

In running these exercises as part of university field classes in Australasia, groups are designated on the basis of the mix of available skill sets and experience (e.g., familiarity in geology, soil science, hydrology, ecology, vegetation analysis, GIS, etc.). Efforts are made to enhance individual experiences through group-based learning. Each group is designated a 5–7 km section of river to walk and analyze. The selected section of river typically has two or three reaches. Contiguous sections are chosen for differing groups, typically including both the trunk stream and tributaries. Depending on the nature of the reach, a determination is made as to the appropriateness of the students to walk that section of river (in terms of terrain, fitness, outdoor experience, and safety concerns).

Each group is tasked to identify reach boundaries on the basis of downstream changes to the assemblage of geomorphic units (channel and/or floodplain features) for their section of river. They are required to collect data and complete a proforma for each reach along their designated section of river. By the end of the day, the entire length of a river has been covered and analysis of downstream patterns along the length of river can be undertaken (see Exercise 2 below). Groups take photographs and videos throughout the day and present this to the class in the evening so that each student has an appreciation of the range of river diversity and geomorphic characteristics of each section (and reach) of river. As a bit of fun in setting up this exercise, a geomorphic unit quiz is frequently held prior to the field excursion, and the group that performs least successfully in the quiz is given a stuffed toy animal to incorporate into photographs of geomorphic units in the field exercise. Some truly memorable images result (what happens on field trips stays on field trips)!

During the evening session, the class as a whole builds a picture of the entire catchment (or section of catchment) under investigation. Emphasis is placed upon determination of downstream transitions along tributaries and the trunk stream, discussing how river morphology and associated process relationships have changed (and why). Each group is tasked to assess and reconsider how many reaches they identified in the field. Some groups are "splitters"; others are "clumpers." Within-group discussions about "what is changing along the river, and why" are key learning elements of this exercise. Particular emphasis is placed upon identification and verification of field boundaries. Once these determinations have been made, the group then formalizes information on each reach on the River Styles proforma and generates a representative planform map and valley scale cross-section for that reach (i.e., a completed version of the proforma shown in Fig. 5.3.4).

An overview of primary attributes of each reach is presented to the class as a whole, working from upstream to downstream groups. Tributary reach analyses are incorporated along the way. Presentations provide an overview of the range of river types, their key characteristics, and their downstream patterns. In the subsequent assignment, students are tasked to compile information across the catchment, integrating information from all groups. This entails synthesis of reach-scale information into differing river types, for which a representative proforma is constructed for each River Style. Particular attention is given to analysis of the behavioral regime of each type of river. Presentations during the evening session must be carefully orchestrated to ensure that the overall class has a coherent synthesis of catchment-wide types and patterns of rivers (the baseline for analyses of river types along longitudinal profiles that is outlined as Exercise 2 below).

Results sections for reports are framed around the catchment-wide map of River Styles. A summary table identifies the distinguishing attributes of each River Style. An accompanying set of tables systematically overviews key attributes of each river type, including full documentation of their character and behavior, an annotated planform map (alongside an image from Google Earth©), and a schematic valley-scale cross-section (both the map and cross-section should incorporate the distinguishing geomorphic units for that reach of river).

5.3.3.2 Exercise 2: Using Longitudinal Profiles as a Teaching and Interpretation Tool: Assessing Downstream Patterns of River Styles and Controls on Their Character and Behavior

To develop an understanding of how and why each reach looks and behaves in the manner that it does, a summary assessment of controls on the distribution of River Styles is developed. Key controls on river behavior may vary from reach to reach. Initial insights are gained by plotting downstream patterns of River Styles onto longitudinal profiles (Figs. 5.3.5 and 5.3.6). The diagram showing reach boundaries atop the longitudinal profile is accompanied by plots of downstream trends in channel geometry, grain size, instream, and floodplain geomorphic unit assemblage, and so forth. (Fig. 5.3.5). Tributary inputs are represented on these diagrams using the catchment area plot (Fig. 5.3.6). Slope and discharge relationships and associated ranges of gross stream-power conditions under which differing River Styles are observed are also shown in Fig. 5.3.6. This presents a basis to examine linkages along river courses, placing each reach in its catchment context. When linked to the nature of the sediment transport regime (i.e., whether the river is bedload, suspended load, or mixed load dominated), process zone functioning is interpreted (i.e., whether the reach operates as a sediment source, transfer, or accumulation zone; see Schumm, 1977).

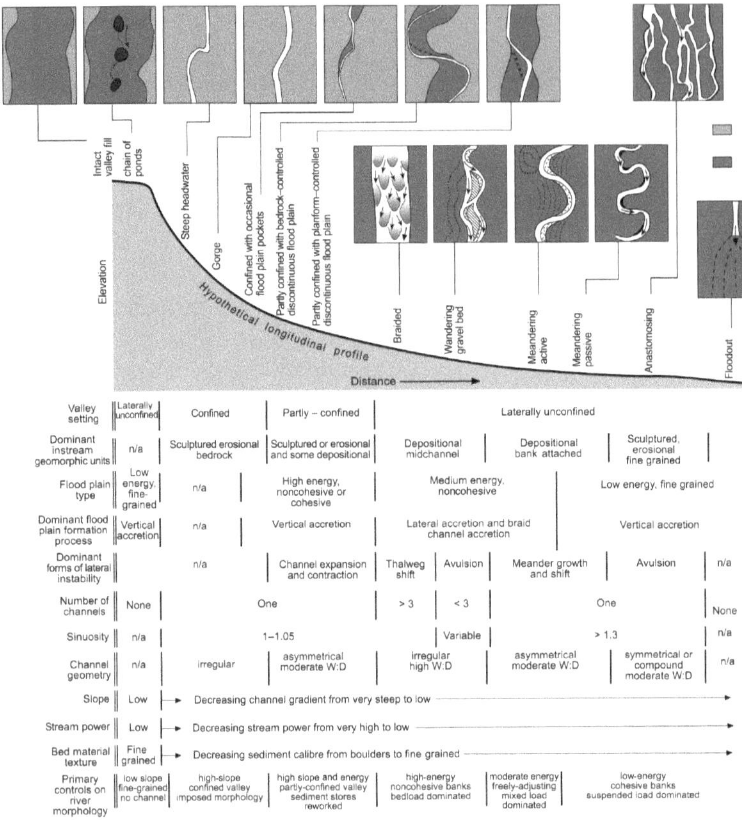

FIGURE 5.3.5 Text-book representation of downstream transitions in primary morphological and behavioral attributes of a river system shown atop a hypothetical longitudinal profile (modified from Fryirs and Brierley, 2013).

FIGURE 5.3.6 Downstream changes in river character and behavior for the Bemboka-Bega River, South Coast, New South Wales, Australia (modified from Brierley and Fryirs, 2005). The upper image shows the cumulative contributing area, with abrupt increases marking tributary inputs. The longitudinal profile portrays the prominence of the escarpment along this part of the eastern seaboard of Australia. The jagged peaks reflect slope (and area) induced increases in gross stream power (2-year-recurrence interval event) at differing positions along the longitudinal profile. The uplands landscape unit drains small but relatively steep areas characterized by the Steep headwater River Style. A gorge occurs at the distinct step in the longitudinal profile coincident with the escarpment zone. Beyond the escarpment, the longitudinal profile has a relatively smooth, concave-upwards form. The progressive decline in slope is accompanied by progressive widening of both the channel and the valley through the base of the escarpment, the rounded foothills and the lowland plain landscape units. A Low sinuosity boulder bed River Style with fan deposits occurs under high stream power conditions at the base of the escarpment. The Confined valley with occasional floodplain pockets River Style is evident along most of the moderate slopes of the Bega-Bemboka trunk stream. Instream areas are dominated by sculpted, erosional bedrock geomorphic units, whereas small floodplain pockets are regularly stripped in this setting. Gross stream powers are moderate, increasing markedly at steeper bedrock steps along the longitudinal profile (e.g., at ~60km).

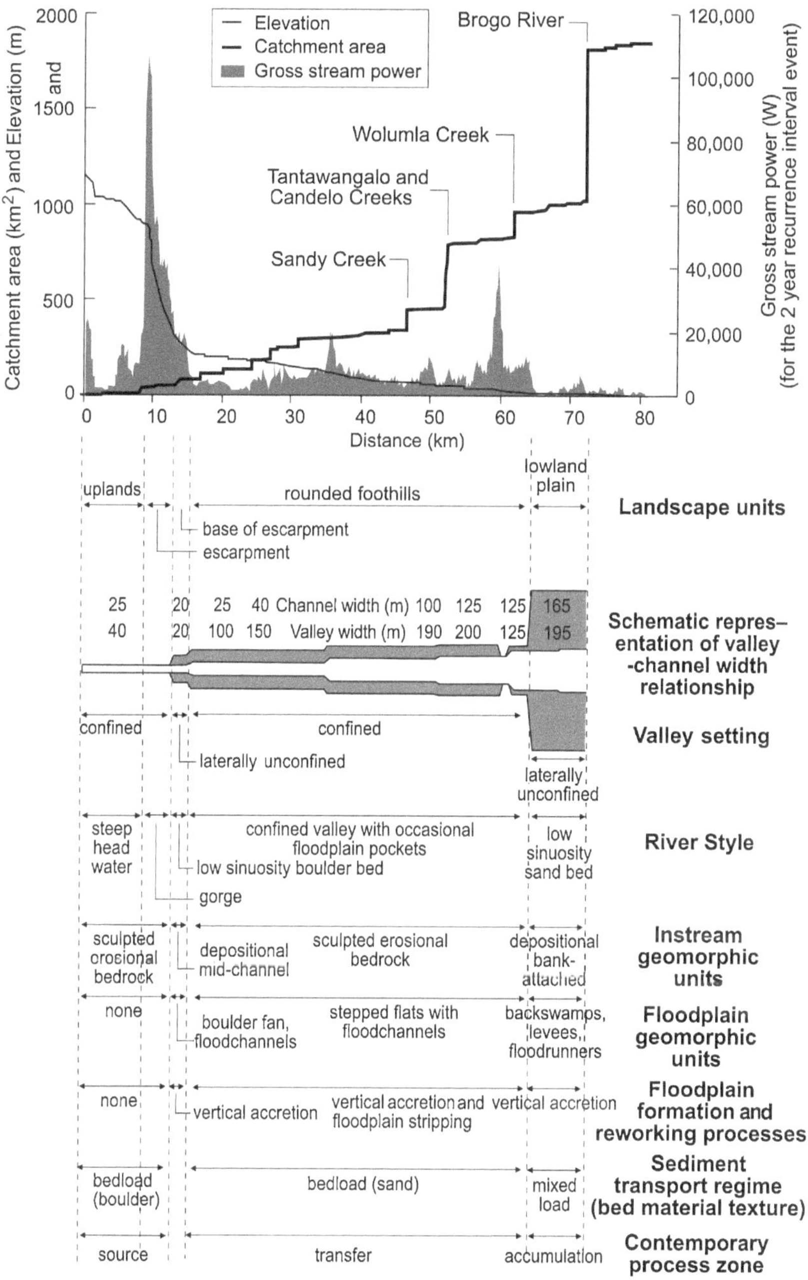

FIGURE 5.3.6 ***(cont.)*** It is not until the valley widens and slope decreases beyond Wolumla Creek confluence that the transition to a Low sinuosity sand bed River Style occurs. This laterally-unconfined river of the lowland plain is dominated by depositional, bank-attached instream geomorphic units with extensive, vertically accreted floodplains with levees and backswamps. The decline in slope and gross stream power induces a marked decrease in bed material texture to sandy, mixed load sediments.

The first step in these analyses entails placement of River Styles boundaries atop the longitudinal profile plots. Downstream patterns of River Styles are determined for all river courses in the catchment, differentiating those that have similar downstream patterns from those that are distinct. This provides a basis to assess the similarity in downstream controls on patterns of River Styles, helping to explain the distribution of river character and behavior in different subcatchments.

Adding data derived from air photograph and field interpretation to a longitudinal profile is a valuable teaching and learning exercise. This can initially be done on a white board in the field with groups of students adding information from their river reach analysis (Exercise 1) underneath the longitudinal profile of the river (as per Figs. 5.3.5 and 5.3.6). These diagrams facilitate understandings of a wide range of cross-scalar data, highlighting linkages between differing patterns, controls and process relationships. As a starting point, determinations can be made as to what has (or has not) changed at the boundaries between River Styles. A typical question asked at this stage of analysis is: What attributes have changed upstream and downstream of the boundary, and why has that change occurred?

Isolating controls on river behavior is often challenging, and a process of elimination is suggested. Differentiating imposed-boundary conditions from flux boundary conditions helps in this assessment (Fig. 5.3.3). Considerable overlap may exist in imposed-boundary conditions, as a range of River Styles may form under similar sets of these controls. The types of questions asked in these analyses include:

- Can discernible changes in imposed boundary conditions be isolated in the transitional zones between River Styles?
- Do transition zones coincide with changes in valley confinement, slope, and base-level controls along the longitudinal profile, or geological (lithologic and/or structural) changes?
- Do transition zones coincide with tributary-trunk stream confluences?
- Do certain River Styles occur in certain landscape units at particular positions in the catchment?
- Have particular geologic or climatic events influenced the distribution of river types (e.g., tectonic uplift, volcanic inputs, patterns of terraces)?

To make an interpretation of flux boundary conditions the type, position, and pattern of rivers is compared to a range of controls upon flow and sediment fluxes. Stream power provides a guide to the energy regime at differing positions along a river (see Jain et al., 2006, 2008). Outputs required to generate stream power include contributing area and a running average of slope. These can be plotted as a continuous data set along the longitudinal profile as gross stream power. If possible, unit stream power estimates are derived for each River Style, recognizing explicitly that the range of estimates may vary markedly for differing reaches. Generally, only one recurrence interval relationship (e.g., 1 in 2 years or 1 in 5 years event) is depicted, but all values are presented in a summary table of controls on the character, behavior, and pattern of River Styles. The distribution of geomorphic process zones (i.e., source, transfer, and accumulation zones) and associated sediment transport regime (i.e., bedload, mixed load, or suspended load system) are plotted beneath these curves. An estimate is made as to whether each reach is sediment supply or transport limited. In general, rivers in confined and partly-confined valley settings act as sediment transfer reaches, as sediments are readily flushed through these parts of the catchment (see Fryirs and Brierley, 2001; Macnab et al., 2006; Kuo and Brierley, 2013). Rivers in laterally unconfined valley settings tend to be accumulation zones, unless the reach responds to disturbance by reworking its sediment stores, thereby acting as a sediment source zone. The types of questions asked at this stage of analysis include:

- How do downstream changes to flow and sediment fluxes affect transitions in source, transfer, and accumulation zones along the longitudinal profile, and resulting patterns of river types?
- Do downstream changes in River Style reflect a change in stream power (if so, give ranges of stream power for each type)?
- How have human factors such as dams or river management applications affected flux boundary conditions and the distribution of river types?

Interpretation of controls on river character and behavior typically results in assessments that are catchment (or region) specific.

5.3.3.3 Exercise 3: Scenario Building to Interpret River Evolution and Future Trajectory of Change

This exercise extends the analysis of contemporary river adjustment (behavioral regime) completed during Exercise 1. Geomorphic and sedimentologic analyses of bank exposures and floodplain units are used to assess river evolution for a valley cross-section. This can be complemented by seismic or radar methods to analyze subsurface sediments, but is most commonly conducted using pits and auger holes. Continuous bank exposures are especially important as they allow interpretations of the geometry of depositional units, and the nature of the bounding surface that separates differing features. Textural attributes and bedform structures are assessed for each bedded unit to define facies units. Commonly, a range of facies units can be identified within broader-scale features such as channel fills. The latter scale features are variously referred to as elements, architectural elements, or morphostratigraphic units (Brierley, 1996). Essentially, these features are defined on the basis of their geometry and their sediment composition – they are the depositional record of geomorphic units. These sediment attributes provide insight into phases of river activity. Sediment reworking is marked by erosional contacts (bounding surfaces). Analysis of the package or stack of elements guides interpretations of the lateral and vertical adjustment of the channel, and its relation to floodplain features. These field-based geomorphic and sedimentologic analyses are typically supplemented by interpretation of historical maps and air photographs, satellite images and local environmental histories to derive evolutionary scenarios for the reach.

Descriptions and interpretations of the sedimentology of differing channel and floodplain geomorphic units (their juxtaposition, stacking arrangement, and indications of primary depositional processes and reworking) are used to reconstruct schematic valley cross-sections and planform maps of river landforms for various timeslices. The contemporary cross-section is used as the starting point. Sediment columns for the various geomorphic units are superimposed atop this cross-section. This aids reconstruction of former surfaces, such as terrace levels, differing channel positions, or adjustments to channel geometry. From this, inferences can be made about the former character and behavior of the river. Often, sedimentologic insights can be used alongside vegetation attributes (position and inferred age of trees) to provide analysis of channel and floodplain conditions around 1950. These analyses are aided by assessment of air photograph and Google Earth© images.

In making assessments of river adjustment/evolution, river behavior is differentiated from river change. A wholesale shift to a different type of river with a different behavioral regime is referred to as *river change*, whereas adjustments within the natural range of variability for that river type are referred to as *river behavior* (Brierley and Fryirs, 2005; Fryirs and Brierley, 2013). Making this distinction is an important interpretative skill. Simple, field-based observational assessments can provide critical insight for such analyses. For example, abandoned channels on the floodplain may provide guidance into former channel geometry and planform conditions. Terraces or incised channel conditions provide guidance into altered bed-level conditions. Bench or ledge features provide insight into channel contraction or expansion. The lack of indicators of change is also a critical field-interpretation skill. For example, fine-grained floodplains and a lack of palaeochannels can be a key indicator of resistance to adjustment. Essentially, these analyses entail interpretation of what has gone on from what is left, interpreting whether the structure and function of the contemporary river is different to that of the past.

Endeavors to unravel sequences of events and the nature and timing of formative events often entail interpretation of what's missing (which events stripped and reworked parts of the depositional record). Typically the depositional record is quite patchy across most of a catchment. Prospect for more complete records occur in downstream accumulation zones. In many instances, intriguing insights such as an erosional contact or a definitive geomorphic unit provide confirmatory evidence that "nails" a story. Harden (2013) refers to this as "aha" fieldwork moments. All too often, however, frustrations are encountered in the quest for confirmatory evidence. Hence, an open mind and commitment to multiple working hypotheses is required.

As a starting point for these interpretations, the question is asked: In what ways can this type of river adjust under the prevailing set of flow, sediment, and vegetation characteristics? Three degrees of freedom are appraised to assess the capacity for adjustment, namely, bed character, channel character, and river planform (Brierley and Fryirs, 2005). These are directly comparable to scales of analysis described in Exercise 1. In some cases, reaches have significant potential to adjust, whereas in others only localized or minimal adjustment can take place.

Building upon assessments of process–form relationships that characterize the contemporary reach, practitioners are asked to draw schematic cross-sections for different timeslices and to provide field-observed evidence for the character and behavior of the river at these times. Historical sources and field evidence provide insights that can guide the construction of timeslices for recent decades (say, 1950). In an Australasian context, much can be gained by extending analyses back to the time of European settlement (say 1800). In this instance, field evidence is used to support inferences made about changes to flow and sediment conditions in the reach, and associated adjustments to channel geometry and planform (was the channel smaller or larger; is there any indication of adjustments to bed level (aggradation or degradation); are alterations to channel planform likely to have occurred, with associated adjustments to channel-floodplain interactions?). Subsequent analyses (and timeslice construction) can be performed for much longer timeframes, invoking adjustment to sea level or known periods of climate change, or specific geological events that may have occurred in that region (in most instances students are asked to reconstruct conditions at 5 ka and the Last Glacial Maximum (15 ka), on the basis of features such as terraces at valley margins that provide inferential insights into former bed level conditions, and associated sedimentologic attributes that provide insight into the sediment regime of the river).

Presentation and explanation of the surveyed cross-section with sediment columns superimposed atop it, along with the schematic representations of valley cross-sections and channel planform for each timeslice, provide a platform to interpret adjustments to river character and behavior over time. Observational evidence from geomorphic unit assemblages (reading the landscape) and their sediment attributes are related to "textbook" understandings of geologic, climatic, and land-use conditions (including flood history and river management activities) to support inferences. Fig. 5.3.7, based on Johnston and Brierley (2006), presents a simplified version of these analyses. To support these analyses, ergodic reasoning is also used to provide students with a visual sense of the range of river character that may have occurred for that river type elsewhere within the region (Fryirs et al., 2012).

For a bit of fun, students are asked to use their evolutionary trajectory to make predictions of likely future adjustments over future decades (say, 2050), commenting in presentations on their optimistic or pessimistic worldview. The range of future scenarios is based on the flux boundary conditions operating in

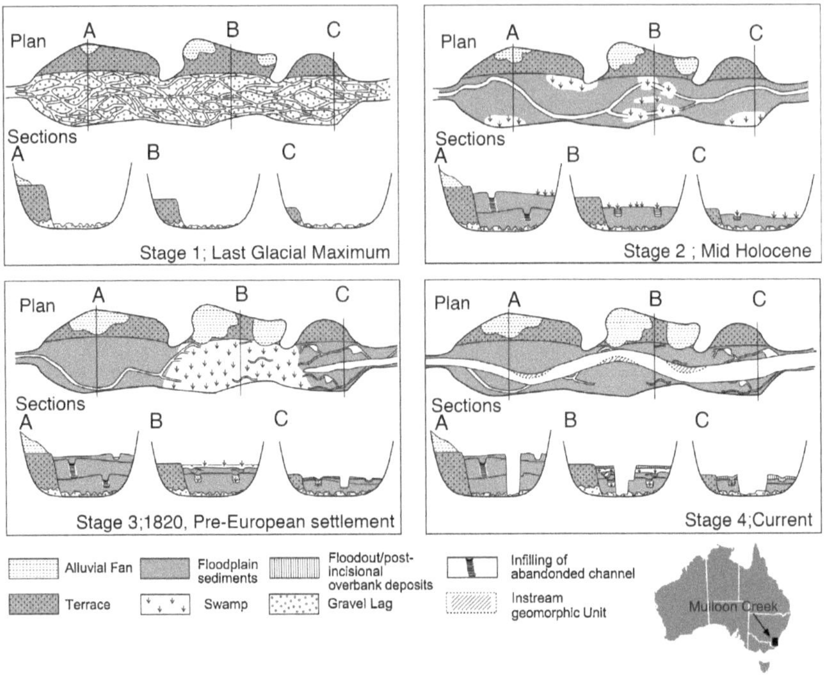

FIGURE 5.3.7 Schematic representation of evolutionary timeslices showing changes to channel planform and valley cross-sections (channel-floodplain relationships) for upper-mid catchment reaches of Mulloon Creek, New South Wales, Australia (modified from Johnston and Brierley, 2006; Fryirs and Brierley, 2013). Valley confinement restricts floodplain development to a series of distinct pockets in this non-glaciated landscape. These pockets comprise a downstream-thinning wedge of sediment packages. Stage 1 depicts a bedload-dominated braided river with valley marginal fan and terrace sequences that are inferred to have occurred at the last glacial maximum. Some time before 12 500 years ago the river was transformed into a suspended load system. By mid-Holocene, vertically accreted swampy, valley-fill deposits extended down pocket as sediment lobes with perched channels (Stage 2). Sedimentation at the head and center of the swamp starved the downstream areas of sediment, forming a wedge-shaped deposit that decreases in thickness down pocket (Stage 3). A channel extended through the pocket shortly after European settlement (Stage 4). Today, a wedge-shaped unit of post-incisional alluvium thickens down pocket, commencing in mid-pocket atop the swamp facies.

the system (from Exercise 2) and the range of potential pressures and limiting factors operating in the catchment. Common factors considered are the impact of climate change, changes to land use, and the impact of river management activities.

5.3.4 DISCUSSION AND CONCLUDING COMMENTS

"The intuitive mind is a sacred gift and the rational mind is a faithful servant. We have created a society that honours the servant and has forgotten the gift."

Albert Einstein, quoted in Davis (2009, p. 116)

Geomorphic inquiry is a critical component in resource and environmental management, addressing key concerns for the sustainability agenda, climate change, management of risk, ecosystem services, and so forth. In the river management arena, systematic catchment-framed understandings of landscapes provide a platform to develop coherent cross-disciplinary conceptual models that articulate how a particular river system works (see Mika et al., 2010). Such toolkits can be used to predict future scenarios and appraise system responses to disturbance events and management interventions. Key issues to be addressed in such endeavors include: What changes are likely to occur, what may trigger them, and can critical thresholds of potential concern be identified? Importantly, field investigations prompt progressive refinement of conceptual models, improving experimental designs and procedures while supporting efforts to make more effective use of emerging technologies and measurement techniques.

Principles outlined in this chapter emphasize two key points. First, fieldwork analyses in fluvial geomorphology must be appropriately contextualized in space and time, relating coherent place-based (catchment-framed) information bases and understandings to theoretical principles (commonly expressed as "text book knowledge"). Second, the procedures and exercises outlined here apply bottom-up, constructivist principles that endeavor to allow a landscape to "speak for itself," rather than conforming to a prescribed "type." Our approach to reading the landscape entails identification, analysis and interpretation of geodiversity (form), behavioral traits (process), spatial dynamics and linkages (pattern), and evolutionary trajectory. Collectively, these understandings provide a geomorphic template that can inform detailed measurement, monitoring and modeling applications that support quantitative predictions (Small and Doyle, 2012). By design and intent, procedures outlined here develop place-based understandings that extend the empirical, ground-based traditions of geomorphic inquiry, while embracing a theory-driven, conceptually intensive future that is underpinned by increasingly sophisticated mapping, measurement and dating techniques (Legleiter and Marston, 2013). In these investigations, sufficient space must be retained to allow "aha" moments to emerge (Harden, 2013).

Although fieldwork is framed here as a bottom-up exercise, emphasis is placed upon generation of systematic, conceptually sound information bases and understandings that are appropriately contextualized. Site-specific observations and analyses are viewed in their "big picture" context. The various underlying themes and threads brought together in these analyses are captured in a memorable acronym: CRAPPER.

- **C**atchment-scale considerations because **C**ontext is vital (How does the configuration of the catchment affect patterns and rates of gravitationally-induced biophysical interactions?) If working in a part of the world where the term watershed is used in place of catchment, the acronym simply becomes WRAPPER.
- **R**iver diversity at the **R**each scale (What is the type of river under consideration, and where does that type of river fit in the spectrum of river diversity?)

- **A**djustment (How does that type of river adjust; what is its behavioral regime?)
- **P**rocess (What are the key formative processes for that type of river; how frequently do those formative events occur (i.e., magnitude-frequency relations and geomorphic effectiveness)?)
- **P**attern (How does the part of the river system under consideration link with other parts of the catchment; how effective are connectivity linkages (upstream-downstream, tributary-trunk stream, hillslope-valley floor, channel-floodplain, surface-subsurface)?)
- **E**volution (How has that landscape adjusted over time; how do contemporary conditions relate to the imprint from the past (landscape memory); what events determined the "preserved" record?)
- **R**ecovery (How is the landscape likely to adjust into the future; how can insights into likely future trajectories inform prospects for **R**ehabilitation?)

A discerning eye is the key to interpretative skills that underpin effective practice in field-based, detective style analyses of landscapes. Unraveling causality may be relatively straightforward in some instances, but elsewhere an instinct (intuition) for relevant details, often accompanied by a healthy dose of good fortune, accompanies critical thinking. Serendipity is largely underplayed in efforts to allow landscapes to "speak for themselves," as availability of information, and access to it, is patchy for any given landscape. Open-ended thinking recognizes prospects for multiple possible causes. Processes of elimination through abduction (inference) engage meaningfully with notions of geomorphic convergence (equifinality – similar form, different formative process) and complex response (similar formative process, divergent responses). Experience and training may help, but in reality some people seem to have an intuitive knack for these endeavors, whereas others do not (Sauer, 1956). Working with as many practitioners as possible, with differing backgrounds and skill sets, may prove to be invaluable. Such is the wonder of geomorphic fieldwork, as opportunities to build our knowledge base and apply our craft in a range of landscapes are boundless (indeed, for many practitioners it is virtually impossible to turn-off such thinking processes, whether engaged in professional activities or otherwise). In making these various assertions, it is important to remember that key insights are not just provided by interpretations of the landscape itself. Key guidance may be provided by local residents – it surely pays to ask with respect and humility. Indeed, incorporation of local knowledge is increasingly integral to teaching and management practices (e.g., Brierley et al., 2002; Lane et al., 2011; McEwen et al., 2014; Wilcock et al., 2013).

Appraising what is similar from what is different is an experiential construct that changes with time and experience. Effective practice builds upon carefully targeted systematic analyses that ask the right questions, ensuring that methods are appropriately documented, the rationale for site selection and field procedures is provided, observations and measurements (description) are clearly separated from analyses and interpretations (explanation), and the evidence base that supports the narratives that emerge is conveyed in a transparent manner. Important considerations must be addressed in appraising reliable transferability of understandings from one situation to another. Differing combinations of roles support these exercises. Moments of inspiration are but a small part of systematic data gathering and analytical phases, wherein appropriately conceptually framed analyses (e.g., CRAPPER) are undertaken with considerable hard work and graft. Why would anyone expect this to be simple? Immense fun is to be had in fashioning encounters for the next field campaign.

ACKNOWLEDGEMENTS

We thank the editor and publisher for the opportunity to reflect upon the importance of fieldwork in geomorphology. Writing this chapter has prompted many memories of amazing experiences and engaging conversations. Practice is about people, and we have been blessed in sharing great times in the field with countless great colleagues and classes. Long may it continue.

REFERENCES

Baker, V.R., 1999. Geosemiosis. Bull. Geol. Soc. Am. 111, 633–645.

Baker, V.R., Twidale, C.R., 1991. The re-enchantment of geomorphology. Geomorphology 4, 73–100.

Brierley, G.J., 1996. Channel morphology and element assemblages: a constructivist approach to facies modelling. In: Carling, P., Dawson, M. (Eds.), Advances in Fluvial Dynamics and Stratigraphy. Wiley Interscience, Chichester, UK, pp. 263–298.

Brierley, G.J., 2010. Landscape memory: the imprint of the past on contemporary landscape forms and processes. Area 42, 76–85.

Brierley, G.J., Fryirs, K.A., 2005. Geomorphology and River Management: Applications of the River Styles Framework. Blackwell Publications, Oxford, UK, pp. 398.

Brierley, G.J., Fryirs, K.A., 2008. River Futures: An Integrative Scientific Approach to River Repair. Island Press, Washington, DC, pp. 303.

Brierley, G.J., Fryirs, K.A., Cullum, C., Tadaki, M., Huang, H.Q., Blue, B., 2013. Reading the landscape: integrating the theory and practice of geomorphology to develop place-based understandings of river systems. Prog. Phys. Geogr. 37, 601–621.

Brierley, G.J., Hillman, M., Devonshire, L., 2002. Learning to participate: responding to challenges in Australian land and water management policy and practice. Aust. J. Environ. Educ. 18, 7–13.

Carbonneau, P., Fonstad, M.A., Marcus, W.A., Dugdale, S.J., 2012. Making riverscapes real. Geomorphology 137, 74–86.

Chorley, R.J., 1978. Bases for theory in geomorphology. In: Embleton, C., Brunsden, D., Jones, D.K.C. (Eds.), Geomorphology: Present Problems and Future Prospects. Oxford University Press, Oxford, pp. 1–24.

Church, M., 2011. Observations and experiments. In: Gregory, K.J., Goudie, A. (Eds.), The SAGE Handbook of Geomorphology. SAGE, London, pp. 121–141.

Church, M., 2013. Refocusing geomorphology: field work in four acts. Geomorphology 200, 184–192.

Davis, W., 2009. The Wayfinders. Why Ancient Wisdom Matters in the Modern World. Anansi, Scarborough, Canada.

Fisher, S.G., 1997. Creativity, idea generation, and the functional morphology of streams. J. N. Am. Benthol. Soc. 16, 305–318.

Frissell, C.A., Liss, W.J., Warren, C.E., Hurley, M.D., 1986. A hierarchical framework for stream habitat classification: viewing streams in a watershed context. Environ. Manage. 10, 199–214.

Frodeman, R., 1995. Geological reasoning: geology as an interpretive and historical science. Bull. Geol. Soc. Am. 107, 960–968.

Fryirs, K.A., Brierley, G.J., 2001. Variability in sediment delivery and storage along river courses in Bega catchment, NSW. Australia: implications for geomorphic river recovery. Geomorphology 38, 237–265.

Fryirs, K., Brierley, G.J., 2013. Geomorphic Analysis of River Systems. An Approach to Reading the Landscape. Blackwell, Oxford, UK.

Fryirs, K.A., Brierley, G.J., Erskine, W.D., 2012. Use of ergodic reasoning to reconstruct the historical range of variability and evolutionary trajectory of rivers. Earth Surf. Proc. Land. 37, 763–773.

Grant, G.E., O'Connor, J.E., Wolman, M.G., 2013. A river runs through it. Conceptual models in fluvial geomorphology. Schroder, J., Switzer, A.D., Kennedy, D.M. (Eds.), Treatise on Geomorphology9Academic Press, San Diego, CA pp. 6–21, Fluvial Geomorphology.

Harden, C.P., 2013. Geomorphology in context: dispatches from the field. Geomorphology 200, 34–41.

Jain, V., Preston, N., Fryirs, K.A., Brierley, G.J., 2006. Comparative assessment of three approaches for deriving stream power plots along long profiles in the upper Hunter River catchment, New South Wales, Australia. Geomorphology 74, 297–317.

Jain, V., Fryirs, K.A., Brierley, G.J., 2008. Where do floodplains begin? The role of total stream power and longitudinal profile form on floodplain initiation processes. Geol. Soc.Am. Bull. 120, 127–141.

Jennings, J.N., 1973. Any Millenniums Today, Lady? The geomorphic bandwagon parade. Aust. Geogr. Stud. 11, 115–133.

Johnston, P., Brierley, G.J., 2006. Late Quaternary river evolution of floodplain pockets along Mulloon Creek, New South Wales, Australia. The Holocene 16, 661–674.

Kellerhals, R., Church, M., Bray, D.I., 1976. Classification and analysis of river processes, Journal of the Hydraulics Division. Proceedings of the American Society of Civil Engineers 102(HY7), 813–829.

Korup, O., 2013. Landslides in the fluvial system. In: Schroder, J.F., Wohl, E. (Eds.), Treatise on Geomorphology. Academic Press, San Diego, pp. 475–510.

Kuo, C.W., Brierley, G.J., 2013. The influence of landscape configuration upon patterns of sediment storage in a highly connected river system. Geomorphology 180, 255–266.

Lane, S.N., Odoni, N., Landström, C., Whatmore, S.J., Ward, N., Bradley, S., 2011. Doing flood risk science differently: an experiment in radical scientific method. T. I. Brit. Geogr. 36, 15–36.

Legleiter, C.J., Marston, R.A., 2013. Introduction to the special issue: the field tradition in geomorphology. Geomorphology 200, 1–8.

Macnab, K., Jacobson, C., Brierley, G.J., 2006. Spatial variability of controls on downstream patterns of sediment storage: a case study in the Lane Cove Catchment, New South Wales, Australia. Geogr. Res. 44, 255–271.

Magnani, L., 2001. Abduction. Reason and ScienceKluwer, New York, pp. 205.

May, C., Roering, J.J., Eaton, L.S., Burnett, K.M., 2013. Controls on valley width in mountainous landscapes: the role of landsliding and implications for salmonid habitat. Geology 41, 503–506.

McEwen, L., Stokes, A., Crowley, K., Roberts, C., 2014. Using role-play for expert science communication with professional stakeholders in flood risk management. J. Geogr. Higher Educ. 38, 277–300.

Mika, S.J., Hoyle, J.T., Kyle, G., Howell, T., Wolfenden, B.J., Ryder, D., Keating, A., Boulton, A., Brierley, G., Brooks, A.P., Fryirs, K., Leishman, M.R., Sanders, M., Arthington, A., Creese, R., Dahm, M., Miller, C., Pusey, B., Spink, A., 2010. Inside the "black box" of river restoration: using catchment history to identify disturbance and response mechanisms to set targets for process-based restoration. Ecol. Soc. 15 (4), 8.

Phillips, J.D., 2007. The perfect landscape. Geomorphology 84, 159–169.

Phillips, J.D., 2012. Storytelling in Earth sciences: the eight basic plots. Earth Sci. Rev. 115, 153–162.

Preston, N., Brierley, G.J., Fryirs, K.A., 2011. The geographic basis of geomorphic enquiry. Geography Compass 5, 21–34.

Rhoads, B.L., Thorn, C.E., 2011. The role and character of theory in geomorphology. In: Gregory, K.J, Goudie, A. (Eds.), The SAGE Handbook of Geomorphology. SAGE, London, pp. 59–77.

Richards, K., Clifford, N.J., 2011. The nature of explanation in geomorphology. In: Gregory, K.J., Goudie, A. (Eds.), The SAGE Handbook of Geomorphology. SAGE, London, pp. 36–58.

Roering, J.J., Mackey, B.H., Marshall, J.A., Sweeney, K.E., Deligne, N.I., Booth, A.M., Handweerger, A.L., Cerovski-Darriau, C., 2013. "You are HERE": connecting the dots with airborne lidar for geomorphic fieldwork. Geomorphology 200, 172–183.

Sauer, C.O., 1956. The education of a geographer. Ann. Assoc. Am. Geogr. 46, 287–299.

Schumm, S.A., 1977. The Fluvial System. John Wiley & Sons, New York.

Schumm, S.A., 1991. To Interpret the Earth. Ten ways to be wrong. Cambridge, New York.

Seijmonsbergen, A.C., 2013. The modern geomorphological map. Schroder, J., Switzer, A.D., Kennedy, D.M. (Eds.), Treatise on Geomorphology14Academic Press, San Diego, CA pp. 35–52, Methods in Geomorphology.

Slaymaker, O., 1997. A pluralist, problem-focused geomorphology. In: Stoddart, D.R. (Ed.), Process and Form in Geomorphology. Routledge, London and New York, pp. 328–339.

Small, M.J., Doyle, M.W., 2012. Historical perspectives on river restoration design in the USA. Prog. Phys. Geogr. 36, 138–153.

Stoddart, D.R., Adams, W.M., 2004. Fieldwork and unity in geography. In: Matthews, J.A., Herbert, D.T. (Eds.), Unifying Geography: Common Heritage, Shared Future. Routledge, pp. 46–61.

Tooth, S., 2013. Google Earth™ in geomorphology: Re-enchanting, revolutionising, or just another resource? Schroder, J., Switzer, A.D., Kennedy, D.M. (Eds.), Treatise on Geomorphology 14Academic Press, San Diego, CA pp. 53–64, Methods in Geomorphology.

Wilcock, D., Brierley, G.J., Howitt, R.I., 2013. Ethnogeomorphology. Prog. Phys. Geogr. 37, 573–600.

Wohl, E., 2013. The complexity of the real world in the context of the field tradition in geomorphology. Geomorphology 200, 50–58.

Yu, G., Brierley, G.J., Huang, H.Q., Wang, Z.Y., Blue, B., Ma, Y., 2014. An environmental gradient of vegetative controls upon channel planform in the source region of the Yangtze and Yellow Rivers. Catena 119, 143–153.

Chapter 6

Conclusion

Mary J. Thornbush*[,1], Casey D. Allen**, Faith A. Fitzpatrick[†]
*School of Geography, Earth and Environmental Sciences University of Birmingham, Edgbaston, Birmingham, United Kingdom; **Department of Geography and Environmental Sciences, University of Colorado Denver, Colorado, USA; [†]U.S. Geological Survey, Wisconsin Water Science Center, Middleton, Wisconsin, USA
[1]Corresponding author: E-mail: m.thornbush@bham.ac.uk

This volume addresses geomorphological fieldwork from a predominantly pedagogical approach. From the justification of "*Why Fieldwork?*" to preparing for fieldwork and getting into the field – including field safety – the emphasis has been on the fieldwork trainee (the learner) as well as already established fieldworkers. This remains particularly evident in the second section of the book, where consideration was given to different learning styles, modern (digital) technology, and field-based learning projects for (undergraduate) learners. In a final section on geomorphological methods for the field learner and practitioner, a classic (fluvial) physical landscape was presented along with field-based case studies of soil geomorphology for investigations at the landscape-scale.

Unlike previous volumes, this volume centers on learning fieldwork and includes material that is relevant at different levels of training for field-based research. Its emphasis is on using fieldwork as an active (experiential) form of learning, from the undergraduate to postgraduate levels, and beyond. Focusing on active and experiential learning involves a deep processing of information, which is possible and straightforward to accomplish in the field, where practitioners can readily observe landforms, track processes, discern trends, and verify models directly where they occur (*in situ*). Fieldwork makes it possible to observe natural landforms and processes, but also enables the researcher to take measurements and quantify the landscape. Rather than it being a question of "why fieldwork?" then, the real question is, "why *not* fieldwork?"

This volume conveys that, although field-based learning can be challenging, especially for younger undergraduates and underfunded programs, students respond positively (in both quantitative and qualitative assessments of their fieldwork experience) to geomorphological field-based learning. In addition to providing further justification for the necessity to conduct fieldwork in learning geomorphology, the volume also provides ample preparation tips for actually performing the work. From the preparatory stages, knowing the preferred

Developments in Earth Surface Processes, Vol. 18. http://dx.doi.org/10.1016/B978-0-444-63402-3.00014-5

learning styles of field practitioners helps as part of an educational geomorphology, where learning styles affect students' perception of preparation and performance in the field. Some of the research presented in this volume conveys a generally positive perception of fieldwork as a learning experience. Nevertheless, suggestions are made with ways to improve the reception of fieldwork depending on learning type. For instance, R-learners (generally physical geographers) benefit from a preparatory unit and need to be monitored in the field, whereas, K-learners (generally human geographers) do not agree with the need to have a preparatory unit and summary (report or field notebook), yet still need support and immediate feedback. Knowing the different learning styles allows instructors to better prepare and execute fieldwork, improving the overall student experience and rewards.

Preparation is key, and understanding that things can and will go wrong has been the stipulated approach here, as is assuming that what can go wrong, may in fact, lead to the formulation of backup plans for different aspects of doing fieldwork. This volume also addresses the specific needs regarding field assistance, instruments, finances, transportation, accommodation, and health and safety concerns (support systems, fitness, emergency planning, authorization, etc.). It is crucial that individuals participating in fieldwork are able to adjust to the relevant (different physical and human) environments. A critical part of preparing for any fieldtrip remains field safety, and in this case, this volume gives attention to field operations, transportation, equipment and instruments, wildlife, personnel, and setting diversity. Identifying risks and hazards of fieldwork is also a common preparatory event, as mishaps are to be expected and appear in the field. However, with continued mitigation efforts throughout the fieldwork experience, the intention of risk and task hazards can be minimized. Still, critical thinking is key here, as is adopting a culture of safety where ample feedback provides for improvement of the overall field experience. Although the "Field Safety" chapter is written from a Canadian perspective, there are approaches therein that can be applied more broadly. For example, while going beyond the bare minimum safety requirements is advocated, particularly with supportive leadership in place, it is necessary to adopt at least a minimum field safety standard.

In addition to the Canadian (North American) perspective, other settings are considered here, including New Zealand and Europe, in terms of using digital technology for the field-teaching of physical geography. For these case studies, video clips are encouraged in order to summarize landscape features and processes, as well as management. An experimental approach is also advocated, with the deployment of digital video in experiments to capture process geomorphology occurring directly in the field. Digital video can also be used to train students for research, such as when introducing them to methods and instruments as well as data collection (field-sampling procedures), which has been known to enhance student engagement, group interactions, and overall enjoyment of the field experience. An experimental approach was also taken in field studies of soil erosion. Field experiments, some involving simulation, are considered to be ideal in a realistic setting and with necessary adjustments made, as in coping with instrumentation problems. They allow for updating of research questions as well as for testing what is relevant in a field situation.

Adjustments are a necessary part of any field-based learning and study, and pivotal for reading the landscape. To read the landscape, students need to examine relationships between processes and forms and consider various scales where phenomena are at work. In fluvial studies, at the catchment scale, mechanisms can be inferred based on discernible patterns that enable fieldworkers to interpret their landscape, be it fluvial or otherwise. In this volume, the preparation for this work has been examined within a karst landscape specifically, but the fluvial landscape is also considered, including classic methods such as measuring channels in planform and their morphological cross-temporal changes through surveying and coring. Sediment budgets, for instance, can be approached from a historical framework, with consideration given to both fluvial features as well as glacial landforms.

The overarching theme running through this volume is that of fieldwork as an active and experiential pedagogical approach to learning. Whether through running field experiments, making observations (as in a field notebook) and measurements, or recording field phenomena, engaging learners in fieldwork early in their education is crucial for field verification of geomorphological knowledge, for application, and relevance to the "real world." Fieldwork remains a recognized part of doing geomorphology that is still incorporated in (and, arguably, at the forefront of) educational geomorphology. It stimulates participation in learners and an engagement that fosters deep learning as part of active and experiential learning. The field offers opportunities for "hands-on," active learning, as well as a connection to the environment, other students, and instructors. Finally, fieldwork provides the novice with an impression of what it is to be an environmental researcher beyond the classroom walls. It allows for casual interactions between fieldworkers, and the likelihood of gaining further connections with the subject-matter and existing field practitioners. It also grants the experienced fieldworker reality checking of accepted theories as well as their own inputs to the discipline. As such, and more, it ultimately remains an important part of their identity as geomorphologists.

Index

A

Academics, 45, 124, 131, 135
Acclimatization, 52
Accommodation, 37, 47, 49, 51, 86, 260
Agriculture, 192, 201, 202
  clearing of upland vegetation for, 201
  development following Euro-American settlement, 201
  related erosion in uplands, 202
Amazonian camp, 19
Analytical skills, 15
Antarctic mountains, 35
Anthropocene, 54
Antimalarial prophylaxis, 53
Approach, mitigating risks
  flexible enough to accommodate a range of field activities, 67
  standardization, 67
  supporting mechanisms, 67
Aquatic environments, 204
Archaeogeophysical survey and mapping, 135
Association of American Geographers, 35
ATMs malfunction, 56
Attitudes, 2, 56, 66, 158
Audio-visual resources (AVRs), 131
Aural learning style, 101
Authorization, 46
AVRs. *See* Audio-visual resources (AVRs)

B

Backscatter microscopes, 12
Behavior, 16, 56, 84
Biological considerations, 53
Biome-specific research, 19
Biotic interactions, 237
"Black box" scenario, 176
Bretz's field maps, 21
British Society for Geomorphology, 35
Budgets, 40, 46, 47, 214

C

Carbon footprint, 48
Case studies
  mobile devices to residential fieldwork, application of, 125
  rainfall alter erosion rates on arable land, influence of wind on, 182–187
  runoff and infiltration experiments on terracettes, using simulated rainfall, 188–196
  testing new and measuring known processes, 181
Catastrophism hypothesis, 21
Centers for Disease Control and Prevention, 53
Certification, 75, 77–79, 83
Chernobyl, 66, 68, 84
Cholera, 53
"Citizen scientists", 43
Classroom size, 133
Climate, 19, 48, 52, 142, 204, 240, 253
  Holocene climate change, 201, 204
  role in the decay process, 22
Clinometers, 40
Cloud-based data-storage facilities, 125
Cognitive exercises, 232
Cognitive novelty, 94
Communications, 68, 82, 147
  emergency, 82
  emergency communication devices, 73
  field safety, 82
Compasses, 40
Computer-based "field" exercises, 132
Consciousness, of safety issues, 66
Controls, 74
  allow for accountability, 74
  sign of complacency to resort to field-safety controls, 74
Cook's Tour' fieldtrip, 117
  Marks analysis, 121
  methods of evaluation, 120
Coring devices for inundated sediment, 216
Cross-contamination, 42
Cross-slope variations, 40
Cultural heritage, 135
Cultural stigmas, 51
Culture, 19
  and conscience, 66
Cyber world, 35

D

Data archival, 42
DataPad, 42
Data security system, 42
Dating, 203, 221, 232, 253
Declarative knowledge, 94
Degradation (erosional) process, 20
Delegation, 75
 as elements of a dynamic field-safety culture, 84
 of responsibility, 75
Dengue, 53
Diastrophism (tectonic) process, 20
Digital cameras, 40, 123
Digital elevation models, 35
Digital files and data, 42
Digital maps, 35
Digital technology, 117
 to enhance field teaching in geomorphology and physical geography, 127
Digital video, 121, 127, 260
 assessment, 120
  and understanding of geomorphology, 120
 assignments, 91, 117
Dress (codes), 56
Drinking water, 52
Dropbox, 125
Dual-purpose vehicles, 49
Dyes, 42

E

Earth sciences, 135
Ebola, 53
Economies, 17, 56, 77
Education, 16
 geomorphology, 261
 implications, 99–100
 literature in relation to fieldwork, 94
 platforms, 124
 post-secondary outdoor, 131
 traditional, 16
Electric power, 41
Electronic devices, 35
Emergency communication devices, 73
Emergency planning, 73
Emotional responses, 132
Emotional stability, 40
Empowerment, 16
Entomologist, 19
Environmental geomorphology course, 140, 155–157
 feedback from students in study, 157
 final reservations about standard assessments, 157
 frequency counts of responses on six-point scale for, 151
 gaining useful skills through applied learning, 156
 grade allocations for, 139
 importance of fieldwork defended, 156
 learning through the affective domain, 156
 PBL approach and responses, 155
 positive perception of fieldwork, 155
 problem with assessments, 156
 questions for Part A in changed survey administered in 2009 for, 141
 questions on six-point scale in student satisfaction survey for, 143
 specific commentary on course improvement
  and to recognize student complaints, 164
 student-evaluation forms, 142
 students assessment, 142
 understanding of actual field-based (modern) instruments, 156
 use of "on-street" teaching alongside visual sources, 156
Environmental hydrology, 133
Environmental sciences, 135
Epistemology, 23
Ethical dimensions of fieldwork, 33
Evaluation methods, for fieldtrips, 120
Evidence-based process, 24
Exotic locations, 134
Expeditions, 19
Expenses, 46
Experiential (or constructivist) learning, 134
Experimental Advanced Airborne Research LiDAR (EAARL), 213
Exxon Valdez Alaskan oil spill, 68

F

Field-based "applied learning", 6
Field-based learning
 throughout undergraduate curriculum, 131
Field courses, 6
Field data, 59
Field data security, 42

Field equipment, 83
custom-made or modified instruments, 83
first aid kits, 83
personal protective equipment, 83
portable automated external defibrillators (AEDs), 83
respiratory equipment, 83
snow machines, 83
Field equipment security, 41
Field experiments, as a tool to explore surface processes, 196–197
Field experiments, in geomorphology, 175–177
aims of, 175–176
Field experiments in process geomorphology, 122
approach to teaching and assessment, 122–123
learning evaluation and discussion, 123
Field methodologies, 8
Field observation, 33
Field parties, 80
Balance of Expertise and Compatibility, 80
Joint Fieldwork, 81
Medical Health Evaluations and Fitness for Fieldwork, 80
Threats to Health and Safety Within Field Parties, 80
Working Alone, 81
Field projects, 5
Field safety, 7
Field school, 144–145
counts of tick marks in responses on a six-point scale for, 153
field-based assessment, 144
group-based learning and for the development of social interaction, 158
improvement of teacher-student rapport, 159
informal mentoring by graduate students, 159
lectures, providing students, 144
list of questions in the undergraduate satisfaction survey, 146
PBL approach found to be challenging, 159
positive remarks about field experience, 165
prerequisite, 144
purpose of the final fieldtrip, 145
qualitative open-ended questions, 158
sex-, gender-, and cultured-related conflicts, 159
small-group dynamics provided opportunities for interactions, 159
standard student-evaluation forms, 145
textbook for the course, 144
textbook to be unhelpful, 159
upper-level (also at third-year) undergraduates enrollment, 144
Field techniques, and equipment, 40–42
Field Techniques in Glaciology and Glacial Geomorphology, 144
Fieldtrips, 93, 118
briefings from staff, 118
digital story, 118
digital video, 118
field notebook, 118
hands-on technical skills, 118
illustrative material, 118
knowledgeable expert, 118
methods of evaluation, 120
report critically appraising management at a site, 118
road trip, 118
video diary, 118
Fieldwork
in the Canadian Arctic, boreal forest, and Rocky Mountains, 72
challenges established paradigmst, 20–23
enhancing subfield and interdisciplinary relationships, 18–19
in geomorphology, benefits, 12
pedagogic considerations, 34
potential range of activities, venues, and timeframes, 34
transdisciplinary considerations, 33
Fieldwork exercises, 240
assessing downstream patterns of river styles
and controls on character and behavior, 245–249
interpret river evolution and future trajectory of change, 249–251
CRAPPER acronym, 253
river reach analysis, 240–245
Fieldwork personnel, animals pose threats to, 54
Fieldwork transportation, 48
Financial resources, 35
Fish habitat improvements, 203
Floods, 39
Fluvial geomorphologists, 237

Fluvial geomorphology studies, 203
geologic approach to field methods, 209
age determinations of field sediment samples, 221
calculations and interpreting field data, 221
characterizing fluvial sedimentary sequences, 214–218
describing cores and exposures in field or laboratory, 219
historical floodplain sedimentation depths and volumes and rates, 223
historical watershed sediment budgets, 225
interpreting river behavior and floodplain dynamics, 222–223
topographic mapping and surveying, 209–214
prefield activities, 204–207
reconnaissance trip, 208–209
Fluvial landscapes, 233
Fluvial responses, in channel behavior and sediment dynamics, 201
Food and beverages, 51
Funding, 46

G

Geographical expeditions, 4
Geographic information system (GIS), 213
Geography, 131
Geomorphologic models, 11
Geomorphology, 7, 11
Geomorphology and Global Environmental Change, 142
GeoPDF maps, 41
Georeferencing via GPS, 41
Geosciences, 1
GIScience data, 23
GIS geodatabase, 41
Global Positioning System (GPS), 122, 208, 212
Google Earth, 125
Google Scholar, 35, 124
Gray- and black-market economies, 56
Ground penetrating radar (GPR), 214
Ground truthing, 11, 12
Groundwater, 52
Gully erosion in the Buffalo River, 202

H

Hepatitis A, 53
Holocene climate change, 201, 204
Homesickness, 45
Hot-air balloons, 48
Human-accelerated erosion, 201
Human environment, 55
"Humans as geological agents", 42
Hydraulic engineering, 203
Hydrography maps, 204
Hydrology, 19
Hypothermia, 12
Hypothetico-deductive method, 35

I

"Ideal" academic fieldwork program, 47
Image, from Google Maps, 13
Influenza, 53
Information technology, 35
Inquiry-based learning approach (IBL), 135, 136
benefits, 136
Institutional and personal support, 39
Institutional review boards (IRBs), 47
Institutional transparency, 47
Interconnectivity in ecology and biology, 18
Interdisciplinarity issue, 135
International Association of Geomorphologists, 35
International currency, 56
International Nuclear Safety Advisory Group (INSAG), 66
International shipping permits, 58
Internet, 41, 56, 126, 205
Interpretations
of field-safety requirements, 74
resources for, 75
Interpreters, 44
Introductory geomorphology course, 138, 154–155
field-based learning, provide opportunities for selfdiscovery, 155
grade allocations for, 139
lectures covere topics, 138
most frequent (modal) responses
for Part A in April 2008, 147
for Part A in November 2008, 148
for Part A in November 2009, 149
PBL approach appreciated by, 155
research-led learning, trigger deeper learning, 154
student satisfaction survey, 139
written comments in open-ended questions, 162
iPads, 126

K

K-learners, 101, 259
Knowledge, 24
  and decision-making, 50
"Know your Catchment" mantra, 233

L

Landform assemblages, 33
Landform development, 23
Landforms (or geomorphic units), 236
  analyze and interpret package and assemblage, 237–238
  controls on the package and assemblage of, 238
  identify individual landforms, 236
  integrate understandings of geomorphic relationships at the catchment scale, 239
  process-form relationships that determine behavior, 236
Landscape
  analysis, 232
  questions, making field-based interpretations, 235
  reading, 234
    approach to, 236
Language, 55
Laptop computers, 40
Leadership, of safety culture, 84–87
  collaborate efforts, 85
  key element of organizational leadership, 84
  leadership authority, decentralized, 86
  OHS experts, role, 84
  O-ring issue, 86
Learning cost
  fieldwork learning *vs.* classroom teaching, 133
Learning cycles, 3
Learning styles
  affecting students' perception, 259
  types, 95
Learning through fieldwork, 15–18
Learning tool, 100
Lecture-based classroom, 133
LIDAR, 12, 173, 213
Local contacts, 50
Local customs, and traditions, 56
Local guides, 44
Local tolerances, 51
Local water treatment, 52
Logistic advantages, 43

M

Management of research funding, 47
Mapping fluvial landforms, 204
Mendeley, 124
Mental illnesses, 45
Metadata analysis, 46
Method of Multiple Working Hypotheses, 21
Mobile technologies
  and fieldwork, 124
  for personal learning environment, 127
Monetary woes, 17

N

NASA Challenger space shuttle tragedy, 68
NASA's Landsats 1 and 2, 21
Nuclear accidents, 68

O

Occupational health and safety (OHS), 66
Off-the-grid electricity generation, 41
Online banking, 56
Ontology, 23
Open system, 14
Organizations, 35, 73
  culture, 87
  leadership, 85
  listing of potential hazards during fieldwork, 73

P

Paid workers, 44
(Paleo)flooding, 19
Paleohydrology, 203
PaperShip, 124
Partial landforms, 33
Particle size, 17
Perceptions, 48
"Personal chemistry", 44
Personal health considerations, 45
Personalities, 44
Personal lifestyles, and preferences, 51
Personal safety, 55
Personnel considerations, 43
Personnel safety, 54
Physical environment, 52, 59
Positive culture, of safety, 66
Post fieldtrip report, 121
Potential fieldwork sites, 57
Practices, 68
  to identify hazards and mitigate risk, 68

Preparatory unit, aim of, 94
Preparing for problems, 42
Preservation, 65
Principle investigators, 43
Problem-based learning (PBL), 135, 136
- deployed at second-year level in an introductory field-based course, 137
- disliked workload associated with PBL, 136
- integral part of, 136
- poses various concerns for instructors, 136
- positive experience with group work, 136

Processes, as risk management tools, 69
- mitigation strategies, 71
  - potential hazards, 71
- risk management planning, 70
- seismic survey in a downtown environment, 70
- task-hazard analysis, 70, 71

Professional arrogance, 50
Professional ethics, 47
Professional geomorphological organizations, 35
Professional support, 40
Provisions, 51
Psychological novelty, 94
Public transportation, 48
p -value, 121

Q

Qualitative information, 43
Questionnaire, 79, 95, 103, 151, 154, 157
- gauge student satisfaction with, 134

R

Radiocarbon (carbon-14) dating, 221
Rainfall alter erosion rates on arable land, influence of wind on, 182–187
- background and research question, 182
- overview of results, 186, 187
- plot dimensions and scale considerations, 185
- study site and date of experiments, 185
- supplemental parameters, 186
- test sequence and of sediment collection, 186
- wind-driven rain simulation, 183–184

Rainfall simulations, 175
Rainfall simulator setup, for a microscale erosion experiment, 176
Reading, the Landscape, 233
Real field trips, 5
Real Time Kinematic survey (RTK-dGPS), 122
Recommendations, 126, 147
- to improve safety, 85
- limitations and cautions, 126–127
- transferable skills development for geomorphologists, 126

Reconciliation
- academic and logistic considerations, 39
- academic and logistic preparation, 47

Referees, 11
Richard Foster Flint, 21
Risk-management tools, 66
River behavior, 201, 203, 222, 231, 241, 248, 250
River conditions, 203
"River dance", 234
River-management applications, 233
River patterns, 205, 232, 233
River reach analysis, 232
River Styles framework, 233
R-learners, 101
"Rock decay" science, 22
Rock samples, 58
Rocky desertification, 54
Runoff and infiltration experiments on terracettes using simulated rainfall, 188–196
- antecedent soil moisture conditions, 193
- background information about terracettes and series of experiments, 188, 189
- expectations of a field experiment, 189–190
- field experiments: dealing with unexpected findings, 195
- key resources and logistical constraints, 190–191
- potential relevance of observed infiltration on terracettes for hillslope hydrology, 194
- rainfall simulation experiments, 192, 193
- study sites, 191–192
- surface hydrology and volumetric moisture content
  - related to runoff generation, 194

S

Safety culture, 66, 68, 82, 86. *See also* Leadership, of safety culture
Safety issues, 65
Safety protocols, 39
Safety reporting system, 73
Sailing to islands, 48
Satellite images, 35

Scanning electron microscopy, 12
SCUBA equipment, 39
Sedimentary records, in fluvial landforms, 203
Sedimentation, 202
from Holocene climate change and, 201
processes to nutrient sequestration in Upper Mississippi River bottomlands, 201
Sediment budgets, 203, 260
Sedimentology, 203
Sediment samples, 58, 221
Sediment yield, from wind-driven rainfall, 187
Seminars, 138
Sensory-intensive approach to learning, 131
Shipment
rock samples, 58
of soil samples, 58
"Signature pedagogy", 117
Skills, 24
analytical, 15
interpersonal, 134
Sleeping bag, 49
Smartphones, 124
Socialization, 19
Social sciences, 93
Soil conservation, 202
Soil-erosion research, 54, 178, 202
advantages/disadvantages of field experiments, 178
experiments in geomorphology, 175–177
field experiments *vs.* field monitoring, 178–179
field *vs.* laboratory experiments, 180
preparation of field experiments
general considerations, 180–181
Soil geomorphology, 203
Soil import, 58
Soil science, 16, 23
Standardization, 67
Standards, for field safety, 68
benchmarking, 69
organizational, 68
regulatory, 68
training, 69
Stratigraphy, 203
Structural geology, 23
Students
disliked workload associated with PBL, 136
field-based learning throughout the undergraduate curriculum, 131
"hands-on" learning, 135
inquiry-based approach to, 133
learning styles in education, 95
methodology, 95–96
questionnaire, 95
results, 97
VARK-model study, 96
Marks analysis, for student performance, 121
perceptions of "novelty spaces", 97–99
perceptions of their performance, 95
reality-based part of exercises, 134
taking up the module, 95
Subject-specific information, 121
Supervisors role, 38
SURRGO soils data base, 205
Swimming, 48
Symbiosis, 11

T

Teacher-training program, 140
Teaching, 95
through fieldwork, 15–18
Teaching tool, in environmental geology, 133
Tetanus, 53
Theory of continental drift, 20
The Physiography of Southern Ontario (third edition), 144
Time-lapse photography, 12
Timing, 35
as critical consideration in fieldwork, 36
fieldwork preparation, at temporal scale of days to months, 36
location coincides with extreme conditions, 37
minimal data needs of statistical analysis, 39
20-point preparatory checklist, 37
practical constraints, to take into account, 38
realistic assessment, 38
Training coordination, 77–78
Training delivery, and expertise recognition, 78–79
Training evaluation, 79
Training, examples of, 79
Training need analysis, 75
components, 76
Training requirements, 75
Training standards, 76–77
Transportation, 48
Travel agencies, 47
Travelers' checks, 56
t -test, 121
Twitter, 125
Typhoid, 53

U

Ultra-light aircraft, 48
Unanticipated qualitative information, 43
Union Carbide toxic gas leak, 68
Urbanization, 17
US Bureau of Land Management, 207
US Soil Conservation Service, 201

V

Vaccinations, 53
VARK test, 96
Verbal agreements, 57
Video diaries, 120
Video productions, 123
Virtual fieldtrip experience, 132
Virtual fieldtrips (VFTs), 5
Virtual learning, 133
Visualization
  physical models for three-dimensional visualization, 132
V-learners, 97, 99, 100
Volcanoes, 42

W

Walking lectures, 93
Water, 52
Water-borne diseases, 53
Water samples, 58
Watershed-wide erosion, 203
Weathering science, 22, 135
WEB 3.0 technology, 124, 126
Wind-driven rain, 181
  simulation, 183
Wind influence, on erosion rates, 176
Wind turbines, 41
Wind velocity, 184, 185
Working relationships, 57
Workplace Health and Safety Committees, 76
WorldCat, 35
World Health Authority, 53

Kolibri 360 GmbH
Ferdinand-Jühlke-Straße 7
99095 Erfurt, Deutschland
produktsicherheit@kolibri360.de